Elastic analysis of slab structures

To the memory of my father
Prof. Nicolae Negruţiu

MECHANICS OF SURFACE STRUCTURES
Editors: W. A. Nash and G. Æ, Oravas

P. Seide, Small elastic deformations of thin shells. 1975. ISBN 90-286-0064-7
V. Panc, Theories of elastic plates. 1975. ISBN 90-286-0104-x
T. L. Nowinski, Theory of thermoelasticity with applications. 1978. ISBN 90-286-0457-x
S. Lukasiewics, Local loads in plates and shells. 1979. ISBN 90-286-0047-7
V. Firt, Statics, formfinding and dynamics of air-supported membrane structures. 1983. ISBN 90-247-2672-7
Yeh Kai-yuan, Progress in applied mechanics. 1986. ISBN 90-247-3249-2
R. Negruţiu, Elastic analysis of slab structures. 1986 ISBN 90-247-3367-7

Elastic analysis of slab structures

by

Radu Negruțiu

EDITURA ACADEMIEI
București, Romania

1987 **MARTINUS NIJHOFF PUBLISHERS**
a member of the KLUWER ACADEMIC PUBLISHERS GROUP
DORDRECHT / BOSTON / LANCASTER

Distributors

for the United States and Canada: Kluwer Academic Publishers, 190 Old Derby Street, Hingham, MA 02043, USA
for the UK and Ireland: Kluwer Academic Publishers, MTP Press Limited, Falcon House, Queen Square, Lancaster LA1 1RN, UK
for all other countries: Kluwer Academic Publishers Group, Distribution Center, P.O. Box 322, 3300 AH Dordrecht, The Netherlands
for socialist countries: Editura Academiei, Calea Victoriei 125, R-79717, Bucharest, Romania

Library of Congress Cataloging in Publication Data

Negruțiu, Radu.
  Elastic analysis of slab structures.

  (Mechanics of surface structures; 7)
  Rev. translation of: Analiza elastică a structurilor cu dale.
  Bibliography: p.
  Includes index.
  1. Slabs. I. Title. II. Series: Mechanics of surface structures; v. 7.
TA660.S6N4313 1987   624.1'772   86-12538

ISBN 90-247-3367-7

Book information

This book is the revised and updated English version of the Romanian book *Analiza elastică a structurilor cu dale*, published in 1976 by Editura Academiei, Calea Victoriei 125, București 79717.

Copyright

© 1987 by Martinus Nijhoff Publishers, Dordrecht and Editura Academiei, Bucharest.

All rights reserved. No part of this publication may be reproduced, stored in a retrieval system, or transmitted in any form or by any means, mechanical, photocopying, recording, or otherwise, without the prior written permission of the publishers,
Martinus Nijhoff Publishers, P.O.Box 163, 3300 AD Dordrecht,
The Netherlands and
Editura Academiei, Calea Victoriei 125, R-79717, Bucharest, Romania.

PRINTED IN ROMANIA

Contents

Preface ix

1. Introduction 1
  1.1 *Aim of the book* 1
  1.2 *State of the art. Main research trends* 6
    1.2.1 Theoretical approaches of plates 6
    1.2.2 Classical approaches to flat slabs with capitals 12
    1.2.3 Modern approaches to flat slabs without capitals 30
    1.2.4 Early studies on surface couples. The effective plate width 54
    1.2.5 Early approaches to the elastic analysis of slab structures 70
  1.3 *Scope of the book* 84

2. New mathematical models for the deflected middle surface of plates 89
  2.1 *Preliminaries* 89
    2.1.1 Basic assumptions 89
    2.1.2 General computation formulae 91
    2.1.3 Characteristics of the categories of plates investigated 92
    2.1.4 Characteristics of the types of design loads . . 94
  2.2 *Equations of the middle surface of plates deflected by loads of type P, M and q* 95
    2.2.1 The infinitely long plate 95
      2.2.1.1 The locally distributed load P 95
      2.2.1.2 The surface couple of vector $\vec{M} \parallel Oy$. 97
      2.2.1.3 The surface couple of vector $\vec{M} \parallel Ox$. 98
      2.2.1.4 The uniformly distributed load q . . . 100
    2.2.2 The rectangular plate with two parallel free edges 101
      2.2.2.1 The locally distributed load P 101

	2.2.2.2 The surface couple of vector $\vec{M} \parallel Oy$.	104
	2.2.2.3 The surface couple of vector $\vec{M} \parallel Ox$.	105
	2.2.2.4 The uniformly distributed load q	108
2.2.3	The rectangular plate simply supported along the entire boundary	110
	2.2.3.1 The locally distributed load P	110
	2.2.3.2 The surface couple of vector $\vec{M} \parallel Oy$.	111
	2.2.3.3 The surface couple of vector $\vec{M} \parallel Ox$	113
	2.2.3.4 The uniformly distributed load q	115
2.3	*Algorithms for computing elastic displacements, sectional stresses and boundary reactions*	115
2.3.1	Systematization of numerical computations	115
2.3.2	Expressions for dimensionless coefficients	118
	2.3.2.1 The infinitely long plate	119
	2.3.2.2 The rectangular plate with two parallel free edges	124
	2.3.2.3 The rectangular plate simply supported along the entire boundary	132

3. Behaviour of plates subjected to a surface couple ... 137

3.1 *Aim of the numerical study. Computation parameters and grid* ... 137

3.2 *Factors influencing the magnitude and distribution of elastic displacements of plates subjected to a surface couple* ... 141

    3.2.1 Magnitude and distribution of elastic displacements along the plate width ... 142

    3.2.2 Magnitude and distribution of elastic displacements along the plate length ... 152

3.3 *Factors influencing the magnitude and distribution of stresses in plates subjected to a surface couple* ... 157

    3.3.1 Magnitude and distribution of bending moments along the plate width ... 158

    3.3.2 Magnitude and distribution of bending moments along the plate length ... 174

    3.3.3 Magnitude and distribution of torsional moments ... 180

    3.3.4 Influence of Poisson's ratio on the values of bending and torsional moments ... 185

3.4 *Accuracy of numerical computations* ... 190

    3.4.1 Influence of the number of summed terms in the series on the accuracy of numerical computations ... 190

    3.4.2 Checking the accuracy of computations by use of the reciprocity theorem of unit displacements in the case of plates 196
    3.4.3 Checking the accuracy of computations by reference to the results obtained by other authors 200

4. **Elastic plates as structural members** 203
  4.1 *Slab structures subjected to transverse loads. Effect of M- and P-type loads* 203
    4.1.1 Symmetrical and antisymmetrical sets of surface couples M 205
      4.1.1.1 The infinitely long plate 206
      4.1.1.2 The square plate 219
    4.1.2 Symmetrical and antisymmetrical sets of locally distributed forces P 224
      4.1.2.1 The infinitely long plate 225
      4.1.2.2 The square plate 233
    4.1.3 Associated loads of type M and P applied on partially symmetrical or unsymmetrical slab structures 236
      4.1.3.1 Unsymmetrical sets of couples M and of locally distributed forces P 237
      4.1.3.2 Sets of couples νM and forces νP of different value and sign 244
  4.2 *Slab structures subjected to lateral loads. Evaluation of the effective slab width* 248
    4.2.1 The effective width of the plate subjected to only one surface couple M 249
      4.2.1.1 Definition. Calculation procedure . . . 249
      4.2.1.2 Comparative values of the effective plate width 252
    4.2.2 The effective width of the plate subjected to sets of surface couples M associated in a simple row 256

5. **A new method for the elastic analysis of slab structures** 261
  5.1 *Use of the general force method. Transverse and in-plane loads* 264
    5.1.1 General scheme 264
      5.1.1.1 Basic systems and condition equations 264
      5.1.1.2 Calculation of the elements included in the matrices **d** and **D** in the general case 268
      5.1.1.3 Systematization recommended in the case of associated unknowns 270

Contents

    5.1.1.4 Expressions for the elastic displacements and stresses in the initial statically indeterminate slab structure 272
  5.1.2 Influence of the relative slab-column stiffness on the value of displacements and stresses in slabs and columns 274
    5.1.2.1 Transverse loads 277
    5.1.2.2 In-plane loads 298
  5.1.3 Effect of failure of supports on slab structures 302
 5.2 *Use of the general displacement method. In-plane loads* 312
  5.2.1 Basic scheme 312
  5.2.2 Use of the design procedure in the analysis of symmetrical multistoried slab structures . . . 314

6. Introduction to the dynamic analysis of slab structures 319
 6.1 *Calculation elements for the dynamic analysis of slab structures* 319
  6.1.1 Elements for the study of free vibrations . . . 320
  6.1.2 Elements for the study of forced vibrations . . 327
 6.2 *Behaviour of slab structures subjected to dynamic action. Comparative numerical studies* 335
  6.2.1 Determination of natural circular frequencies 335
  6.2.2 Dynamic deflected shape and stresses induced by disturbing forces 341
  6.2.3 Comparative dynamic analysis of a multistoried slab structure 346

Conclusions 354

Appendix 357

References 403

Author index 413

Subject index 417

Preface

Any practitioner who takes his profession in earnest, such that daily work is not a heavy duty but part of their life, will recognize in this book the rigorousness of the analysis and the comprehensive presentation of the problems. This professional attitude is solely able to make the research and design engineer deal with strength structures and their behaviour.

Indeed, the computational means that are nowadays available permit the numerical computation of whatever problem; the program libraries are extremely rich and programs themselves have developed intensively. However, though computers are available at any moment without restrictions on the frequency with which they are employed, they finally impoverish the creative competency of the civil engineer. Thus, he will calculate increasingly more while devising increasingly less. He will draw less and less on the experience gained in devising and implementing bearing structures because the computational process can be repeated as often as desired over a minimum time-period by means of the available programs. We note that nowadays structures are no longer investigated or economically designed to comply with the requirements of the topic of interest. Much to the contrary, the solutions are chosen so as to comply with the capabilities of the programs.

A bearing structure lives as is prescribed by its initial constructive data. However, the numerical computations that hold pre-eminence today, are only a later confirmation by static analysis means of the fact that the sizes, stiffness and other items have been well chosen. A correctly built structure can be obtained only if one acquires full knowledge of the distribution of stresses in the over-all structures and in each of the strength members. Of course, it is essential to have a clear idea of the structural behaviour under stress and strain and of the structural response to loads.

Preface

It is in this sense that, Dr. Negruțiu's work should be understood. So far, this is the most comprehensive research of the behaviour of plates subjected to locally distributed couples, defined as surface couples.

Let us note the preliminary theoretical approach which gives solutions to the fundamental equation of plates under this new type of load. Owing to the rigid column-slab connection, the study of surface couples is of outstanding importance whenever slab structures having an arbitrary location of columns are subjected to uniformly or locally distributed loads or to horizontal in-plane forces.

An independent chapter revolves about the dynamic behaviour of slab structures, i.e. the response of such structures to dynamic loads as may be induced, for instance, in seismic zones.

To the familiarized reader, the pertinent diagrams for the variation of the static quantities and the numerical data tabulated for the sectional stresses and elastic displacements furnish a clear image of the behaviour of slab structures under load.

It is my hope that this book will find wide audience among engineers, particularly among the youngest.

Karlsruhe, March 1985

KLAUS STIGLAT

1

Introduction

1.1 Aim of the book

For many decades flat slabs have been employed in reinforced concrete structures. The use of flat slabs with capitals — as they were first introduced — and the more recent adoption of flat slabs without capitals has witnessed a remarkable extension for reasons of functional and technological efficiency. Moreover, the great versatility of slab structures is also favoured by economic advantages which stem from their simplified geometry and the combined working of their structural members, i.e. slabs and columns.

The analysis of flat slabs taken individually or as structural members has been the object of a relatively rich literature. Fewer approaches have been made to the global analysis of these categories of structures. Indeed, the behaviour of slab structures as a three-dimensional problem could not have been rigorously modelled so far. Thus, most of the present studies are confined to the design of flat slabs alone, considering only the effect of gravity loads. These approaches fail to attain a rigorous static analysis, which could be obtained in the case of reticular systems consisting of bars alone. Worth mentioning is also the dearth of theoretical studies dealing with the lateral load effect on slab structures. Additionally, the effect induced by the failure of supports and the dynamic analysis of this category of structures are not dealt with in the literature.

★

The emergence of the *elastic analysis* of structures consisting of bars and plates without beams — hereinafter referred to as mixed structures or slab structures — as a distinct field of Structural Mechanics is of a recent date. Given the high topological complexity of these structures, it has become conservative in their investigation, even in recently published studies, to use

1 Introduction

models of a common characteristic, i.e. schematic diagrams that reduce the three-dimensional problem to a two-dimensional one. By this treatment, the effects of space interaction are neglected either in the vertical or in the horizontal plane. The solutions thus obtained are restricted either to the study of the plate-like behaviour or of the framed structure treated as a plane reticular structure.

Although the solutions obtained for the first category pertain to the Theory of Elasticity, they are still approximate because the effect of combined working is practically neglected. The bars of the slab structure, i.e. the columns, are assumed to be equivalent to intermediate supports of the plate under study. With rare exceptions, these discrete supports are introduced in the analysis as point-like supports, i.e. as hinges.

The solutions pertaining to the second category are also approaximate although the structural effect is studied by means of the Theory of Structures. The approximate character of these solutions derives from the fact that the schematic diagram of the plane structure is obtained by assimilating the slab with a frame beam.

In the research underlying the ideas contained in this book (1969—1985), these simplified models were abandoned. The elastic analysis of slab structures is performed in the three-dimensional context, assuming that the bar-plate connections are rigid non-point-like connections.

The interaction between plates — defined as in-plane continuous members — and bars — discrete elements located in planes normal to the plate — has forced a global treatment of the analysis. To this end, use was made of the original solutions derived from the development of the Theory of Plates and the general methods of the Theory of Hyperstatic Structures. The unified method of analysis presented in this book allowed the consideration of the vertical loads acting transverse to the slab, the horizontal loads acting in the plane of the slabs and the effect due to the failure of supports. Additionally, the new approach to the static analysis of structures consisting of bars and plates and the results thereby obtained allowed us to approach the dynamic analysis of this category of structures.

★

Throughout the book, *slab structures* are used to designate mixed structures with beamless floors, for civil and industrial constructions. The floors consist of slabs resting on columns that can be with or without capitals. This work deals mainly with reinforced concrete structures having one or more flat slabs (Fig. 1.1). The

1.1 Aim of the book

results obtained are also valid in the investigation of structures with bearing walls, e.g. shear-wall rigidly connected by discrete beams, cellular structures, foundation caissons, caissons for piers and docks, etc. From the topological standpoint, flat slabs without capitals are a modern structural solution derived from

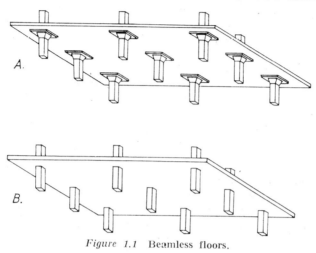

Figure 1.1 Beamless floors.
A. Flat slabs with capitals. B. Flat slabs without capitals.

flat slabs with capitals by eliminating the capitals at the upper end of the column. Because of the geometric and mechanical similarity of these two categories of flat slabs, the main investigations and current code requirements [154], [155], [158], [160] — including the Romanian code of practice in preparation [162] — give a similar treatment in their analysis. This similarity will be taken into account in the rigorous elastic analysis that is the object of this book. Thus, the range of applicability of our method extends from structures having flat slabs without capitals to flat slabs with capitals, of course under certain conditions.*

For all these categories of structures, the connections between slabs and columns (shear walls and beams, respectively) forbid — partially or totally — the relative rotations in the bar-plate

---

* From the results obtained (Chaps. 4—6), we can derive the following basic conditions for extending the range of applicability covered by our method of analysis: a) limitation of the relative sizes of the capitals with respect to the cross-section of the column and the span of the slab; b) consideration in calculation of the sizes of the contact area between the capital and the slab; and c) accurate evaluation of the relative bending rigidity of the slab and column of varying cross-section.

1 Introduction

nodes. The mathematical modelling of the effect of these *rigid connections*, which represent the basic characteristics of reinforced concrete slab structures and shear-wall structures, has been intensively studied by the author over the last 15 years [64], [65], [68], [69], [70], [73], [74], [75], [77], [78], [82], [83], [86]. The main results and the conclusions thereby derived are given in a unified view in this book.

Throughout this work, we have attempted to cover a wide range of applications. Thus, the expressions of the static magnitudes which occur in the structural analysis — transverse displacements, rotations, stresses and boundary reactions — have been derived with no constraint on the physical and mechanical characteristics of the material employed or on Poisson's ratio in particular. In this way, the conclusions of this book can be extended to the analysis of topologically equivalent metal structures (such as installation and machine elements) whose form mimics the geometry of slab structures, assuming that the bar-plate connections are rigid non-point-like connections. Additionally, the numerical data obtained for the value $\mu = 1/3$ of Poisson's ratio could be used as reference values in the tests carried out on reduced-scale steel models [77].

★

Basically, this work deals with the *distribution of the elastic displacements and sectional stresses* in slabs and throughout the structure, assuming various loading diagrams and boundary conditions. The investigation of the state of stress and deformation leads to the construction of original mathematical models which allow a representation of the behaviour of the structure that is closer to the real state of stress. Whence a more rational distribution of material throughout the structure, especially of the reinforcement in slabs and columns.*

As the main object of this book is to place at the disposal of researchers and design engineers a basic tool for analysis, the results obtained are systematically presented so as to allow their easy handling in current practice. Thus, the algorithms derived

---

* As concerns a rational distribution of concrete, it is well known that the slab thickness, hence the overall volume of concrete, is mainly determined by the condition that the shearing stresses are carried over. Shear — a highly complex phenomenon — was the object of numerous testings on full-scale models carried out in several countries under the umbrella of Comité Européen du Béton [161]. Similar testings were performed in Romania. The program of CEB for experimental research was essential for the progress achieved in this field. The results obtained resulted in codes of practice for the design against shear, which have been recently introduced in the above mentioned design requirements. Hence, aspects pertinent to shear will not be approached here.

1.1 Aim of the book

from the author's solutions to the fundamental equation of plates are presented in a simplified formulation. The expressions suggested for these algorithms have the same structure for any boundary condition or loading type of the plate. In this way, the direct computation of the elastic displacements and the stresses is more convenient. In addition, the numerical data to be used in practice are presented in tables and diagrams.

The new method for analysis must be considered to be rigorous within the limits given by the basic hypotheses of the Theory of Elasticity [4], [5], [6], [7], [9], [16], [20], [21], [23] and of the Theory of Structures [8], [15], [22], etc. The use of this method in the analysis of reinforced concrete slab structures requires the adoption of the hypothesis of the linear elastic behaviour of the material. In the case of statically indeterminate structures, this hypothesis is accepted by most of the codes of practice including the Romanian one [159], [162], etc. In the practice of the new analysis method, we must examine the value of the real rigidity of columns and slabs.

As was already shown, the proposed method leads to a new treatment of slab structures, the design of which relies at present on approximate methods. However, the present method of analysis is not intended to supercede the approximate methods in design. The *empirical method* outlined in the American code [160] and the *equivalent frame method* recommended by the American, French [154], German [155], Swiss [156], British [158] and Romanian [162] codes are of peculiar advantages. The author's rigorous method of elastic analysis can be used in parallel for slab structures that are incompatible with approximate methods (unsymmetrical structures, oblique structures, decks of bridges, etc.). This new method is also recommended in the case of structures under heavy service loads or having large spans, where the design elements furnished by the above mentioned approximate methods are scanty. In all these cases, the author's method of analysis and the conclusions drawn from comparative studies lead to a more exact and comprehensive representation of the distribution of stresses throughout the structure and, hence, to a more economical design of slab structures.

From the investigations underlying the ideas presented in this book, more than 95,000 items of numerical data for plates (transverse displacements, rotations, bending moments and

1 Introduction

torsional moments) were obtained. These data can be used in experimental research, including testings on full-scale models, as reference values in the elastic range for the displacements and stresses determined experimentally.

1.2 State of the art. Main research trends

1.2.1 Theoretical approaches of plates

Classical and recent literature on the analysis of plates resorts to the Theory of Plates and to the solutions of the Lagrange-Sophie Germain equation in Cartesian co-ordinates

$$\nabla^2 \nabla^2 w = \frac{p(x, y)}{D}, \qquad (1.1)$$

where $w(x, y)$ represents the deflected mid-surface of the plate and $p(x, y)$ denotes the load function, i.e. the load normal to the mid-plane, taken per unit surface (Fig. 1.2).

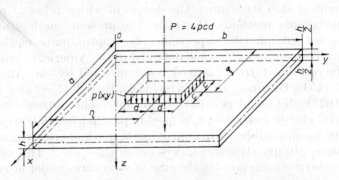

Figure 1.2 Rectangular plate under locally distributed transverse load P.

In the theoretical developments deriving from the modelling of the boundary conditions, frequent use is made of the homogeneous differential equation

$$\nabla^2 \nabla^2 w = 0 \qquad (1.1a)$$

corresponding to the transverse load $p(x, y)$ which is zero throughout the plate, except for the edges.

1.2 State of the art. Main research trends

In Eqn. (1.1) the cylindrical stiffness of the plate is denoted by

$$D = \frac{Eh^3}{12(1-\mu^2)}, \qquad (1.2)$$

where E is the modulus of elasticity of the material; h represents the plate thickness, which is assumed to be constant; and μ denotes Poisson's ratio.

For a given type of load $p(x, y)$ and for given boundary conditions, knowledge of the function

$$w(x, y) = w(p, E, a, b, c, d, h, \xi, \eta, \mu, x, y), \qquad (1.3)$$

as a solution of the fundamental equation (1.1), permits the determination of the state of stress and deformation at any point (x, y) in the plane of the plate (Fig. 1.3).

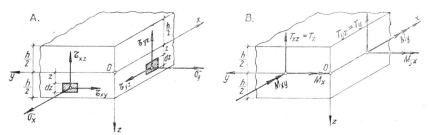

Figure 1.3 Rectangular plate in Cartesian co-ordinates.
A. Stresses induced by transverse load. B. Equivalent sectional stresses.

Beside the above-mentioned parameters, the expression for $w(x, y)$ given by Eqn. (1.3) includes the following geometrical parameters (Fig. 1.2):

- a, b — in-plane sizes of the rectangular plate under examination;
- $2c, 2d$ — dimensions of the area ($2c < a$; $2d < b$) loaded with the locally distributed transverse load $p(x, y)$;
- ξ, η — the co-ordinates of the centroid of the area loaded with $P = 4\,pcd$.

The main types of solutions of the fundamental equation of plates (1.1) differ either by the method of numerical analysis employed or by the schematic diagram used for the bearing element under examination. Thus, we have

1 Introduction

a. Rigorous solutions, within the limits imposed by the fundamental hypotheses of the *Theory of Plates* (see Sub-sec. 2.1.1), which represent expansions into simple or double series of trigonometric and exponential functions of Navier, Lewe, Timoshenko and Girkmann types. Usually, these solutions are obtained by expressing the load function $p(x, y)$ as one or several series of trigonometric functions and by identifying the general term satisfying the partial differential equation of fourth order (1.1), i.e. by integration of this equation [1], [3], [7], [9], [16], [20], [21], [27], [75], etc. In several particular cases expansion into series could sometimes be replaced by "closed" expressions, which represent the sum of the respective series (Sub-secs. 1.2.2.3 and 2.2.1.4).

Further solutions to the fundamental equation (1.1), which pertain to the category of rigorous solutions, are developed herein for the elastic analysis of slab structures (Chap. 2).

b. Approximate solutions of the Marcus or Westergaard type, which are obtained using the *finite difference method* as an approximate procedure for integrating the equation of Lagrange. This means that we must replace differential operators in the known relations between the stress tensor and the strain tensor (Sub-sec. 2.1.2 — Table 2.1) by finite difference operators. The boundary conditions can be established easily and the computation programs are conveniently written. These advantages explain the wide use of this method [2], [7], [9], [16], [20], [21], [25], [50], [72], [103], [112], [135], etc.

c. Approximate solutions of the Rayleigh, Galerkin or Ritz-Timoshenko type, which are obtained with the aid of the so-called *energy methods*. These methods are developments of the energy variational principle, which states that of all the possible fields of displacements of a structure, the displacements produced by external forces are those for which the total energy Π of the loads-structure system is constant:

$$\Pi = U + W \tag{1.4}$$

where U represents the deformation energy and W denotes the potential energy of the external forces.

A particular application of this principle is the minimum total energy theorem

$$\delta(U + W) = 0, \tag{1.4a}$$

which holds for a wide range of problems in the elastic analysis of structures, including the static and dynamic analysis of plates. This theorem allows us to write the differential or algebraic

1.2 State of the art. Main research trends

equations of equilibrium using variational operations. The solutions thus obtained are not rigorous because the functions defining the phenomenon under study in topological and rheological terms (for instance, the displacements of the deflected mid-surface of the plate) are only approximate in most cases and the boundary conditions are seldom fully satisfied [3], [7], [9], [16], [20], [21], etc.

d. Approximate solutions of the Marcus type [2], where the approximation derives both from the method of numerical analysis and the schematic representation of the plate. These solutions were obtained by use of the *membrane analogy*, and replacing the membrane with an *elastic network*. In his method of analysis, Marcus replaces the partial differential equation of fourth order (1.1) by two second-order equations, which represent the deformed surfaces of two membranes. These membranes are next treated in calculations as two equivalent networks consisting of elastic bars, and this facilitates the use of the finite difference method as an approximate integration procedure.

Other solutions pertaining to this category are those in which the plate is replaced by a network of integrally built orthogonal beams. The twisting and bending rigidities of the beams thus defined are given by the condition that the deformation energy of the network is equivalent to that of the real plate [50].

e. Approximate solutions recently obtained by use of the *finite element method* which is conveniently adapted to work on computers. This modern method allows us to solve field problems that can be expressed in variational form and can be considered to be an extension of Rayleigh-type methods. Many such problems, e.g. force, tension, deformation and displacement fields, arise in structural mechanics. Instead of using this variational method for the entire field of the domain under study, the finite element method is applied to several small but finite elements of this domain. By assembling these elements we can reconstitute the entire field. For the elements in which the domain was divided in order to study the field, most frequently the displacement field under examination, we can choose simpler functions than those used for the description of the entire field. This method can be easily adapted to deal with complex structural forms, particularly with boundary field problems, as the domain is divided up into a series of elements. Several major difficulties pertaining to the energy method, e.g. the selection of the function to approximate the entire field and the fulfilment of the boundary conditions, are avoided in the finite element method.

The variational problem defined by relation (1.4a) is approximated in the finite element method through the limit of a steady

1 Introduction

value problem of a function $f = f(\alpha_i)$ with respect to a finite number n of parameters α_i, which are expressed by the equations

$$\frac{\partial f}{\partial \alpha_i} = 0 \quad (i = 1, 2, \ldots, n). \tag{1.5}$$

The variables α_j define the partial fields u, confined to small ranges

$$u = \sum_1^m \alpha_j W_j \quad (j = 1, 2, \ldots, m), \tag{1.6}$$

where W_j represents m known functions.

From the topological and rheological data of the problem, we derive the set of external forces, the manner in which the structure is idealized and discretized, the method of analysis (usually the general displacement method) used in dealing with the idealized structure, the boundary condition and the constitutive laws (of Hooke's type) of the material. Since n is finite, the set of equations (1.5) leads to an approximate solution. In the linear elastic analysis of structures, the set (1.5) is linear and the solution of this problem means to determine a finite number n of parameters α_i related by a set of algebraic equations.

As a physical interpretation of this method, the structure under study is considered to consist of an assembly of individual structural members, i.e. finite elements. These elements are related by a finite number of continuity conditions for a finite number of points connecting several individual elements. These conditions express the equality at the connecting node of the parameters defining the partial elements. The parameters α_i represent local values of the field under examination, e.g. the displacements at the nodes.

The approximation made between the real and the idealized structure is of a physical nature as the real structure is replaced by a structurally modified system. However, no approximation appears in the analysis of the idealized structure. This is unlike the finite difference method, where the exact equation of the real system is solved with the aid of an approximate mathematical method [11], [14], [85].

★

All the categories of known — rigorous or approximate — solutions to the Lagrange equation (1.1) admit the assumption

1.2 State of the art. Main research trends

of the linear-elastic behaviour of the material *. This restriction could not prevent the wide use of these solutions in the design of plates, including the design of reinforced plates. Nor did it mar the concerns for finding new solutions for the static analysis of this category of flat bearing elements. This is illustrated by the numerous studies, published since 1955, which extend or particularize known solutions in the Theory of Plates [87]—[143], etc.

Further developments of the classical solutions in the elastic range have been extensively approached over the last two decades, with special reference to the topology of plates. From the bulk of recent contributions to the solution of engineering practice problems, special mention should be made of studies on:
— continuous plates [87]—[100]; cantilever plates; clamped plates [101]—[113]; elastic boundary plates; free boundary plates [114]—[124]; plates of variable thickness; plates with openings [125]—[138]; effect of the surface couple on plates [60], [64], [65], [66], [68], [70], [71], [73], [79], [80], [82].

The classical and contemporary works and the above-mentioned solutions and analysis methods rely on the linear elastic Theory of Plates. This theory underlies also the solutions obtained in this book for modelling the effect of the rigid connections at the bar-plate nodes (column-slab or beam-shearwall nodes).

The author's preference for the elastic analysis of reinforced concrete slab structures is firstly justified by the lack of this kind of treatment in the literature. Secondly, the main design prescriptions in force recommend the use of the elastic analysis in the determination of the state of stress in hyperstatic structures, including slab structures [154]—[162]. Thirdly, it is generally acknowledged that the analysis of the behaviour of plates treated as structural members and loaded in the plastic range, requires prior knowledge of the distribution of displacements and stresses in the elastic range.

Johansen's *yield-line theory* [144]—[147], which underlines the ultimate state design of plates, brought about a new conception in this domain. Numerous theoretical and experimental studies dealing with the behaviour of plane reinforced concrete bearing members in the post-elastic range have been carried out in Romania [151], [153] and abroad [148], [149], [150], [152]. The conclusions of these studies only reinforce the idea that investigations must be performed first in the elastic range. Additionally, the

---

* Investigations of the elasto-plastic or plastic ranges, including the simulation of the effect of cracks, which characterizes reinforced concrete structures, are theoretically possible by use of the finite element method.

1 Introduction

need for mathematical models in this range increases with higher topological and rheological complexity of the structures under examination, including reinforced concrete structures.

1.2.2 Classical approaches to flat slabs with capitals

The first attempts to approach design stresses in beamless reinforced concrete floors resting on columns were made seventy-five years ago. These attempts consider only the case of columns with capitals. Because of the empirical nature of these methods of analysis, including the procedures suggested by American authors (Turner, Maurer, Mensch a.o.) in the early century, the reinforcement consumption varied with the method employed by 400 per cent [29], [40].

From the classical theoretical works on the analysis of flat slabs with capitals, which proceed from the fundamental equation of plates (1.1), special mention deserve those by Nádai [1], [26]; Marcus [2], [32]; Galerkin [3]; Timoshenko [7]; Girkmann [9]; Lewe [27], [28]; Tölke [30]; and Grein [33]. In the schematic diagrams first used by these authors, support on the columns with capitals was assumed to be concentrated at a point and the analysis was confined to only one interior panel. Later on, the edge panels were introduced in the analysis under the assumption that the edges are simply supported. Some of the analysis procedures suggested by these authors were later extended to the analysis of flat slabs without capitals. These procedures will be briefly presented below. Details will be given only for those procedures that are closely related to the author's concept of analysis.

The *elastic network method* was established by Marcus [2], [32] and used in the study of both plates in general and flat slabs in particular. As an approximate method, it can be used in the analysis of simply supported or edge-restrained flat slabs, under distributed or concentrated, symmetrical or nonsymmetrical loads. The interior columns of the slabs are assumed to be discrete and concentrated at a point. The *compatibility conditions* prescribe only that the transverse displacements of the slab is cancelled in the column axis:

$$w(x, y) = 0. \tag{1.7}$$

This schematic treatment simplifies the analysis notwithstanding the inconvenience of operating with two networks. However, the effect of the column stiffness is neglected in this approach.

The *finite difference method* is employed by Marcus and many others, beginning with Westergaard-Slater [25], down to a recent comparative study [50]. The analysis of flat slabs

1.2 State of the art. Main research trends

with or without capitals through the finite difference method can be made with the aid of fourth-order differential equations, discarding the membrane analogy or Marcus' elastic network method. In calculations, the intensity of the load $p(x, y)$ occurs in each point of the calculation grid. Hence, a uniformly distributed load of type q can be expressed in a straightforward manner. A concentrated force P — including the reaction in the axis of the intermediate column — corresponds to a zero intensity throughout the grid, except for the closest node, where we apply the intensity obtained by referring the force P to the area of a mesh of the grid.

In the analysis, account can be taken of given displacements across the plate, or of elastic supports, for instance, in the axes of the intermediate columns. As was also observed in the case of the elastic network method, the use of the finite difference method requires that the intermediate columns be assimilated with point-like supports. The effect of the transverse dimensions of the columns could be introduced in calculations by prescribing zero displacement conditions of type (1.7) in several points of the computation grid. However, this procedure prohibits the evaluation of stresses in the immediate vicinity of the discrete supports, i.e. in the regions where the bending and torsional moments reach extreme values. Similar inconveniences might arise if in order that the effect of the rigid connections be considered in the analysis, we would attempt to express the zero relative rotation of the slab-column node using the finite difference method.

1.2.2.1 *Nádai-Timoshenko's procedure.* From the classical procedures of analysis that may be considered to be rigorous inasmuch as they rely on the solutions of Lagrange's equation (1.1), mention should be made of the procedure proposed by Nádai [1], [26], who gave the stress distribution for an infinitely extended slab with concentrated column reactions. Both Galerkin [3] and Nádai consider only one panel which pertains to a flat slab, consisting of an infinite number of identical panels in both directions of the slab plane and neglect the dimensions of the column cross-section.

Using the results obtained by Nádai, Timoshenko and Woinowsky-Krieger [7] suggested another procedure, which subsequently became the source of other recent approaches [9], [42] etc. This analysis relies on three simplifying assumptions, which particularize the schematic diagram:

a. The in-plane sizes of the flat slab are large as compared to the sizes a and b of an *interior panel*, which represent the equal distance between the rows of successive columns, measured

1 *Introduction*

in the two directions of the plane (Fig. 1.4). Additionally, the panel under examination is assumed to be sufficiently far away from the indefinite edges of the flat slab.

    b. The load q is uniformly distributed throughout the slab.

    c. The interior supports of the slab are *point-like supports*, the effect of the column stiffness being neglected.

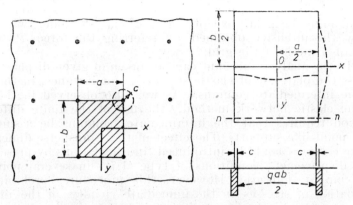

Figure 1.4 Flat slabs with capitals. Schematic diagram for Nádai-Timoshenko's procedure.

Since the states of stress and strain can be assumed to be identical in all the interior panels which satisfy these assumptions, it is sufficient to study the behaviour of only one corner-supported panel under the load q.

From the topological symmetry, it can be deduced that in the case of the panel thus defined, the slope of the deflected mid-surface, measured normal to the boundary, is zero. Hence, the torsional moment M_{xy} vanishes also on the entire boundary of the panel. The shearing force is also equal to zero at any point on the boundary, except for the four support points:

$$(\varphi_x)_{x=\pm\frac{a}{2}} = (\varphi_y)_{y=\pm\frac{b}{2}} = 0 \,;\, (T_x)_{x=\pm\frac{a}{2}} = (T_y)_{y=\pm\frac{b}{2}} = 0. \qquad (1.8)$$

Hence, the continuity of the slab on the boundary of the panel under study is equivalent with the full clamping effect. Consequently, the only compatibility condition used by Timoshenko is the condition of type (1.7) of *zero transverse displacement* of the slab at the slab-column nodes. The expression of the deflected mid-surface of the slab, which is determined as a solution of the differential equation (1.1), results as a simple series of

1.2 State of the art. Main research trends

hyperbolic functions:

$$w(x, y) = \frac{qb^4}{384D}\left(1 - \frac{4y^2}{b^2}\right)^2 - \frac{qa^3b}{2\pi^3 D}\left\{\sum_{m=2,4,6,\ldots}^{\infty}\frac{1}{m^3}\left(\alpha_m - \frac{\alpha_m + \operatorname{th}\alpha_m}{\operatorname{th}^2\alpha_m}\right) - \sum_{m=2,4,6,\ldots}^{\infty}\frac{(-1)^{m/2}\cos\frac{m\pi x}{a}}{m^3 \operatorname{sh}\alpha_m \operatorname{th}\alpha_m} \times\right.$$

$$\left.\times\left[\operatorname{th}\alpha_m \frac{m\pi y}{a}\operatorname{sh}\frac{m\pi y}{a} - (\alpha_m + \operatorname{th}\alpha_m)\operatorname{ch}\frac{m\pi y}{a}\right]\right\}, \qquad (1.9)$$

where $\alpha_m = m\pi b/2a$.

At the center ($x = 0$, $y = 0$) of the interior panel under examination (Fig. 1.4), Timoshenko calculates for different values of the parameter b/a and taking the value $\mu = 0.2$ for Poisson's ratio, the transverse displacement w and the corresponding sectional bending moments M_x and M_y (Table 1.1).

Table 1.1

Flat slabs with capitals under the uniformly distributed load q

Values of w, M_x and M_y at ($x = 0$, $y = 0$), after Timoshenko

$\mu = 0.2$

b/a	$w / \dfrac{qb^4}{D}$	M_x/qb^2	M_y/qb^2
1.0	0.00581 (0.00835)	0.0331 (0.0428)	0.0331 (0.0428)
⋮			
2.0	0.00292	0.0092	0.0411
∞	0.00260	0.0083	0.0417

Expression (1.9) is of type

$$w(x, y) = w(p, a, b, \mu, x, y), \qquad (1.10)$$

i.e. it is independent of the sizes ($2c \times 2d$) of the slab-column contact area which was assumed to be concentrated at a point ($c = 0$; $d = 0$). The values of the bending moments M_x and M_y at the support points ($x = \pm a/2$; $y = \pm b/2$) (Fig. 1.4), calculated from solution (1.9), are hence infinite.

Considering the pronounced influence of the parameters defining the contact area on the value of the moments at the

1 Introduction

intermediate supports, Timoshenko gives also expressions for the sectional bending moments M_x and M_y in the column axes. These expressions were given by Nádai [1] for columns of circular cross-section and by Woinowsky-Krieger [31] for columns of rectangular cross-section. If in the latter case (Fig. 1.5) he assumes that

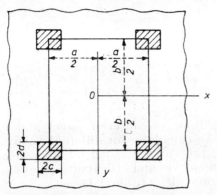

$$\frac{2c}{a} = \frac{2d}{b} = k, \qquad (1.11)$$

then the following expressions are obtained for the bending moments $M_x = M_y$, which hold only in the case of square $(a=b)$ interior panels and square $(c=d)$ columns (capitals):

— in the column axis $\left(x = y = \dfrac{a}{2}\right)$

Figure 1.5 Schematic diagram for Woinowsky-Krieger's procedure. Sizes of slab-column contact area incorporated in the analysis.

$$M_x = M_y = -\frac{(1+\mu)\,qa^2}{4}\left[\frac{(1-k)(2-k)}{12} + \right.$$
$$\left. + \frac{1}{\pi^3 k^2}\sum_{m=1}^{\infty}\frac{2}{m^3 \operatorname{sh} m\pi}\operatorname{sh}\frac{m\pi k}{2}\operatorname{ch}\frac{m\pi(2-k)}{2}\sin m\pi k\right]; \quad (1.12a)$$

— at the centre of the panel $(x = y = 0)$

$$M_x = M_y = \frac{(1+\mu)\,qa^2}{4}\left[\frac{1-k^2}{12} + \right.$$
$$\left. + \frac{1}{\pi^3 k^2}\sum_{m=1}^{\infty}(-1)^{m+1}\frac{\operatorname{sh} m\pi k}{m^3 \operatorname{sh} m\pi}\sin m\pi k\right]. \quad (1.12b)$$

Taking the value $\mu = 0.2$ for Poisson's ratio, Woinowsky-Krieger calculates, using expressions (1.12), the value of the bending moments M_x and M_y at three characteristic points of the square panel under study as a function of the value of parameter k (1.11) (Table 1.2).

In the case of the static diagram adopted by Timoshenko and Woinowsky-Krieger (Figs. 1.4 and 1.5) the values of the transverse elastic displacements w and those of the bending moments M_x and M_y are significantly influenced by the full clamping effect assumed on the boundary of the panel $a \times b$ under examination. Compared to the corresponding values given in the present

1.2 State of the art. Main research trends

Table 1.2

Flat slabs with capitals under the uniformly distributed load q

Values M_x and M_y, after Timoshenko and Woinowsky-Krieger

$a = b$; $c = d$; $\mu = 0.2$

$k = \dfrac{2c}{a}$	$x = y = \dfrac{a}{2}$	$x = y = 0$	$x = \dfrac{a}{2}$; $y = 0$	
	M_x/qa^2 $M_y = M_x$	M_x/qa^2 $M_x = M_y$	M_x/qa^2	M_y/qa^2
0	$-\infty$	0.0331	-0.0185	0.0512
0.10	-0.1960	0.0329	-0.0182	0.0508
	(-0.0852)	(0.0428)	(-0.0080)	(0.0560)
0.20	-0.1310	0.0321	-0.0178	0.0489
.				
.				
0.50	-0.0487	0.0265	-0.0140	0.0361

book for a similar loading case but using a static diagram that represents more faithfully a real slab structure (Sub-sec. 5.1.2.1), large differences appear in compliance with expectations (the bracketed values in Tables 1.1 and 1.2). Indeed, if we consider the case of the square panel, which was investigated by the American authors, the values obtained at the center of the panel are 44 per cent smaller for the displacement w and 30—33 per cent smaller for the moments $M_x = M_y$. However, the values of $M_x = M_y$ in the column axes are substantially higher in absolute value than those obtained by the rigorous method presented herein [75], [78].

In establishing solution (1.9), Timoshenko dispensed with the continuity condition for rotations and thus, the effect of the relative slab-column stiffness against bending (Sub-sec. 5.1.2) was neglected in calculations. Considering the importance of this effect, Timoshenko developed several local corrections for the slab-column region, recalling the solutions of the circular plate theory [7]. Expressions (1.12), which are given only for the case of the interior square panel and for some of its characteristic points, should be also regarded as corrections to solution (1.9). An approach to the analysis of stresses in the edge panels is out of the question. Because of the simplifying assumptions used in the analysis, the Nádai-Timoshenko procedure is restrictive in practice. However, it is the merit of these authors to have approached by theoretical means the study of flat slabs with capitals, which was subsequently developed by other authors in the analysis of flat slabs.

1 Introduction

1.2.2.2 Timoshenko's procedure. In the simplified calculation scheme adopted by Timoshenko in order to establish a solution (1.9), one of the conditions used for determining the integration constants was to cancel the shearing forces T_x and T_y (1.8) on the boundary of the interior panel under study. Under this particular assumption, Timoshenko [7] approaches the study of the infinitely long plate with the *free parallel edges* resting on two rows of equidistant columns which are assumed to be *point-like supports* (Fig. 1.6).

In order to represent the bending moments M_{y0} along the two free edges ($y = \pm b/2$), he suggests the general expression

$$M_{y0} = M_0 + \sum_{m=2,4,6}^{\infty} M_m \cos \frac{m\pi x}{a}, \qquad (1.13)$$

where a denotes the equal distance between the columns of a row. As the edge moments M_{y0} (1.13) are the only load applied on the slab in the first step of analysis, the deformed mid-surface

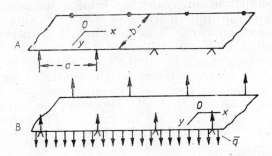

Figure 1.6 Schematic diagram for Timoshenko's procedure. The infinitely long plate with free parallel edges resting on two rows of assumedly point-like equidistant supports.
A. Uniformly distributed load q. B. Linear load $\bar{q}$ uniformly distributed along the two free edges.

of the slab can be expressed by means of a symmetric solution of the homogeneous differential equation (1.1a):

$$w(x, y) = A_0 + A_1 \left(y^2 - \frac{b^2}{4} \right) + \sum_{m=2,4,6,\ldots}^{\infty} \left(A_m \operatorname{ch} \frac{m\pi y}{a} + B_m \frac{m\pi y}{a} \operatorname{sh} \frac{m\pi y}{a} \right) \cos \frac{m\pi x}{a}. \qquad (1.14)$$

1.2 State of the art. Main research trends

The integration constants are determined from the boundary conditions:

$$(M_y)_{y=\pm \frac{b}{2}} = M_{y0}; \quad (R_y)_{y=\pm \frac{b}{2}} = 0. \tag{1.15}$$

The *only continuity condition* at the slab-column nodes is of type (1.7) and can be used to determine A_0. Hence,

$$A_1 = -\frac{M_0}{2D}; \tag{1.16a}$$

$$A_m = -\frac{a^2 M_m}{\pi^2 m^2 D} \cdot \frac{(1+\mu)\,\text{sh}\,\alpha_m - (1-\mu)\,\alpha_m \text{ch}\,\alpha_m}{(3+\mu)(1-\mu)\,\text{sh}\,\alpha_m\,\text{ch}\,\alpha_m - \alpha_m(1-\mu)^2}; \tag{1.16b}$$

$$B_m = -\frac{a^2 M_m}{\pi^2 m^2 D} \cdot \frac{\text{sh}\,\alpha_m}{(3+\mu)\,\text{sh}\,\alpha_m\,\text{ch}\,\alpha_m - \alpha_m(1-\mu)}, \tag{1.16c}$$

where α_m is given by (1.9a).

The constants M_0 and M_m are determined in the second step of analysis, taking into account the load function $p(x, y) \neq 0$. For the sake of illustration, Timoshenko outlines the solution of the present problem only in the case of the transverse load q uniformly distributed throughout the floor under study (Fig. 1.6A).

For this type of load, the slab is assumed to behave like a sequence of interior panels pertaining to the flat slab with capital previously examined (Fig. 1.4). As the elastic surface of the slab pertaining to one of the panels is modelled by means of (1.9), it becomes possible to express in an analytical manner the bending moments M_y generated along the edges ($y = \pm b/2$) of this panel under the load q. The constants M_0 and M_m in (1.16), which correspond to this type of load, are obtained by equating term by term the expression $(M_y)_{y=\pm \frac{b}{2}}$ and M_{y0} (1.13) but with a change of sign:

$$M_0 = \frac{qb^2}{12}; \tag{1.17a}$$

$$M_m = \frac{qab}{2\pi m}(-1)^{m/2}\left[\frac{1+\mu}{\text{th}\,\alpha_m} - \frac{\alpha_m(1-\mu)}{\text{sh}^2\alpha_m}\right]. \tag{1.17b}$$

The final solution $w(x, y)$ for the particular case under study (Fig. 1.6A) can be obtained by superposition of effects:

$$w(x, y) = w_1 + w_2, \tag{1.18}$$

1 *Introduction*

where w_1 is given by (1.9) and w_2 represents (1.14) where we introduced the constants A_1, A_m, B_m, M_0 and M_m, given by (1.16) and (1.17). Although Timoshenko did not give the final expressions, he made a brief exposition of the manner in which the solution $w(x, y)$ can be obtained in the case of the same flat slab with capitals, assuming that the transverse load reduces to a linear load of type $\bar{q}$ uniformly distributed along the two free parallel edges (Fig. 1.6B).

1.2.2.3 *Lewe's procedure*. This procedure [27], [28] belongs to the category of analysis procedures for flat slabs with capitals which rely on the Theory of Plates and use the condition of zero transverse displacement in the column axis, of type (1.7), as the *only continuity equation*. Both Nádai-Timoshenko (Sub-sec. 1.2.2.1) and Lewe consider only the case of a *central panel* that is sufficiently far away from the edges of the floor. However, unlike Timoshenko who assumes point-like supports, Lewe introduces in the analysis the *sizes of the column cross-section* (*ABab*). The connecting forces are considered to be continuously distributed throughout the cross-sectional area, according to the rule $r(x, y)$ (1.20). The columns are identical and spaced at an equal distance, and the number of spans is infinite in both directions of the plane (Fig. 1.7).

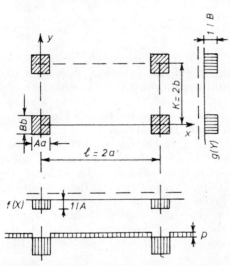

Figure 1.7 Schematic diagram for Lewe's procedure.

Assuming that the load q is uniformly distributed throughout the surface of the floor and superposing the effects of column reactions, Lewe expresses the load function $p(x, y)$, occurring in the Lagrange equation (1.1) in the form

$$p(x, y) = q - r(x, y), \qquad (1.19)$$

where

$$r(x, y) = q \cdot f\left(\frac{\pi x}{a}\right) \cdot g\left(\frac{\pi y}{b}\right). \qquad (1.20)$$

1.2 State of the art. Main research trends

Developing the functions $f\left(\dfrac{\pi x}{a}\right)$ and $g\left(\dfrac{\pi y}{b}\right)$ into Fourier series, he obtains the expression for the deflected mid-surface of the plate $w(x, y)$ for the interior panel ($lk = 4ab$) under consideration (Fig. 1.7):

$$w(x, y) = \frac{1}{D}\left\{C - \frac{2qa^4}{\pi^5}\left[\frac{1}{A}\Sigma\frac{1}{m^5}\sin mA\pi\cos\frac{m\pi x}{a} + \right.\right.$$
$$\left.+\frac{b^4}{Ba^4}\Sigma\frac{1}{n^5}\sin nB\pi\cos\frac{n\pi y}{b} + \frac{2}{AB\pi}\Sigma\Sigma\frac{b^4}{m\,n\,(b^2m^2+a^2n^2)^2}\times\right.$$
$$\left.\left.\times\sin mA\pi\sin nB\pi\cos\frac{m\pi x}{a}\cos\frac{n\pi y}{b}\right]\right\}, \tag{1.21}$$

where $Aa = 2c$; $Bb = 2d$.

Given the geometrical and mechanical symmetry of the schematic diagram, the continuity condition (1.7) is prescribed only once in the axis of the column ($x = 0$, $y = 0$) in order to determine the integration constant C.

The numerical results obtained hold for various types (uniformly distributed, in strips or in rectangles) of loads, but assuming that the cross-sections of the columns are of well-defined

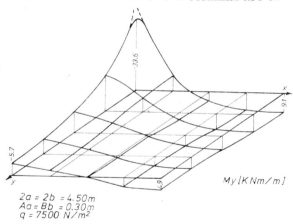

Figure 1.8 Example of analysis illustrating Lewe's procedure. Distribution of sectional bending moments M_y over panel quarter under uniformly distributed load q.

form and size. For the sake of illustration, Lewe considers a square panel ($2a = 2b$), a square cross-sectional area ($Aa = Bb$) of the columns, and a uniformly distributed load q. The image of the distribution of the moments M_y (Fig. 1.8), derived from the solution $w(x, y)$ given by (1.21), resembles the diagrams plotted in this book for the case of the square panel (Sub-sec.

1 *Introduction*

5.1.2.1 — Fig. 5.7). This similitude holds also in the case of the positive moments M_y in the symmetry axis ($x = a$) of the panel, where higher values appear along the column row ($y = 0$), than at the centre ($y = b$) of the panel.

The difference between these values appears in a predetermined way and is mainly due to the simplifying assumption adopted by Timoshenko and Lewe. According to this assumption, the rotations φ_x and φ_y of the slab are zero on the boundary of the panel under consideration. The main contradiction in Lewe's procedure is that the effect of the nonrigid slab-column connection discards the assumptions of nonpoint-like intermediate supports. Nor can Lewe's procedure be used in the analysis of edge panels.

★

In his fundamental treatise [9], Girkmann reverts to Lewe's solution (1.21) in a beneficial attempt [101] to transform the rather complicated expression — in simple and double trigonometric series — of the deflected mid-surface $w(x, y)$ into an expression in simple series of exponential functions (Fig. 1.9):

$$w(x, y) = \frac{qb^4}{384D} \left(1 - \frac{4y^2}{b^2}\right) \left[\left(1 - \frac{4y^2}{b^2}\right) - \frac{8d^2}{b^2}\right] -$$

$$- \frac{qa^5 b}{64\pi^5 Dcd} \sum_m \frac{(-1)^m \sin \alpha_m c \cos \alpha_m x}{m^5 \operatorname{sh} \frac{\alpha_m b}{2}} \left\{\left[\left(2 + \frac{\alpha_m b}{2} \operatorname{cth} \frac{\alpha_m b}{2}\right) \times \right.\right.$$

$$\left.\left. \times \operatorname{sh} \alpha_m d - \alpha_m d \operatorname{ch} \alpha_m d \right] \operatorname{ch} \alpha_m y - \alpha_m y \operatorname{sh} \alpha_m d \operatorname{sh} \alpha_m y \right\}, \quad (1.22)$$

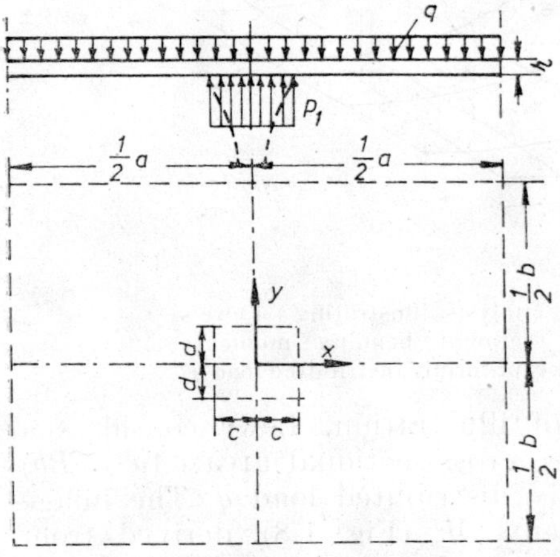

where $\alpha_m = 2m\pi/a$.

Owing to this transformation, the time required for work on computers is substantially shortened (see Sec. 3.4).

1.2.2.4 *Grein-Girkmann's procedure.* Because of the use of a simple static diagram, where the interior panel is sufficiently far away from the outer boundary of the floor, none of the classical procedures save for Timoshenko's (Sub-sec. 1.2.2.2) permits the evaluation of the displacement and stress distribution

Figure 1.9 Schematic diagram for Lewe-Girkmann's procedure.

1.2 State of the art. Main research trends

in the edge panels of the floor. In order to bridge this gap, Girkmann extends the study of the floor with an indefinite number of panels in both directions of the plane to the case of a floor with only one infinite size, *simply supported on the longitudinal edges* and resting on one or several rows of intermediate supports parallel to these edges (Fig. 1.10). The equidistant columns are *point-like intermediate supports* and the effect of their stiffness does not occur in calculations. The present problem was first approached by Grein [33] who, assuming that the load function $p(x, y)$ in Eqn. (1.1) is constant or can vary only in the direction of the Ox axis, obtains also expanded solutions into simple series of hyperbolic functions.

First, Girkmann [9] considers the diagram of a floor with only one longitudinal row of columns, assuming that the load q is uniformly distributed on one of the strips framed by this column row (Fig. 1.10). Two distinct groups of solutions $w(x, y)$ are obtained as follows:
— for the loaded panel $u \times b$

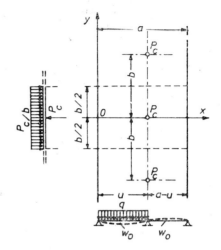

Figure 1.10 Schematic diagram for Grein-Girkmann's procedure. Infinitely long plate with simply supported parallel edges and only one intermediate row of equidistant columns. Load q on one of the two spans.

$$w(x, y) = w_0 + w_1, \quad (0 \leqslant x \leqslant u), \tag{1.23a}$$

and
— for the unloaded panel $(a - u) \times b$

$$w'(x, y) = w'_0 + w'_1 \quad (u \leqslant x \leqslant a). \tag{1.23b}$$

The analytical calculations develop in two steps. Under the partial load q, the row of intermediate columns is first assumed to be the continuous linear support of the floor slab. Assuming that the slab behaves like a continuous beam along two spans, and taking $EI = D$, the corresponding solutions w_0 and w'_0 are obtained in the form

$$w_0 = \sum_n w_n \sin \alpha_n x; \quad (0 \leqslant x \leqslant u) \tag{1.24a}$$

$$w'_0 = \sum_n w_n \sin \alpha_n x, \quad (u \leqslant x \leqslant a) \tag{1.24b}$$

where $\alpha_n = n\pi/a$.

1 Introduction

Under the same continuous-beam assumption, they determine the linear reaction P_c/b uniformly distributed along the column row. The interacting forces P_c concentrated in the column axis, all of equal value, result from the equilibrium condition over the width b of the two adjacent panels.

In the second step of the analysis, Girkmann establishes the solutions w_1 and w_1', which introduce in Eqn. (1.23) the corrections found through the consideration of the final static diagram, where the columns are assumed to be equidistant point-like supports. Superposing the effect of the linear reactions P_c/b — which is here assumed to be the load applied on the slab — over the effect of concentrated reactions P_c, he obtained the following expression of the linear load $\bar{p}(y)$ of the slab along the column row:

$$\bar{p}(y) = \frac{P_c}{b} - P_c(y) = -\frac{2P_c}{b} \sum_n \cos \beta_n y, \quad (n = 1, 2, 3, \ldots)$$
(1.25)

where $\beta_n = 2n\pi/b$.

The middle surface of the infinitely long plate, deformed under load $\bar{p}(y)$ at $x = u$, is defined by the equation

$$w_1 = \frac{P_c b^2}{8\pi^3 D} \sum_n \frac{\cos \beta_n y}{n^3 \operatorname{sh} \beta_n a} [\beta_n x \operatorname{sh} \beta_n (a-u) \operatorname{ch} \beta_n x + \beta_n (a-u) \times$$
$$\times \operatorname{ch} \beta_n (a-u) \operatorname{sh} \beta_n x - (1 + \beta_n a \operatorname{cth} \beta_n a) \operatorname{sh} \beta_n (a-u) \operatorname{sh} \beta_n x],$$
(1.26)

$(n = 1, 2, 3 \ldots)$

which is valid in the range $(0 \leqslant x \leqslant u)$.

Solution w_1', corresponding to the range $(u \leqslant x \leqslant a)$, is derived from w_1 given by (1.26) by replacing x by u. Because of the symmetry with respect to the Ox axis, solutions w_1 and w_1' require four groups of integration constants: C_n, D_n and C_n' and D_n', respectively. It is interesting to note that the determination of these constants requires the use of both the boundary conditions of the plate

$$x = 0 \begin{cases} w_1 = 0; \\ M_{1x} = 0; \end{cases} \quad x = a \begin{cases} w_1' = 0; \\ M_{1x}' = 0; \end{cases}$$

$$y = \pm \frac{b}{2} \begin{cases} \varphi_{1y} = 0; \; \varphi_{1y}' = 0. \\ T_{1y} = 0; \; T_{1y}' = 0. \end{cases} \tag{1.27a}$$

and the continuity conditions along the intermediate supports

$$x = u \begin{cases} w_1 = w_1'; \; M_{1x} = M_{1x}' \\ \varphi_{1x} = \varphi_{1x}'; \; \bar{q}_x - \bar{q}_x' = \bar{p}(y). \end{cases} \tag{1.27b}$$

1.2 State of the art. Main research trends

The use of Grein-Girkmann's procedure in the case of flat slabs with capitals with two or more rows of equidistant columns is only sketched [9]:
— solutions of the type w_0, given by Eqn. (1.24), are obtained assuming that the slab behaves like a continuous beam over three or more spans and the column rows are continuous linear supports;
— solutions of the type w_1, given by Eqn. (1.26), are established in a similar manner to those obtained under the assumption of an intermediate column row, the linear loads of type $\bar{p}(y)$ given by (1.25) being expressed in a similar manner for each row.

The schematic diagram of a flat slab with capitals with two equidistant rows of intermediate supports is presented by Girkmann in the form of a numerical example (Fig. 1.11). The distribution of the bending moments M_{0x} — calculated with the aid of solution w_0 alone — is identical along the column row

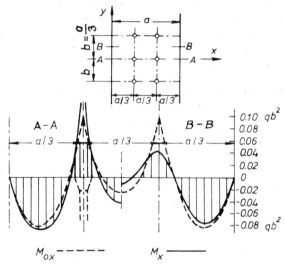

Figure 1.11 Example of analysis illustrating Grein-Girkmann's procedure. Infinitely long plate with two intermediate rows of equidistant columns. Distribution of sectional bending moments M_{0x} (solution w_0) and M_x (solution $w = w_0 + w_1$).

($y = 0$) and in the symmetry axis of the panels ($y = b/2$). Introducing solution w_1, i.e. considering the over-all solution (1.23), the corrected values of the moments M_x increase substantially along the column row ($A - A$) and are significantly smaller in the axis ($B - B$) of the panels.

1 Introduction

It is of interest to compare the values of M_x with those obtained by Timoshenko (Table 1.2) who adopts the same distribution for columns ($a = b$) and neglects in calculation the effect of the column cross-section ($k = 2c/a = 0$). It becomes clear that the edge position of the panels bears on the magnitude of the bending moments in the field and along the column row, and also that the presence of edge panels adjacent to the central panel substantially reduce the values of M_x in the field of this panel (Table 1.3).

Referred to the Nádai-Timoshenko and Lewe procedures, Grein-Girkmann's procedure is more comprehensive as it approaches the *evaluation of stresses in the edge panels*. The approximations of this procedure derive from disregarding of the effect of column cross-section ($c = d = 0$) and the *assimilation of the intermediate supports with hinges*. Like all the classical authors (Sub-secs. 1.2.2.1.−1.2.2.3.), Grein-Girkmann do not treat the slab as a structural element, i.e. they neglect the effect of the column-slab interaction. These approximations are all the more important as all the classical procedures were first established for the analysis of flat slabs with capitals, for which the values of the parameters c and d are relatively high.

1.2.2.5 *Girkmann's procedure.* Using a similar approach (Sub-sec. 1.2.2.4) but resorting to Nádai's solution [1] to the Lagrange equaiton (1.1), Girkmann [9] makes an essential contribution to the study of plates as structural members integrally built with the columns (Fig. 1.12). The *rigid connection* at the slab-column nodes is shown here by writing the condition of *zero relative rotation* for these nodes. Unlike Andersson (Sub-sec. 1.2.3.3), Girkmann has failed to give an explicit formulation of the relative stiffness factor between slab and column, like that of type η given by (1.48) or (5.23). Moreover, he considers that both the interacting forces

Table

Flat slabs with capitals under

Comparative values of M_x/qb^2, after

Square panels

Authors	$y = 0$ (A − A)		
	($x = b/2$) Field 1	($x = b = a/3$) Column axis	($x = a/2$) Field 2
Grein-Girkmann	0.0806	−∞	0.0400
Timoshenko	(0.0512)	−∞	0.0512

1.2 State of the art. Main research trends

P_c and the reaction couples M_c are concentrated. However, by avoiding the artifice of calculation and the empirical formulae used by the Swedish author in introducing into the analysis the parameters defining the column stiffness, Girkmann found a more elegant solution which relies only on the Theory of Plates.

Girkmann's procedure introduces three other simplifying assumptions which specialize the solution and, hence, restrict its range of applicability (Fig. 1.12):

a. the slab is indefinitely long in both directions of the axis Oy and the parallel edges are assumed to be simply supported. The interior supports consist of *only one longitudinal row of equidistant columns* which divides the floor into two strips of unequal width $(u, a - u)$.

b. the load function $p(x)$ can vary only in the transverse direction or is constant of type q. The load is *unsymmetrical*, i.e. it acts only on the width u of one of the strips.

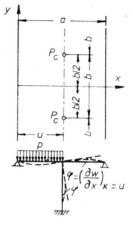

Figure 1.12 Schematic diagram for Girkmann's procedure. Continuity condition set for rotations.

c. the intermediate supports of the slab are *concentrated* at a point. Unlike the schematic diagram considered in the previous procedure (Fig. 1.10), here the supports are not treated as hinges.

The states of stress and deformation are assumed to be identical in all the panels situated on either side of the column row. Hence, all the reactions P_c are of equal value. Thus, here too the problem reduces to the investigation of only one pair of adjacent panels $(u, a - u) \times b$.

1.3
the uniformly distributed load q
Grein-Girkmann and Timoshenko

$b = a/3$

	$y = b/2$ (B − B)			
$(x = b/2)$ Field 1	$(x = b = a/3)$ Column axis	$(x = a/2)$ Field 2	Remarks
0.0750	−0.0400	0.0090	$\mu = 0$
(0.0331)	−0.0185	0.0331	$\mu = 0.2$

1 Introduction

In the first step of the statical analysis, Girkmann superposes the effect of the effective load q over the counter-effect of the concentrated reactions P_c acting on the slab and obtains the following solution to the fundamental equation (1.1)

$$w(x, y) = w_q + w_P \quad (0 \leqslant x \leqslant a). \tag{1.28}$$

The expression of the particular integral w_q, given by Eqn. (1.24a) can be obtained by considering only the partially distributed load q. The solution w_P was established by Nádai [1] and represents the mid-surface of the plate, deflected under the set of equidistant forces P_c acting in the vertical plane $x = u$

$$w_P = \frac{P_c a^2}{2\pi^3 D} \sum_n \frac{\sin \alpha_n u \sin \alpha_n x}{n^3 \operatorname{sh} \frac{\alpha_n b}{2}} \left[\left(1 + \frac{\alpha_n b}{2} \operatorname{cth} \frac{\alpha_n b}{2}\right) \operatorname{ch} \alpha_n y - \alpha_n y \operatorname{sh} \alpha_n y \right]. \tag{1.29}$$

The interacting forces P_c are obtained from a continuity equation of type (1.7). This means that we must prescribe the condition

$$(w_q + w_P)_{\{x=u,\ y=\pm \frac{b}{2}\}} = 0;\ (1.30a) \quad \text{or} \quad (w_q + w_P)_{\{x=u,\ y=\pm \frac{b}{2}\}} = \Delta_c \quad (1.30b)$$

in the column axis. Δ_c above represents the linear elastic displacement of the column under the load P_c. This displacement is generally negligible as compared to the elastic transverse displacement of the slab. From solution $w(x, y)$, given by (1.28), thus determined, the states of stress and deformation are fully defined in the slab at the end of the first step.

In the second step of the analysis, Girkmann considers the rotation φ_c of the column end under the transverse unsymmetrical load q. As was shown above, unlike Girkmann, Nádai, Timoshenko and Lewe assume that the slope of the tangent is zero on the entire boundary of the panel under study. Girkmann takes into account the *bending deformation of columns* due to some reaction-couples M_c at the slab-column nodes. Hence, the idea of rigid connections between these structural elements is implicit.

Dispensing with the details of analysis, Girkmann shows that the magnitude of the moments M_c derives from the condition of equal slab and column rotations, prescribed at the concentrated supports ($x = u$, $y = \pm b/2$) (Fig. 1.12):

$$\varphi_c = (\varphi_x)_{\{x=u,\ y=\pm \frac{b}{2}\}}. \tag{1.31}$$

1.2 State of the art. Main research trends

It may be derived that the angular displacement φ_x of the slab can be obtained only by means of Eqn. (1.28) of the deflected mid-surface determined in the first step.

Next, Girkmann approaches the effect produced on the slab by the equal, equidistant concentrated couples M_c, the moment vector of which is parallel to the line of supports (the Oy axis). The equation of the mid-surface of the infinitely long plate deflected under the concentrated couple $\vec{M_c} \| Oy$ at $x = u$, which was earlier established by him (see Sub-sec. 1.2.4.1 — formula 1.58), is of the type

$$w_M(x, y) = w(M_c, a, u, \mu, x, y). \qquad (1.32)$$

According to Girkmann, the additional stresses generated in the slab by the reaction-couples M_c would be determined from the solution $w_M(x, y)$ given by Eqn. (1.32). However, Girkmann does not give the numerical results that would illustrate the use of his procedure.

From the above-presented classical procedures of analysis for flat slabs with capitals, Girkmann's second procedure is unique in that the effect of the rigid connection between the slabs and the columns is effectively considered. Referred to other modern procedures of analysis, it is of interest to note the limits and approximations of the analysis procedure suggested by Girkmann:

a. The schematic diagram is highly theoretical in character and the generalized interacting forces of type P_c and M_c are assumed to be concentrated. The use of this procedure is restrictive in practice.

b. Nádai's solution w_P, given by Eqn. (1.29) and used by Girkmann in (1.28), (1.30) and (1.31), was established assuming a set of equidistant concentrated forces P_c associated in a simple row (Fig. 1.12). However, Girkmann's solution w_M, given by Eqn. (1.32), is established for the slab subjected to only one concentrated couple M_c in $x = u$. The study of the influence of sets of couples on the states of stress and deformation in the slab, which is assumed to be a structural element (Sub-secs. 4.1.1 and 4.1.3) leads to the conclusion that superposition of the effects of adjacent couples can be neglected only in the case when the columns are spaced at very large equal distances ($b \geqslant 2a$).

c. Finally, an important approximation of Girkmann's procedure derives from the manner in which the magnitude of the statically indeterminate unknowns P_c and M_c is determined. The correct analysis for a given diagram means the superposition of all effects — the given load q and the unknowns P_c and M_c — so

1 Introduction

that the compatibility conditions (1.30) and (1.31) represent a system of condition equations (see Sec. 5.1)

$$w_q + w_P + w_M = 0 ; \tag{1.33a}$$

$$\varphi_{qx} + \varphi_{Px} + \varphi_{Mx} + \varphi_c = 0. \tag{1.33b}$$

Indeed, the transverse displacement w_M given by Eqn. (1.32) is not zero, except for the particular case $u = a/2$, i.e. when the slab is antisymmetrically loaded with the couple M_c (see Sub-secs. 3.2.1 and 4.1.1). In writting the condition of zero relative rotation (1.33b) for $(x = u, y = \pm b/2)$, the slab rotations φ_{Px} and φ_{Mx}, due to the unknowns P_c and M_c, respectively, should not be neglected. Hence, Girkmann fails to deal with a proper structural analysis, the effect of the generalized interacting forces, i.e. reaction couples and reaction forces, being only partially approached.

1.2.3 Modern approaches to flat slabs without capitals

Beginning in the 1960's, the theoretical approaches in the field of slab structures are devoted entirely to flat slabs without capitals. From the numerous contributions published of late, we shall consider only those which can be taken as rigorous within the limits of the basic assumptions of the Theory of Plates (Sub-sec. 2.1.1). Like all the classical authors, except for Girkmann, most of the modern authors — Duddeck [38], Rabe [39], [47], [51], Bretthauer, Seiler [41], [45], Krebs, Baader [44], Franz [47], [94], Pfaffinger, Thürlimann [48], Fuchs [114], Morrison [89] and Appleton [117] — use the equation of transverse displacement of the type (1.7) as the only compatibility condition for the slab-column node. Since the only interacting forces are the reaction forces in the column axes — which are assumed to be hinged supports — the slab is treated as an independent structural member.

A significant development in this field is due to Girkmann (Sub-sec. 1.2.2.5) who inserted in the analysis the effect of concentrated couple reactions. Concentrated interacting couples are also used by Andersson [42] and by Bretthauer and Nötzold in their latest works [49]. However, a structural analysis in its own right would be possible only by the mathematical treatment of the surface couple as a load applied on the slab (Sub-sec. 1.2.4.3).

1.2.3.1 *Rabe's procedure*. The schematic diagram of the flat slab studied by Rabe [51] reduces to the finite boundary plate $(a \times b)$ with nine panels framed by four identical interior columns. The load q is assumed to be uniformly distributed throughout the

1.2 State of the art. Main research trends

floor (Fig. 1.13). The effect of the rigid slab-column connection is neglected in calculations, although the contact area is different from zero.

A similar diagram was treated by Marcus [2] with the finite difference method. However, he approached only the case of the square slab ($b = a$) simply supported on its entire boundary. Marcus' study was extended by Timoshenko [7] and later on by Morrison [89]. What distinguishes this particular analysis from the complex slab case considered by Rabe are the following boundary conditions:

a. The slab of the floor is assumed to be supported on its entire boundary by *elastic edge beams*. In the analysis, we must determine the value of the parameters (Fig. 1.13)

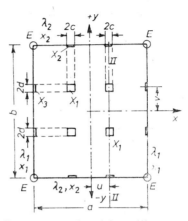

Figure 1.13 Flat slabs without capitals. Schematic diagram for Rabe's procedure. Rectangular plate with elastic edge-beams supported at the corners of the plate.

$$\lambda_1 = \frac{E_1 I_1}{aD}; \quad \lambda_2 = \frac{E_2 I_2}{bD} \quad (1.34)$$

which define the relative slab-beam stiffness against bending. In order to simplify the analysis, the twisting rigidity $G_1 I_{1t}$ and $G_2 I_{2t}$ respectively, of the edge beams, is assumed to be zero in the final step of analytical calculations.

b. The edge beams are assumed to be point-like supported at the four corners of the slab. Additionally, two linear intermediate supports are inserted on each side of the floor, at the junction with the row of columns ($\pm u$, $\pm b/2$, $\pm a/2$, $\pm v$).

c. As a simplifying assumption on the boundary conditions, the zero moment points of the four edge beams are assumed to coincide with the zero moment lines of the plate.

Whereas prior approaches were confined to the study of the interior panel (see Sub-secs. 1.2.2.1 and 1.2.2.3) or to the case of the edge panel pertaining to an indefinitely long floor (see Sub-secs. 1.2.2.2, 1.2.2.4 and 1.2.2.5), Rabe's attempts deal with the stress distribution in the *corner panel* of the flat slab under consideration. Owing to the two-fold symmetry of his schematic diagram, the analysis is reduced to the behaviour of a quarter of the slab ($0 \leqslant x \leqslant +a/2$, $0 \leqslant y \leqslant +b/2$).

The statically indeterminate unknowns are assumed to be the only reaction forces (Fig. 1.13): the set of unknowns X_1 in

1 Introduction

the axes of the interior columns and the sets of unknowns X_2 and X_3, respectively, in the axes of the linear edge supports. In order to avoid infinite values of the bending moments, which arise in the case of concentrated discrete supports, Rabe assumed that the reactions X_1 are uniformly distributed on the slab-column area and that the linear edge reactions X_2 and X_3 are uniformly distributed on the segments $2c$ and $2d$, respectively, of the edge beams. Taken as unit loads, they are inserted in the analysis in the form

$$X_1 = 1 = 4\bar{p}cd; \tag{1.35a}$$

$$X_2 = 1 = 2\bar{\bar{p}}c;\ X_3 = 1 = 2\bar{\bar{p}}d. \tag{1.35b}$$

In determining these three sets of force unknowns, Rabe uses three compatibility conditions for transverse displacements of type (1.7) at the discrete supports.

The global solution $w(x, y)$ is obtained by superposing the effects q, X_1, X_2 and X_3, i.e. by summation of the three distinct solutions

$$w(x, y) = w(q, x, y) + w(X_1, x, y) + w(X_2, X_3, x, y). \tag{1.36}$$

The basic system employed by Rabe for establishing these three solutions (i.e. the sub-ensemble of nine solutions for load $X_1 = 1$) is represented by the plate with only one panel $(a \times b)$, for any boundary conditions and assuming that the plate is point supported at the four corners. The general solution for this basic system was given by Fuchs [114] in the form

$$w(x, y) = \Phi(x, y) + \frac{a^4}{D}\sum_m \left(A_m \text{ch}\frac{m\pi y}{a} + B_m \frac{m\pi y}{a}\text{sh}\frac{m\pi y}{a}\right) \times$$
$$\times \cos\frac{m\pi x}{a} + \frac{b^4}{D}\sum_m \left(C_m \text{ch}\frac{m\pi x}{b} + D_m \frac{m\pi x}{b}\text{sh}\frac{m\pi x}{b}\right)\cos\frac{m\pi y}{b}.$$
$$(m = 1, 3, 5, \ldots) \tag{1.37}$$

$\Phi(x,y)$ above represents the particular solution to the inhomogeneous equation (1.1), which is determined function of the type of load. For load q, Fuchs gave the expression*

$$\Phi(q, x, y) = \frac{2q}{aD}\sum_m (-1)^{(m-1)/2} \cdot \frac{m^5\pi^5}{a^5}\cos\frac{m\pi x}{a} +$$
$$+ \frac{2q}{bD}\sum_m (-1)^{(m-1)/2} \cdot \frac{m^5\pi^5}{b^5}\cos\frac{m\pi y}{b}. \quad (m = 1, 3, 5, \ldots)$$
$$\tag{1.37a}$$

---

* The expression for the particular solution $\Phi(q, x, y)$ given by Rabe is both highly divergent and inhomogeneous. In order to satisfy convergence and homogeneity requirements, the factors of the trigonometric functions should be reversed: $a^5/m^5\pi^5$ and $b^5/m^5\pi^5$, respectively.

1.2 State of the art. Main research trends

In the case of the sets of loads $X_2 = 1$ and $X_3 = 1$ (1.35 b), $\Phi(x, y)$ is zero because these linear loads are applied only on the boundary of the plate.

The last two terms in Eqn. (1.37) represent general solutions of the homogeneous fundamental equation (1.1a). Because of the two-fold symmetry of the schematic diagram, the two known types of boundary conditions for the elastic edge beam (see Timoshenko [7]) are identical on the opposite edges

$$-(R_x)_{x=\pm\frac{a}{2}} = \frac{b}{a} E_1 I_1 \left(\frac{\partial^4 w}{\partial y^4}\right)_{x=\pm\frac{a}{2}};$$

$$-(R_y)_{y=\pm\frac{b}{2}} = \frac{a}{b} E_2 I_2 \left(\frac{\partial^4 w}{\partial x^4}\right)_{y=\pm\frac{b}{2}}; \qquad (1.38a)$$

$$-(M_x)_{x=\pm\frac{a}{2}} = \frac{b}{a} G_1 I_{1t} \left(\frac{\partial^3 w}{\partial x \partial y^2}\right)_{x=\pm\frac{a}{2}};$$

$$-(M_y)_{y=\pm\frac{b}{2}} = \frac{a}{b} G_2 I_{2t} \left(\frac{\partial^3 w}{\partial x^2 \partial y}\right)_{y=\pm\frac{b}{2}}. \qquad (1.38b)$$

Determination of the expressions A_m, B_m, C_m and D_m, which represent integration constants in the general solution (1.37), requires that the four boundary conditions (1.38) are satisfied. Rabe's procedure relies on lengthy mathematical developments and his formulations are *inconveniently complex**.

Moreover, whereas for the loads q and X_2, X_3, solution $w(x, y)$ given by (1.37), was used as such as it holds for the entire domain $(a \times b)$, for the set of loads $X_1 = 1$, given by (1.35a), the domain $(a/2 \times b/2)$ had to be divided into nine distinct subdomains (Fig. 1.14).

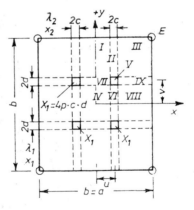

Figure 1.14 Rabe's procedure. Subdomains for which solutions $w(X_1, x, y)$ are valid.

---

* The system of four linear transcendental equations in A_m, B_m, C_m and D_m extends over no less than a review page [51]. The expression of only one "load term" $L_1(X_1 = 1)$ extends on three quarters of a page. Thirty-three substitution expressions are employed, etc.

1 Introduction

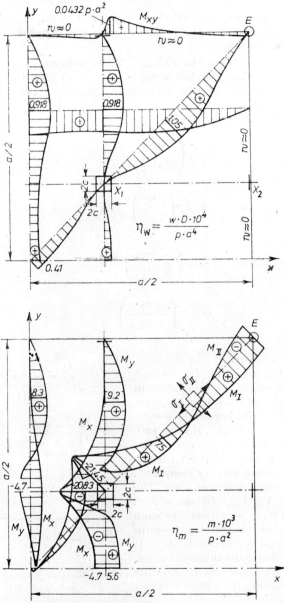

Figure 1.15 Numerical example illustrating Rabe's procedure. Distribution of transverse displacements w and of sectional moments M_x, M_y and M_{xy} ($\lambda_1 = \lambda_2 = EI_b/aD = 1000 \approx \infty$; $I_{1t} = I_{2t} = 0$; $c/a = 0.01667$; $a/b = 1.0$; $\mu = 0$).

1.2 State of the art. Main research trends

Hence, solution $w(X_1, x, y)$, given by (1.36), represents an ensemble of nine distinct solutions, each of which is valid for the respective sub-domain (I, II, ..., IX). So, the use of Rabe-Fuchs' procedure becomes prohibitive in practice, due to possible computation errors.

For illustration, Rabe considers the case of a square ($b = a$) flat slab with columns of square cross-section ($c = d$), assuming that the relative slab-edge beam stiffnes is constant on the entire boundary of the floor. Assuming that the number $m = 23$ of the terms in the series satisfies the required precision, Rabe presents the numerical results (w, M_x, M_y, M_{xy}) under two alternatives: edge-beams of zero stiffness ($\lambda_1 = \lambda_2 = EI_b/a\,D = 0$) and edge-beams of infinite stiffness (Fig. 1.15). As the twisting rigidity of the edge-beams is neglected even in the analytical calculations, the second alternative is equivalent to the computation diagram examined by Marcus [2]. Notwithstanding Rabe's contention that his values differ by only 5 percent from the results obtained by Marcus with the finite difference method, the author's comparison shows that the difference ranges from 12 to 34 per cent (Table 1.4). The order of magnitude of this difference in percentage cannot be explained by different values of Poisson's ratio and distinct relative sizes of the slab-column area with the two authors (Table 1.4).

Despite its inherent limitations, particularly those deriving from *the disregard of the effect of rigid slab-column connection*, Rabe's work makes a noteworthy contribution to the study of flat slabs. He is the first among the modern authors to approach a schematic diagram, which through its boundary conditions reflects the physical reality with the sole means of the Theory of Plates.

1.2.3.2. *Pfaffinger-Thürlimann's procedure*. The latest procedure of analysis for flat slabs without capitals established by modern authors, where the effect of the slab-column interaction is neglected, is due to Pfaffinger and Thürlimann [48]. Because of the characteristics of the schematic diagram employed (Fig. 1.16), the range of applicability of this procedure is substantially wider than that pertaining to Rabe's procedure. Moreover, by the simultaneous use of the known solutions in the Theory of Plates and the Theory of Hyperstatic Structures, Pfaffinger-Thürlimann's procedure gains in simplicity and elegance:

a. The flat slab is assimilated with the rectangular plate ($a \times b$) with *simply supported edges* $x = 0$ and $x = a$. The boundary conditions for the other parallel edges $y = 0$ and $y = b$ can

1 Introduction

Table
Flat slabs under the uniformly
Comparative values of the bending moments
$b = a;\ c = d;\ \lambda_1 = \lambda_2 = \dfrac{EI_b}{aD}$

Authors \ Calculation point	$10^3 M_x/qa^2$			
	$x=0$ $y=0$	$x=a/6$ $y=0$	$x=a/6$ $y=a/6$	$x=a/6$ $y=a/3$
Rabe	2.00	−4.70	−20.83	−0.50
Marcus	2.33	−4.44	−15.56	−0.44
Δ %	−14.2	+5.9	+33.9	+13.6

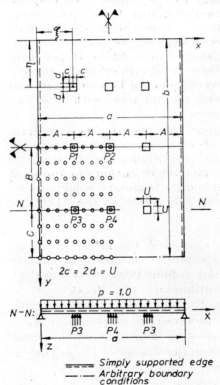

Figure 1.16 Schematic diagram for Pfaffinger-Thürlimann's procedure.
──── Simply supported edge
─·─ Arbitrary boundary conditions

be considered, independently of one another, to be either *clamped, simply supported* or *free edge*[*];

b. The position (ξ, η) of the columns is *arbitrary* in the slab plane. The columns can be of rectangular cross-section ($2c \times 2d$) and can be treated as linear ($2c, 0$) or ($0, 2d$) or as concentrated ($c = 0, d = 0$) supports. The force reactions are assumed to be uniformly distributed on the slab-column line or area, or are concentrated;

c. The uniformly distributed transverse load of type q can be distributed throughout the surface of the floor on strips or in chess board fashion. Discrete loads similar to force reactions — i.e. uniformly distributed or concentrated loads — can also be incorporated in the calculations. However, this pro-

---

[*] In a subsequent program of "Static Analysis of Beamless Floors — SAUD2", the elastic edge beam was also introduced as a boundary condition [53].

1.4
distributed load q
M_x and M_y, after Rabe and Marcus

$= 1000 \approx \infty$; $G_1 I_{1t} = G_2 I_{2t} = 0$

$10^3 M_y/qa^2$				Remarks
$x = 0$ $y = 0$	$x = a/6$ $y = 0$	$x = a/6$ $y = a/6$	$x = a/6$ $y = a/3$	
2.00	5.60	−20.83	9.20	$\mu = 0$ $c/a = 0.01667$
2.33	4.22	−15.56	8.22	$\mu = 0.2$ $c/a = 0.04167$
−14.2	+32.7	+33.9	+11.9	—

cedure *does not allow consideration of the horizontal loads* acting in the slab plane.

d. The effect of the column stiffness, i.e. *the relative slab-column stiffness is not included* in the analysis *.

The elastic analysis of the flat slab under study is carried out in two steps. In the first step, the basic system is assumed to consist of an infinitely long plate subjected to only one locally distributed force $P = 4pcd$ at (ξ, η). The solution $w(x, y)$ to the fundamental equation (1.1) is established by superposition of effects:

$$w(x, y) = w_0 + w_h \qquad (1.39)$$

where w_0 denotes the particular solution given by Girkmann [9] for the basic system under this type of load (see Sub-sec. 2.2.1.1 — formulae 2.11—2.13). Solution w_0 satisfies the boundary conditions for $x = 0$ and $x = a$.

Solution w_h to the homogeneous fundamental equation (1.1a) is expressed as

$$w_h = \sum_n (A_n e^{\alpha y} + B_n y e^{\alpha y} + C_n e^{-\alpha y} + D_n y e^{-\alpha y}) \sin \alpha x, \quad \alpha = \frac{n\pi}{a} \qquad (1.40)$$

---

* The authors draw attention to the eventual errors (up to 20 per cent) stemming from the disregard of the stiff connection between the slab and the columns. This warning refers to the values tabulated in the Appendix, where the authors consider only symmetrical flat slabs with a uniform distribution of columns. However, the comparative studies carried out herein show that the order of magnitude of these errors can be higher (Sub-secs. 5.1.2.1 — Tables 5.6 and 5.7).

1 Introduction

and can be used to correct the particular solution w_0 such that the boundary conditions for $y = 0$ and $y = b$ are also satisfied. Hence, the integration constants A_n, B_n, C_n and D_n are derived from a system of $4n$ linear equations, the final solution assuming the form (Fig. 1.16):

$$w_i(x, y) = w(p, a, b, c, d, \xi, \eta, \mu, x, y). \tag{1.41}$$

No explicit formulation was given for solution (1.41).

The equation of the mid-surface of the plate deflected under only one locally distributed load of type $P_i = 4 p_i c_i d_i$, which is valid for a certain combination of boundary conditions at $y = 0$ and $y = b$, can be written [48], [53] as

$$w_i(x, y) = p_i w(1, a, b, c_i, d_i, \xi_i, \eta_i, \mu, x, y). \tag{1.41a}$$

In this formulation, the equation was used by Pfaffinger and Thürlimann in the second step of analysis. Taking now a set of given loads L_k and a set of force reactions $P_i(i = 1, 2, 3, \ldots m)$ corresponding to one of the schematic diagrams examined, the unknowns p_i are determined from the condition of zero total transverse displacement in the axis of each intermediate support:

$$\mathbf{Dp} + \mathbf{l}_k = \mathbf{0}, \tag{1.42}$$

where $\mathbf{D}$ represents the matrix of the transverse displacements d_{ij} of the slab in the axis of the columns i, induced under the unit load $p_j = 1$ applied at $j(j = 1,2, \ldots m)$; $\mathbf{p}$ denotes the vector containing the locally and uniformly distributed reactions p_i, considered as statically indeterminate unknowns and $\mathbf{l}_k$ is the column vector of the load matrix $\mathbf{L}$, including the transverse displacements l_{ik} of the slab in the axes of the intermediate supports i, due to the given set of loads L_k.

In [48], Pfaffinger and Thürlimann gave a great number of values for the static quantities p_i, M_x, M_y, M_{xy}, M_1 and M_2. These values were calculated by means of the above-presented procedure but using a simplified schematic diagram (Fig. 1.16), where the boundary conditions for $y = 0$ and $y = b$ are identical, the nine interior columns are identical and of square cross-section ($2c = 2d = U$) and the load $q = 1$ is assumed to be uniformly distributed throughout the surface of the floor. The value $\mu = 1/6$, which is adopted in the case of reinforced concrete structures, is uniquely chosen for Poisson's ratio. The accuracy of computations for all the static quantities is presumably satisfied by limiting the summation of series to the order $n = 60$ of the last summed term (30 effectively summed terms) (see Sub-sec. 3.4.1). For illustration, we present the vuales of the bending moments

1.2 State of the art. Main research trends

Table 1.5

Flat slabs under the uniformly distributed load q
Comparative values of the bending moments M_x and M_y, after Pfaffinger-Thürlimann and Marcus
$B/A = 1$; $C/B = 1$; $C/A = 1$; $\mu = 1/6$

Calculation point	M_x/qA^2				M_y/qA^2			
	Pfaffinger-Thürlimann		Marcus ($\mu = 1/5$)	$\Delta_{2/3}$ %	Pfaffinger-Thürlimann		Marcus ($\mu = 1/5$)	$\Delta_{6/7}$ %
	$U/A = 0.080$	$U/A = 0.240$	$U/A = 0.250$		$U/A = 0.080$	$U/A = 0.240$	$U/A = 0.250$	
	1	2	3	4	5	6	7	8
1	0.0268	0.0250	0.021	+19.0%	0.0268	0.0250	0.021	+19.0%
2	−0.0311	−0.0317	−0.040	−20.8%	−0.0303	−0.0317	−0.040	−20.8%
3	0.0681	0.0664	0.069	− 3.8%	0.0249	0.0239	0.025	− 4.4%
4	0.0552	0.0487	0.038	+28.2%				
5	−0.2617	−0.1460	−0.140	+ 4.3%	−0.2653	−0.1464	−0.140	+ 4.6%
6	0.0802	0.0757	0.074	+ 2.3%	−0.0032	−0.0044	−0.004	+10.0%

Note. The values M_x and M_y obtained by Pfaffinger-Thürlimann (columns 1, 2, 5 and 6) correspond to an equidistance $A = a/4$ between the columns, whereas those calculated by Marcus (columns 3 and 7) correspond to an equidistance $A = a/3$.

1 Introduction

M_x and M_y obtained from the schematic diagram which is more faithful to the diagram [2] investigated by Marcus (Fig. 1.17). Pfaffinger-Thürlimann's results are given for two extreme variants of the parameter U/A which defines the sizes of the slab-column

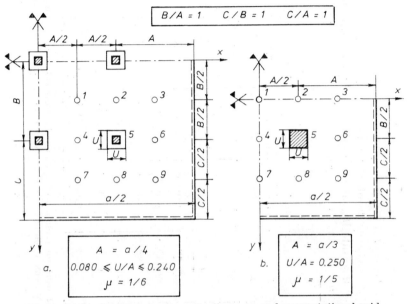

Fig. 1.17 Comparison of schematic diagrams and computational grid.
a. Pfaffinger-Thürlimann's procedure. b. Marcus' procedure.

contact area ($0.080 \leqslant U/A \leqslant 0.240$). The latter value differs moderately from that ($U/A = 0.250$) adopted by Marcus (Table 1.5)*.

The differences Δ in percentage highlights the influence of the interaction of the fourth row of panels on the values of stresses (see Fig. 1.17). This effect is stronger in the central panel (at points 1, 2, 4) and, in compliance with expectations, considerably smaller in the edge and corner panels (points 3, 5, 6, ... 9).

1.2.3.3 *Andersson's procedure*. Among the first modern authors to include in their studies compatibility conditions other than the continuity of transverse displacements of type (1.7) is Andersson [42]. He resumes Timoshenko's approach (Sub-sec. 1.2.2.2) of infinitely long plates with free parallel edges supported at a point by two rows of equidistant columns. Andersson makes

---

* Generally, the accuracy of the numerical results obtained by Marcus with the finite difference method is remarkable if we consider the rudimentary means of numerical calculation he employed.

1.2 State of the art. Main research trends

an interesting contribution by introducing in the analysis further *conditions of continuity for rotations* of type (1.31) (see Sub-sec. 1.2.2.5). Thus, both Girkmann and the Swedish author take into account the effect of the rigid connection between the slabs and the bars of the structure but they fail to give a comprehensive structural analysis.

To the static diagram analysed by Timoshenko (Fig. 1.18 A) Andersson adds another load. This is the *set of concentrated moments* M_c, which he called "clamping moments", acting on the slab in the column axis. Thus, the slab is simultaneously subjected to a given load q, uniformly distributed throughout the surface of the floor, and to two types of loads applied on the boundary: the edge moments M_{y0}, which are continuously distributed, and the concentrated moments M_c (Figs. 1.18B and C, respectively).

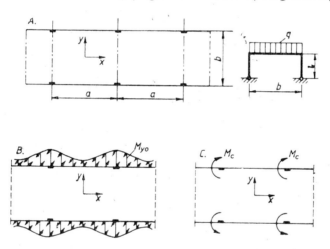

Figure 1.18 Schematic diagram for Andersson's procedure, incorporating the effect of rigid slab-column connection.
A. Uniformly distributed load q. B. Bending moments M_{y0} continuously distributed along the free edges of the plate.
C. Concentrated clamping moments M_c in the column axis.

For the constants M_0 and M_m in Eqn. (1.13) and (1.16), Andersson gives the following expressions corresponding to the additional loads M_c introduced:

$$M_0 = -\frac{M_c}{a}; \quad M_m = -2\frac{M_c}{a}(-1)^{m/2}. \qquad (1.43)$$

1 Introduction

Superposing the effects of these three types of loads, the solution $w(x, y)$ becomes

$$w(x, y) = w_1 + w_2 + w_3, \tag{1.44}$$

where w_1 given by (1.9) and w_2 given by (1.14) are established by Timoshenko (Sub-secs. 1.2.2.1 and 1.2.2.2). The expression of w_3 derives also from Eqn. (1.14), where the constants A_1, A_m and B_m, given by Eqn. (1.16), are defined by M_0 and M_m, given by (1.43). Once Eqn. (1.44) is defined, Andersson gives the expression for the angular displacement $(\varphi_y)_{y=\frac{b}{2}}$ in a direction normal to the free edge of the slab:

$$(\varphi_y)_{y=\frac{b}{2}} = -\frac{qb^3}{24D} \cdot \Upsilon_{M_{y_0}} + \frac{M_c}{2D} \cdot \Upsilon_{M_c}, \tag{1.45}$$

where

$$\Upsilon_{M_{y_0}} = 1 + \frac{24a^2}{\pi^2 b^2} \sum_{2,4,6} \frac{1}{m^2} \left(\frac{1}{\operatorname{th}\alpha_m} - \frac{\alpha_m}{\operatorname{sh}^2\alpha_m} \right) \times \frac{(-1)^{\frac{m}{2}} \cos \frac{m\pi x}{a}}{\left(\frac{3}{\operatorname{th}\alpha_m} - \frac{\alpha_m}{\operatorname{sh}^2\alpha_m} \right)}; \tag{1.46a}$$

$$\Upsilon_{M_c} = \frac{b}{a} + \frac{8}{\pi} \sum_{2,4,6} \frac{1}{m} \cdot \frac{(-1)^{\frac{m}{2}} \cos \frac{m\pi x}{a}}{\left(\frac{3}{\operatorname{th}\alpha_m} - \frac{\alpha_m}{\operatorname{sh}^2\alpha_m} \right)} \tag{1.46b}$$

In the expression of the rotation given by (1.45) as well as all the subsequent formulae, the value of Poisson's ratio was assumed to be zero.

Admitting that $\mu = 0$, the value of the "clamping moment" M_c in the column axis is obtained by Andersson assuming that the slab rotation is equal to the rotation of the upper end of the column, under the load M_c:

$$(\varphi_y)_{\substack{x=a/2 \\ y=b/2}} = -\frac{M_c l}{3\alpha E I_c}, \tag{1.47}$$

where l represents the length of a column; I_c denotes the moment of inertia of the column; α denotes the coefficient characterizing

1.2 State of the art. Main research trends

the clamping mode of the column ends ($\alpha = 4/3$ when the lower end is restrained).

The numerical calculations are carried out by Andersson for various ratios a/b of the in-plane dimensions of a panel, as a function of the parameter

$$\eta = \frac{3\alpha EI_c}{2lEI_{sl}} \cdot \frac{b}{a}, \qquad (1.48)$$

defining the relative slab-column stiffness (Fig. 1.19). In the absence of comparative numerical data, several remarks should be made. First, the series which appears in the expression for γ_{M_c} (1.46 b) is *strongly divergent*. This leads to a substantial decrease of the value of M_c with the increasing number of terms in the series; for the square panel ($b = a$) and for $\eta = 10$, M_c decreases in the range $10 \leqslant m \leqslant 50$ by 28.6 percent *. Second, the smooth-

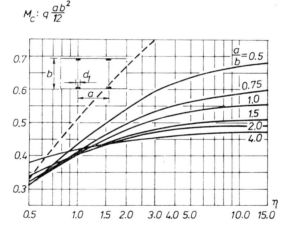

Figure 1.19 Numerical example illustrating Andersson's procedure. Variation of "clamping moment" M_c with parameters a/b and η.

ing of the curves $M_c/q \dfrac{ab^2}{12}$ for ordinary values of the parameter η would lead to an erroneous idea that the dependence of the values of the "clamping moments" M_c on the value of the relative column-

---

* Verifications show that the values given by Andersson were calculated for $m = 20$, which generally does not yield sufficient accuracy (see also Chaps. 3 and 4).

43

1 Introduction

-slab stiffness is weak. Indeed, according to Andersson, for the same panel $b = a$ and for the same number of terms ($m = 20$), to an eight-fold increase of the parameter η ($5 \leqslant \eta \leqslant 40$) there corresponds an increase of only 7.7 percent of M_c (Fig. 1.19). For $m = 50$, this difference reduces to 6.6 percent. However, in the case of the square panel investigated by the author (Sub-secs. 5.1.2.1 — Fig. 5.7), to a similar increase of the relative slab-column stiffness ($10 \leqslant 4EI/lD \leqslant 80$), there corresponds an increase of 51.6 per cent of the value of the couple reaction.

Finally, in order to remove the contradiction between the manner in which the "clamping moment" M_c, which was assumed to be concentrated, is applied and the effect of the slab-column interaction, which was inserted in the analysis, Andersson uses further simplifying assumptions, artifices of calculations and empirical formulae, which reduce the accuracy of his procedure.

1.2.3.4 Bretthauer's procedure. A remarkable theoretical study of the behaviour of flat slabs in the elastic range is due to a team of German researchers headed by Bretthauer [41], [45], [49]. Using the theoretical scheme of the plate of indefinite sizes — at least in one direction — Bretthauer approaches mainly the response of the plate to a load due to concentrated forces. Bretthauer considers either sets of equidistant forces P_k (Fig. 1.20) or successive loadings with the live unit load $P = 1$, the latter being used to determine some influence surfaces. Linear loads of the type $\bar{q} = $ constant are also considered (Figs. 1.21 and 1.23).

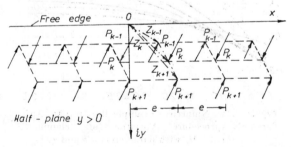

Fig. 1.20 Basic schematic diagram for Bretthauer-Seiler's procedure. Half-plane $y > 0$ subjected to sets of concentrated equidistant forces P_k.

Through its peculiar approach to the static analysis, Bretthauer's procedure is far from the other modern treatments. Its principal contribution is the mathematical modelling of the less investigated effect of the free edge on the stress distribution in the plate, under the uniformly distributed load q. The loads of

1.2 State of the art. Main research trends

type P_k or $\bar{q}$ occur here only as interacting forces. That is, they appear either as force reactions concentrated at the columns, which are treated as point-like supports, or as linear reactions along the row of columns, which are assumed to be continuous supports of the plate.

Throughout his investigations Bretthauer adopts the following simplifying assumptions which limit the applied range of his procedure.

a. All the schematic diagrams employed refer to *one-storey structures*, i.e. structures consisting of only one floor resting on columns.

b. All the columns are assumed to be *point-like supports* (Figs. 1.20, 1.21, 1.23, etc.). Except for one schematic diagram (Fig. 1.27), *the slab-column connection is treated as a hinge*. Hence, the only continuity condition used by Bretthauer is of the type (1.7).

c. The columns are assumed to be arranged as a network of parallel rows spaced at an equal distance. The columns of each row are also assumed to be *equidistant*.

d. The plate is infinitely long in the direction of the Ox axis. In the direction of the orthogonal Oy axis, Bretthauer considers either the scheme of the plate of indefinite sizes ($-\infty \leqslant y \leqslant +\infty$) or the diagram of the half-plane ($y > 0$). The infinitely long strip is treated as a particular case.

e. The boundary conditions for the edge $y = 0$ of the half-plane and on the edges $y =$ constant of the infinitely long plate are those corresponding to the *free edge and simply supported edge* *.

f. The given load is assumed to be a *load q uniformly distributed* throughout the surface of the floor. The linear load $\bar{q}$, which is treated in the analysis as a linear reaction, is assumed to be constant along the column row.

In their early articles [41], Bretthauer and Seiler give a rigorous solution to the problem of the half-plane with free edge, subjected to the action of an arbitrary concentrated force P_k. The solution to the homogeneous equation of plates (1.1a) — which is known as Goursat's formula — as well as the conditions to the limit are expressed by functions of a complex variable. Using the "force-reflection" procedure and the superposition of effects, the authors establish the expressions for the moments M_x, M_y and M_{xy} in the case of one row of equidistant concentrated forces (Fig. 1.20). For illustration, Bretthauer and Seiler consider the case of the plate with simply supported longitudinal edges and free transverse edges. However, the study is here confined to the

---

* The case of the half-plane with the edge $y = 0$ clamped is treated in only one example of analysis.

1 *Introduction*

determination of some influence surfaces for the values of these moments *.

Starting from earlier results, Bretthauer and Seiler establish in two subsequent papers [45] a procedure of analysis for plates with equidistant point-like discrete supports, under a uniformly distributed load q. As was noticed in the approach to the Grein-Girkmann procedure (Sub-sec. 1.2.2.4), here too the analysis is carried out in two steps but using a more complex schematic diagram (Fig. 1.21). Thus,

a. The row of columns is first treated as a continuous support line along which the linear reactions $\bar{q}$ = constant are induced by the load q.

b. In the second step, by superposition of effects, to the stress in the slab, which is assimilated with a continuous beam, add the moments produced by the concentrated forces P_k, which balance the linear loads $\bar{q}$.

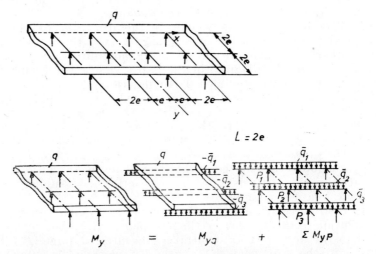

Figure 1.21 Schematic diagram for Bretthauer-Seiler's procedure. Infinitely long plate with parallel free edges resting on three rows of equidistant columns, subjected to uniformly distributed load q.

The examples of analysis include the flat slab of indefinite sizes in both directions of the plane, resting on equidistant point-like supports, under the uniformly distributed load q; the infinitely long plate with two free edges, supported by three rows of equi-

---

* The study of the influence surfaces in the case of the rectangular plate, through the force-reflection procedure, was continued by Bretthauer and Nötzold. Their results are not involved in the analysis of slab structures.

1.2 State of the art. Main research trends

distant columns, under the same type of load (Figs. 1.21 and 1.22); the half-plane with free edge subjected to a linear uniformly distributed load $\bar{q}$ (Figs. 1.23 A and 1.24). For comparison, this is also applied to the infinitely long plate with two parallel free edges (Figs. 1.23 B and 1.24).

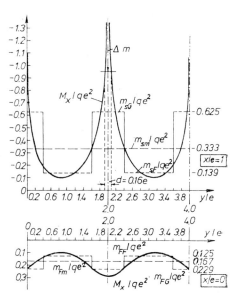

Figure 1.22 Example of analysis illustrating Bretthauer-Seiler's procedure. Distribution of sectional bending moments M_x along column row $(x = e)$ and in axis of panels $(x = 0)$.

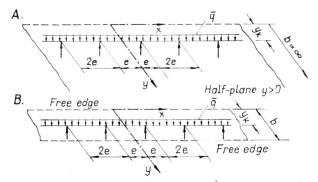

Figure 1.23 Schematic diagram for Bretthauer-Seiler's procedure.
Linear load $\bar{q}$ uniformly distributed on plane $y_k = e$.
A. Half plane $y > 0$. B. Infinitely long plate.

The curves M_x and M_y, plotted for the last three schematic diagrams, show that the values of the bending moments at the

1 Introduction

point-like supports are infinite both when the load q is uniformly distributed throughout the surface of the plate (Fig. 1.22) and under the linear load $\bar{q}$ (Fig. 1.24).

Bretthauer and Seiler's contention that their procedure yields accurate results throughout the slab is open to question.

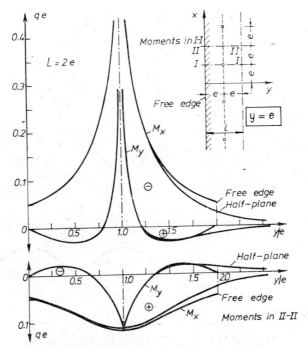

Fig. 1.24 Example of analysis illustrating Bretthauer-Seiler's procedure. Distribution of sectional bending moments M_x and M_y along column axis $x = e\,(I - I)$ and in panel axis $x = 2e\,(II - II)$.

Disregarding the influence of the parameters c and d, which define the sizes of the slab-column contact area, leads to results that reflect only partially the real behaviour of the structure. This is especially true at the vecinity of these supports (Chapters 3 and 4). Equally important is the effect of the interaction between the slab and the column, which was also neglected by Bretthauer, and which substantially changes the distribution and the values of the elastic displacements and stresses in slabs and columns (Chapters 5 and 6) [75], [78].

Several remarks to previous studies are given by Bretthauer and Nötzold in other three papers [49]. For a glimpse of the

1.2 State of the art. Main research trends

limits of the applied range of their procedure, we shall present here two highly complex schematic diagrams dealt with by the German researchers.

First, the authors consider the theoretical scheme of the plate of indefinite sizes in both directions of the plane, under the uni-

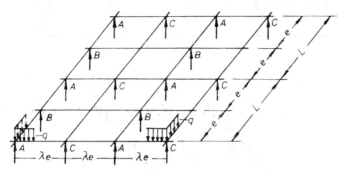

Figure 1.25 Schematic diagram for Bretthauer-Nötzold's procedure. Non-uniform distribution pattern of columns.

formly distributed load q (Fig. 1.25). Unlike the previous schematic diagrams (Fig. 1.21), in this scheme the equidistances e and $2e$ (measured lengthwise the column rows $A + B$ and C, respectively) are different from the distances λe between these rows. Hence, for the determination of the linear statically indeterminate reactions $\bar{q}_A$, $\bar{q}_B$ and $\bar{q}_C$, it is necessary to use a system of three compatibility equations (Fig. 1.26):

$$w_{AA}\bar{q}_A + w_{AB}\bar{q}_B + w_{AC}\bar{q}_C + w_{A,0} + C_0 = 0;$$
$$w_{BA}\bar{q}_A + w_{BB}\bar{q}_B + w_{BC}\bar{q}_C + w_{B,0} + C_0 = 0; \qquad (1.49)$$
$$w_{CA}\bar{q}_A + w_{CB}\bar{q}_B + w_{CC}\bar{q}_C + w_{C,0} + C_0 = 0,$$

where w_{CB} represents the unit transverse displacement at the point C due to the row of forces acting in B. The free terms $w_{A,0}$, $w_{B,0}$ and $w_{C,0}$ are obtained during the first step of analysis, when the slab subjected to the load q is assumed to be in a state of equilibrium under the action of the yet unknown linear reactions $(\bar{q}_A + \bar{q}_B)$ and $\bar{q}_C$ (Fig. 1.26 A):

$$w_{A,0} = w_{B,0} = 0; \qquad (1.50a)$$

$$w_{C,0} = \frac{q\lambda^4 e^4}{24D} - \frac{\bar{q}\lambda^3 e^3}{24D}. \qquad (1.50b)$$

1 Introduction

In order to determine the constant C_0, use is made of the equation of equilibrium for the panels $(2\lambda e \times 2e)$ (Fig. 1.26B):

$$(\bar{q}_A + \bar{q}_B) + \bar{q}_C = 2\lambda eq. \tag{1.51}$$

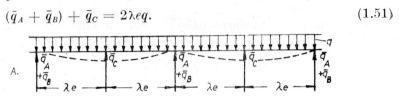

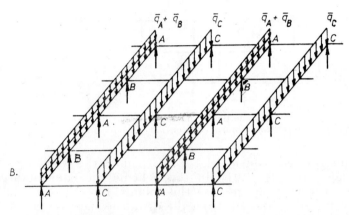

Figura 1.26 Numerical example illustrating Bretthauer-Nötzold's procedure.
A. Step 1. B. Step 2.

Once the values of the linear reactions, which are assumed to be constant, are obtained from Eqns. (1.49)—(1.51), the concentrated force reactions A, B and C are derived as functions of the parameter λ (Table 1.6).

Table 1.6

Flat slabs under the uniformly distributed load q

Value of the reaction forces, after Bretthauer-Nötzold

λ	A/qe^2	B/qe^2	C/qe^2	$(A+B+C)/qe^2$
1/2	0.512	0.952	0.536	2.000
1	1.000	1.500	1.500	4.000
2	2.048	2.144	3.808	8.000

1.2 State of the art. Main research trends

A particular case of the preceding scheme is the infinitely long plate with two simply supported parallel edges and only one row of equidistant columns spaced in the longitudinal axis of symmetry of the plate, under the uniformly distributed load q. The same diagram was treated in a less particularized variant through a simplified procedure by Grein-Girkmann (see Sub-sec. 1.2.2.4).

Following the studies undertaken by Girkmann (Sub-sec. 1.2.2.5) and Andersson (Sub-sec. 1.2.3.3) on the effect of *the rigid slab-column connection*, Bretthauer and Nötzold suggest another procedure for evaluating this effect. The object of this procedure is the analysis of the *concentrated couple-reaction* M_c, as a statically indeterminate unknown, in order to define the state of stress at the slab-column node. The equivalence relation

$$M_c = 2e\overline{m} \tag{1.52}$$

is assumed as a further simplifying assumption, where the linear couple-reaction $\overline{m}$ is considered to be uniformly distributed along the row of columns (Figs. 1.27A and B). The mathematical

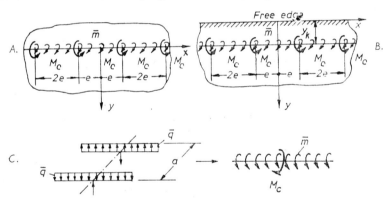

Figure 1.27 Schematic diagram for Bretthauer-Nötzold's procedure. Equivalence between concentrated couple-reaction M_c and linear couple-reaction $\overline{m}$ = constant.
A. Plate of indefinite sizes in both directions of the plane. B. Half-plane $y > 0$. C. Mathematical modelling of linear couple $\overline{m}$ = constant effect on plate.

modelling of the linear load $\overline{m}$ is obtained by superposing the effects of two linear loads $\overline{q}$ and $-\overline{q}$ applied at a distance $a \to 0$ from one another (Fig. 1.27C). The stresses $M(x, y, y_k)$ induced by the load $\overline{m}$ = constant are obtained from the expressions of the corresponding stresses $Q(x, y, y_k)$, induced by the

1 Introduction

load $\bar{q}$ = constant through differentiation with respect to y_k:

$$\overline{m}\, M(x, y, y_k) = \bar{q}\, \frac{\partial}{\partial y_k}\, Q(x, y, y_k). \tag{1.53}$$

The computation formulae of the static quantities w, M_x, M_y, M_{xy}, T_x and T_y from the load $\overline{m}$ are derived with the aid of Eqn. (1.53) for two theoretical schemes: the plate of indefinite sizes in both directions of the plane and the half-plane of free edge $y > 0$.

For illustration, Bretthauer investigates a flat slab diagram of particular characteristics (Fig. 1.28): the slab is of indefinite sizes in the plane, and rests on a network of equidistant columns, which are assumed to be point-like supports under the uniformly distributed load p. By inserting in the analysis the bending rigidity EI_c of the columns, the system is assumed to be simply statically indeterminate. The basic system derived by replacing the rigid nodes by hinges is similar to the previously studied schematic diagrams (Fig. 1.21 and 1.25). Since the only statically indeterminate unknown is the concentrated couple-reaction M_c, the coefficient δ_{11} of the unknown X and the free term Δ_{10} are determined in the two known steps of analysis (Figs. 1.28A and B):

$$\Delta_{10} = \Delta_{10}^{(1)} + \Delta_{10}^{(2)} = -\frac{1}{48} \cdot \frac{pL^3}{D} \tag{1.54}$$

$$\delta_{11} = \delta_{11}^{(1)} + \delta_{11}^{(2)} = \frac{1}{4}\left(\frac{1}{D} + \frac{l_c}{EI_c}\right) + \frac{\varphi}{D} = \frac{1}{4D}(1 + k + 4\varphi), \tag{1.55}$$

where $k = Dl_c/EI_c$ denotes the relative slab-column stiffness and $\dfrac{\partial w}{\partial y} = \dfrac{\varphi}{D}$ represents the slab rotation in the column axis under the load due to the unit couple $X = 1$ ($\overline{m} = 1/L$).

For the schematic diagram under consideration, it follows that

$$M_c = \frac{pL^3}{12}\, \frac{1}{1 + k + 4\varphi}. \tag{1.56}$$

In the numerical example given by Bretthauer-Nötzold, the relative sizes of the column cross-section are $2c = 2d = L/10$ and the relative stiffness factor $k = 0.211$ and hence $M_c = 0.0533\, pL^3$. For a similar schematic diagram of a flat slab of finite dimensions (Sub-sec. 5.1.2.1 — Fig. 5.5), where $2c = 2d = L/16$ and $k = 0.200$, Negruțiu obtained $M_c = 0.0312\, pL^3$. The apparent error of $+ 71$ per cent, induced by the simplifying assumptions adopted

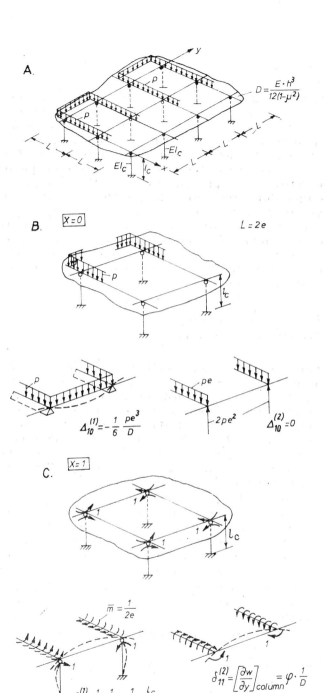

Figure 1.28 Numerical example illustrating Bretthauer-Nötzold's procedure.

A. Static diagram of the assumedly rigid-node structure.
B. Basic system under given load p. C. Basic system under unit-couple $X = 1$.

1 *Introduction*

by the German authors, is in reality higher because the interaction of the adjacent panels should come off as the unloading of the panel under study. Additionally, the parameters defining the slab-column contact area — which is much larger in the case treated with Bretthauer-Nötzold's procedure — should also contribute to the decrease of the value of M_c of the couple-reactions.

The principale approximations of the Bretthauer-Seiler-Nötzold procedure stem from the treatment of the columns of the structure as point-like supports and the assumption that the linear reactions $\bar{q}$ and $\bar{m}$ are uniformly distributed along the row of columns. The limited range of applicability of this procedure is due to the excessively simplified treatment of the real cases. Moreover, all the tabulated values were determined taking Poisson's ratio equal to zero.

1.2.4 Early studies on surface couples. The effective plate width

As we have seen from the presentation of the main works dealing with the analysis of flat slabs using the means of the Theory of Plates, the discrete intermediate supports are usually treated as point-like supports. Several corrections were made to this simplifying assumption by introducing into the analysis either the cross-section of the columns but neglecting their flexural rigidity, or some couple-reactions, which are also assumed to be concentrated. In this situation, the slab of the floors is treated as an independent structural member and, thus, a structural analysis proper is out of the question here.

For a global method of elastic analysis of slab structures, it is strictly necessary to investigate the effect of the rigid non-point-like connection between slabs and columns. Girkmann [9] was the first to give a modelling procedure for this connection but his approach is limitted to the case of a concentrated couple. Further contribution is due to Mohammed and Popov [58] who developed a model for the linear uniformly distributed couple. Relying on the above results, Stiglat [60] was the first to introduce the concept of surface couple, modelling its effect on the infinitely long plate. Stiglat's results became essential to all the subsequent developments in the field, including the author's studies [64], [65], [68], [70], [73], [75], [82], and [86] (Chapts. 2—4).

The results obtained from the study of the surface couple applied on the plate were used by Stiglat in a new procedure for evaluating the effective slab width. This procedure is valid

1.2 State of the art. Main research trends

for a wider variety of cases than is the similar procedure of Mohammed and Popov. From the numerous contributions published of late, mention should be also made of those due to Pecknold [79] and Allen-Darvall [81].

1.2.4.1 Girkmann's procedure. The concentrated couple M_0. For the study of the behaviour of plates under transverse loads, classical treatises give solutions of the Lagrange-Sophie Germain equation (1.1) both in the cases of the concentrated force P_0 and the locally distributed force $P = 4pcd$ [7], [9], [16].

From the equation of the mid-surface of the infinitely long plate, deformed under the load $P_0(\xi, 0)$,

$$w(x, y) = \frac{P_0 a^3}{2\pi^3 D} \sum_m \frac{1}{m^3} \left(1 + \frac{m\pi y}{a}\right) e^{-\frac{m\pi y}{a}} \sin \frac{m\pi \xi}{a} \sin \frac{m\pi x}{a},$$

$$(m = 1, 2, 3, \ldots) \quad (1.57)$$

which holds for $y \geqslant 0$, Ghirkmann [9] established the expressions $w(x, y)$ corresponding to this category of plates subjected to an *exterior concentrated couple* M_0 (Fig. 1.29). The procedure given

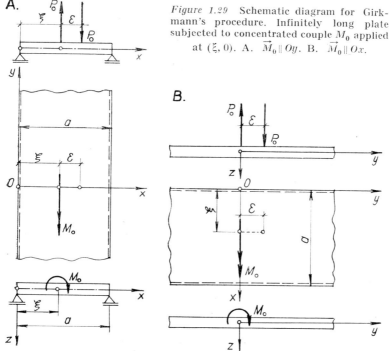

Figure 1.29 Schematic diagram for Girkmann's procedure. Infinitely long plate subjected to concentrated couple M_0 applied at $(\xi, 0)$. A. $\vec{M_0} \parallel Oy$. B. $\vec{M_0} \parallel Ox$.

1 Introduction

by the German scholar is based on the superposition of the effects of two concentrated forces, each acting in opposite directions, the arm ε of the couple $(P_0, -P_0)$ being arbitrarily small. Summating the effects and passing to the limit, for $M_0 = P_0\varepsilon$ and $\varepsilon \to 0$, Girkmann obtains:

— in the case of the moment vector parallel to the line of supports $\vec{M_0} \| Oy$ (Fig. 1.29A)

$$w(x, y) = \frac{M_0 a}{2\pi^2 D} \sum_m \frac{1}{m^2} \left(1 + \frac{m\pi y}{a}\right) e^{-\frac{m\pi y}{a}} \cos\frac{m\pi\xi}{a} \sin\frac{m\pi x}{a};$$

$$(m = 1, 2, 3, \ldots) \qquad (1.58)$$

— in the case of the moment vector normal to the line of supports $\vec{M_0} \| Ox$ (Fig. 1.29B)

$$w(x, y) = \frac{M_0}{2\pi D} \sum_m \frac{1}{m} y e^{-\frac{m\pi y}{a}} \sin\frac{m\pi\xi}{a} \sin\frac{m\pi x}{a}.$$

$$(m = 1, 2, 3, \ldots) \qquad (1.59)$$

Equations (1.58) and (1.59), established by Girkmann for the load due to the concentrated couple M_0 applied at $(\xi, 0)$, are valid for the infinitely long plate. From the investigations underlying the ideas expressed in this book, similar expressions were derived for the finite boundary plate (Sub-secs. 2.2.3.2 and 2.2.3.3). As distinguished from Girkmann's procedure, equations (2.45) and (2.47) were established here assuming that loading of the plate due to a concentrated couple M_0 is a limit case of the loading due to a surface couple M.

1.2.4.2 *Mohammed-Popov's procedure. The linear edge couple $\bar{M}$.* Without knowing of Girkmann's fundamental work [9], the American researchers Mohammed and Popov [58] gave a similar analytical solution for the infinitely long plate subjected to a local edge couple $\bar{M}$, uniformly distributed along the length b_1 of the segment JK (Fig. 1.30). Their aim was to determine the amount of restraint provided by a reinforced bearing wall to a beam framing into it. As this bearing wall is assimilated with a vertical plate, the effective stiffness of this plate is converted into an equivalent beam stiffness. The width b_y of this equivalent

1.2 State of the art. Main research trends

beam — herein referred to as *the effective plate width* (Sec. 4.2) — is expressed as a function of the width b_1 of the actual beam, rigidly connected to the wall. So, the American authors do not consider their results in the investigation of slab structures.

Mohammed and Popov's procedure for modelling the effect of a *local edge couple of moment* $\overline{M}$ is extremely involved. They consider the rectangular plate ($a \times b$) simply supported on the entire boundary, under the locally distributed load $P = p a_1 b_1$ (Fig. 1.30). The solution $w(x, y)$, given by Timoshenko [7] for this type of load, is inconvenient as it requires the determination of 13 integration constants: four constants of type A_m, B_m, C_m and D_m for each of the intervals $(b/2 > y > \eta_2)$, $(\eta_2 > y > \eta_1)$ and $(\eta_1 > y > -b/2)$, and a constant a_m which depends only on the distribution of the load p (Fig. 1.30). For simplification,

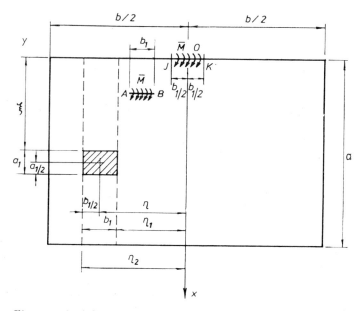

Figure 1.30 Schematic diagram for Mohammed-Popov's procedure. Rectangular plate with all simply supported edges under linear edge couple $\overline{M}$ applied at O ($\xi = 0, \eta = 0$).

the load P distributed on the surface ($a_1 \times b_1$) is replaced by the linear load $\overline{P}$ uniformly distributed along the length b_1 by letting a_1 approach zero. For the loaded range ($\eta_1 \leq y \leq \eta_2$), which is the only range considered by the American authors, the equation

1 Introduction

of the mid-surface of the rectangular plate, deflected under the load $\bar{P} = qb_1$ is

$$w(x, y) = \frac{2\bar{P}a^3}{\pi^4 D b_1} \sum_m \left[\frac{1}{m^4} \sin \frac{m\pi\xi}{a} \sin \frac{m\pi x}{a} \right] \theta_m,$$

$$(m = 1, 2, 3, \ldots), \qquad (1.60)$$

where θ_m is a sum of hyperbolic functions in y and b_1, which is extremely involved [58].

Solution (1.60), established by Timoshenko, is subsequently used for the mathematical modelling of a linear local couple of moment $\bar{M} = \bar{P}\varepsilon$. This procedure is similar to Girkmann's (Sub-sec. 1.2.4.1) and consists in the superposition of two adjacent equal linear loads of magnitude $(\bar{P}, -\bar{P})$, each acting in opposite directions. By letting the distance ε shrink to zero, one obtains an expression for $w(x, y)$ caused by the local linear couple $\bar{M}$, uniformly distributed along the length b_1 of the segment AB (Fig. 1.30):

$$w(x, y) = \frac{2\bar{M}a^2}{\pi^3 D b_1} \sum_m \left[\frac{1}{m^3} \cos \frac{m\pi\xi}{a} \sin \frac{m\pi x}{a} \right] \theta_m.$$

$$(m = 1, 2, 3, \ldots) \qquad (1.61)$$

As was acknowledged by Mohammed and Popov themselves, because of the function θ_m, the resulting equation (1.61) is very involved. However, it greatly simplifies if the length b of the plate approaches infinity ($b \to \infty$) and the couple $\bar{M}$ is applied at O ($\xi = 0, \eta = 0$). For this special case, solution (1.61) reduces to

$$w(x, y) = \frac{\bar{M}a^2}{\pi^3 D b_1} \sum_m \frac{1}{m^3} \sin \frac{m\pi x}{a} \left\{ 2 + \left[\frac{m\pi y}{a} \operatorname{sh} \frac{m\pi y}{a} - \left(2 + \frac{m\pi b_1}{2a} \right) \operatorname{ch} \frac{m\pi y}{a} \right] e^{-\frac{m\pi b_1}{2a}} \right\}. \quad (m = 1, 2, 3, \ldots) \qquad (1.62)$$

Equation (1.62), which holds only over the width $JK = b_1$ of the loaded strip, represents the mid-surface of the infinitely long plate, deformed under a linear edge couple $\bar{M}$ at 0 (Fig. 1.30). This equation is used exclusively by the American authors in the subsequent developments regarding the evaluation of the effective plate width and the numerical applications.

For a local edge couple $\bar{M}$ applied at ($\xi = 0, \eta = 0$), the maximum rotation φ_{x_0} of the plate may be obtained by differentiating equation (1.62) with respect to x, and setting $x = y = 0$. On the other hand, the maximum rotation φ_{b_0} of a simply supported beam loaded with an end couple of equal magnitude $\bar{M}$

1.2 State of the art. Main research trends

is given by the known expression

$$\varphi_{b_0} = \frac{\overline{M}a}{3EI}. \qquad (1.63)$$

The width b_y of the beam equivalent to the plate — i.e. having the same span a and depth h — is obtained by Mohammed and Popov by equating the rotations φ_{r_0} to φ_{b_0}:

$$\left(\frac{\partial w}{\partial x}\right)_{\substack{x=0 \\ y=0}} = \frac{4\overline{M}a}{Eh^3 b_y}. \qquad (1.64)$$

Hence,

$$b_y = k_0 b_1, \qquad (1.65)$$

where

$$\frac{1}{k_0} = \frac{3(1-\mu^2)}{\pi^2} \sum_m \frac{1}{m^2}\left[2 - \left(2 + \frac{m\pi b_1}{2a}\right)e^{-\frac{m\pi b_1}{2a}}\right]. \qquad (1.65a)$$

Observing that the value of the plate rotation $\varphi_x = \partial w/\partial x$ varies in the range $\pm b_1/2$ — the rotation φ_{b_0} of the real beam end being constant — the American authors establish a similar expression for the dimensionless coefficient $k_{b_1/2}$. This expression holds at the extremities of the loaded segment b_1. The results of the numerical computation are presented in the form of a diagram (Fig. 1.31). Mohammed and Popov recommend that the evaluation

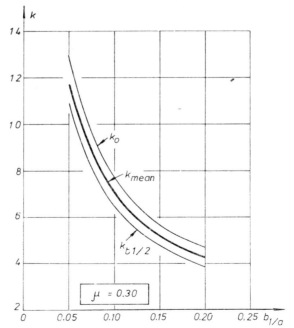

Figure 1.31 Example of analysis illustrating Mohammed-Popov's procedure. Variation of effective plate width with parameter b_1/a.

1 *Introduction*

of the effective plate width b_y, given by Eqn. (1.65), should include the mean value $k_{mean} = (k_0 + k_{b_1/2})/2$ of the dimensionless coefficients (see Sec. 4.2).

The limits of this procedure stem from the treatment of the bar-plate connection as a linear couple and from the fact that the computation algorithms are established for only one position ($\xi = 0$, $\eta = 0$) of this couple. Additionally, the solution obtained is of theoretical interest because the schematic diagram here employed is only the infinitely long plate.

1.2.4.3 *Stiglat's procedure. The surface couple M*. Throughout this book the *surface couple* denotes an exterior couple of moment M, locally applied on a limited area ($2c \times 2d$) of the plate (Fig. 1.32). This concept was introduced by Stiglat ([19], [24], [57], [90], [96], [105], [109], [118], [120]) in 1963 [60]. The mathe-

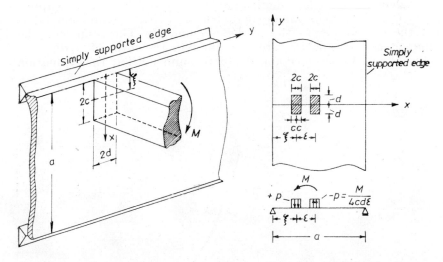

Figure 1.32 The surface couple M as a load acting on the plate, after Stiglat.

Figure 1.33 Schematic diagram for Stiglat's procedure. Infinitely long plate subjected to surface couple M, applied at ($\xi, 0$).

matical modelling procedure of the effect induced by this type of load on plates (Fig. 1.33) is similar to Girkmann's procedure for modelling the concentrated couple M_0 and to Mohammed-Popov's procedure for the case of the linear edge couple $\overline{M}$.

1.2 State of the art. Main research trends

All these three procedures treat only the diagram of the infinitely long plate. Except for one case (Fig. 1.29B), the moment-vector of the exterior couple under study is parallel to the line of supports (Figs. 1.29A, 1.30 and 1.32).

The effect of the force couple $(P, -P)$, of moment $M - P\varepsilon$ (Fig. 1.33) is modelled by Stiglat using the coresponding solution $w(x, y)$ for the infinitely long plate under the locally distributed load $P = 4pcd$ (see Sub-sec. 2.2.1.1 – formulae 2.11 – 2.13). Passing to the limit ($\varepsilon \to 0$), a more elegant general solution than Mohammed-Popov's is obtained for the load due to a surface couple $\vec{M} \parallel Oy$ (see Sub-sec. 2.2.1.2).

From the solution $w(x, y)$ given by Eqn. (2.16), Stiglat derived several numerical results, e.g. computation tables and diagrams showing the variation of the sectional bending moments M_x and M_y (Fig. 1.34). A comparison of the values of φ_x, M_x and M_y obtained by Stiglat and by the author, respectively, is given in Sub-sec. 3.4.3 — Tables 3.24 and 3.25.

The aim pursued by both Mohammed-Popov and Stiglat is to evaluate the relative flexural bar-plate rigidity, which means to determine the effective plate width b_y. As applications, Stiglat suggests both the cases of the bearing wall rigidly connected with a horizontal beam (Fig. 1.32) and of the flat slab supported by vertical columns, under horizontal loads.

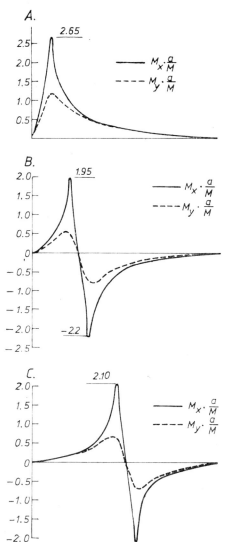

Figure 1.34 Numerical example illustrating Stiglat's procedure. Variation of sectional bending moments M_x and M_y under a surface couple M at $(\xi, 0)$ for $2c = 2d = 0.10\,a$.
A. $\xi = c$. B. $\xi = 0.25a$. C. $\xi = 0.50a$.

1 Introduction

Stiglat's procedure for evaluating the effective plate width relies on equalizing the rotation φ_x of the plate under the load $M(\xi, 0)$ to the rotation φ_ξ of an equivalent beam under a concentrated couple of moment equal to M (Sub-sec. 4.2.1.1 — Fig. 4.28) This procedure represents an extension of Mohammed-Popov's procedure, where only the edge position ($\xi = 0$, $\eta = 0$) of a local linear couple $\overline{M}$ was considered (Sub-sec. 1.2.4.2). Hence, Stiglat's solution holds for the load due to a surface couple whose point of application ($\xi, 0$) can assume any position along the Ox axis. This solution was generalized by the author such that it may cover any position (ξ, η) of the surface couple M in the plane of the finite boundary plate (Sec. 4.2).

As numerical applications, Stiglat gives the values of the ratio b_y/a between the effective width of the infinitely long plate and its span, calculated for two characteristic positions of the surface couple $M : \xi = c$; $\xi = 0.50a$ (Table 1.7).

Table 1.7

The infinitely long plate subjected to a surface couple $\vec{M} \parallel Oy$ applied at $(\xi, 0)$

Comparative values of the effective plate width b_y/a, after Stiglat and Negruțiu

$\mu = 0$

ξ	c/a	d/a					
		0.02		0.04		0.06	
		S	N	S	N	S	N
c	0.02	0.68	0.680	0.70	0.703	0.73	0.729
	0.04	0.83	0.827	0.84	0.840	0.86	0.857
	0.08	1.05	1.048	1.05	1.053	1.05	1.059
0.50a	0.02	0.36	0.355	0.39	0.385	0.42	0.417
	0.04	0.45	0.446	0.47	0.465	0.49	0.489
	0.08	0.62	0.622	0.63	0.634	0.65	0.654

The expression found by Stiglat for evaluating *the effective plate width* b_y (Sec. 4.2.1.1 — formula 4.3) is valid in the case when the plate is subjected to only one surface couple M, i.e. when the superposition of the effects of the neighbouring columns can be neglected in the analysis. In a recent paper [60], Stiglat reverted to this problem, assuming that the infinitely long plate is subjected to a set of equidistant surface couples M grouped in simple row (Fig. 1.35). In the case of the present scheme, which can more faithfully reflect the behaviour of the real structure, the expression of the effective plate width b_y is corrected by inserting the total plate rotation $\Sigma \varphi_{xi}$ given by the

1.2 State of the art. Main research trends

summation of the effects of the couple row M (see Sub-sec. 4.2.2 — formula 4.6).

The numerical applications given by Stiglat include the values of the rotations φ_x at $y = 0$ and $y = e$, determined for the central ($\xi = 0.50a$) and edge ($\xi = c$) positions of the column row (Figs. 1.35A and B). The difference between the alternatives chosen by Stiglat for the equidistance e between columns ($e/a = 0,2; 0.4; \ldots 1.0; 1.2$) and those investigated herein ($e/a = 0.50; 0.75; 1.0; \ldots 2.0$) is such that the values of the rotations can be compared in only one case (Table 1.8).

Using the previously obtained data, Stiglat suggests that the total effect of the set of surface couples disposed in a simple row should be considered in evaluating the effective plate width only for relatively small equal distances between columns ($e \leqslant a$). From the author's investigations it is found that the superposition of effects can be neglected in calculations only when the equidistance between columns is larger ($1.5a \leqslant e \leqslant 2a$) [75].

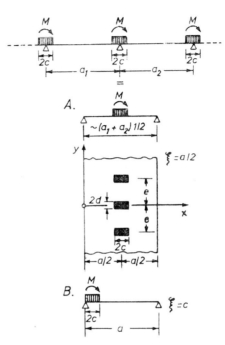

Figure 1.35 Schematic diagram for Stiglat's procedure. Infinitely long plate subjected to a set of surface couples M associated in simple row. A. $\xi = 0.50a$. B. $\xi = c$.

Table 1.8

The infinitely long plate subjected to a surface couple $\vec{M} \| Oy$ associated in a simple row

Comparative values of the total rotations $\Sigma \varphi_{xi}$, after Stiglat and Negruțiu

$\xi = 0.50a; \quad e = a$

$c/a = d/a$	$\dfrac{D}{M}(\varphi_x)_{y=0}$		$\dfrac{D}{M} \Sigma \varphi_{xi}$	
	S	N	S	N
0.02	0.234	0.23409	0.2362	0.23627
0.04	0.179	0.17914	0.1812	0.18131
0.06	0.147	0.14702	0.1492	0.14918

1 *Introduction*

Both Mohammed-Popov and Stiglat show that the value of the rotation φ_x is smaller on the edge $y = \pm d$ of the loaded area than it is at the centre $y = 0$. Inserting in the analysis the value of the rotation would induce higher values of b_y. As distinguished from the American authors, Stiglat does not assume an average value for the effective plate width. For the analysis of slab structures under horizontal loads, he recommends the least value of b_y, calculated at the center of the slab-column area. The author's study of the response of slab structures to the horizontal component of the seismic load leads to the same conclusion (Sub-secs. 5.1.2.2, 5.2.2 and 6.2.2). However, both design values for rotations are presented in this book (Sub-sec. 4.2.2 — Tables 4.8a and 4.8b).

Stiglat's investigations were extended herein in two directions. First, his procedure for evaluating the stiffness of slabs — which are assimilated with equivalent frame beams — was extended to the case of finite boundary plates. Second, and this is the main development of Stiglat's results, the modelling of the effect of the surface couple on the plate allowed its introduction into the structural analysis as a generalized interacting force (Chaps. 4 — 6). This idea underlies the author's method of elastic analysis for slab structures.

1.2.4.4 *Pecknold's procedure*. The determination of the effective plate width, which is required only in the equivalent frame method, is herein a secondary question. It is treated only as a possible illustration of the use of the surface couple model (Sec. 4.2). The first results obtained by Negruțiu in this respect were reported in 1970 [68]. Meanwhile, numerous other contributions have been published by American, Australian and Japanese authors in particular [59], [61], [62], [63], [67], [72], [76], [79], [80], [81], [84], [85]. This wealth of research stemmed from the growing interest in this field of analysis of slab structures which are treated as hyperstatic structures. However, it may be stated that the conclusions of these analytical studies are not essentially different from those presented herein (Secs. 4.2 and 5.2).

Still, for a comprehensive image of the present state of the art, mention should be made of two recent papers. Pecknold's study [79] of 1975 is of interest as it gives *three distinctive modelling procedures* for the effect of a surface couple M on the plate (Figs. 1.36 and 1.37):

a. The "rigid column" model consists of a pair of linear loads $(\bar{P}, -\bar{P})$ and a pair of linear couples $\bar{M}$ which act at

1.2 State of the art. Main research trends

the edges $x = \pm c$ of the contact area of width $2d$. The expressions

$$\bar{P} = \frac{M}{2ad} \quad \text{and} \quad \bar{M} = \frac{M}{4d}\left(1 - \frac{2c}{a}\right) \tag{1.66a}$$

result from the condition that rotation φ_ξ of the equivalent beam is constant along the entire length $2c$ of the contact area (Fig. 1.37A).

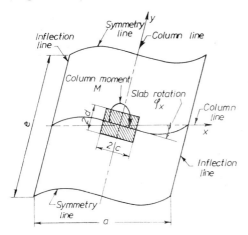
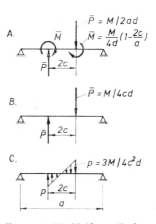

Figure 1.36 Schematic diagram for Pecknold's and Allen-Darvall's procedures. Slab structures of two-fold symmetry, subjected to a system of horizontal forces. Interior panel subjected to a surface couple M.

Figure 1.37 Mathematical modelling of the surface couple M effect (after Pecknold). A. Rigid column. B. Line loads. C. Flexible column.

b. The "line loads" model consists of a pair of linear loads $(\bar{P}, -\bar{P})$ which act at the edges of the contact area of width $2d$. The expression

$$\bar{P} = \frac{M}{4cd} \tag{1.66b}$$

is obtained by taking the effect of the pair of linear loads equivalent to the effect of the surface couple M (Fig. 1.37B).

c. The "flexible column" model is obtained by assuming a triangular pressure distribution as a "lower bound" case along the length $2c$ of the area subjected to the couple M. The expression

$$p = \frac{3M}{4c^2 d} \tag{1.66c}$$

1 Introduction

for the intensity of the triangular load at $x=\pm c$ is obtained using a similar equivalence relation (Fig. 1.37C). In the same manner as Mohammed-Popov (Sub-sec. 1.2.4.2) and Stiglat (Sub-sec. 1.2.4.3), Pecknold assumes that the equivalent loads $\bar{M}, \bar{P}$ and p (1.66a, b and c) are uniformly distributed in the range $-d \leqslant y \leqslant +d$. As distinguished from the procedures employed by the above-mentioned authors and Girkmann (Sub-sec. 1.2.4.1), in Pecknold's procedure the effect of the surface couple M is modelled for a *finite lever arm* $\varepsilon = 2c$ and $\varepsilon = 4c/3$, respectively, and not by passing to the limit ($\varepsilon \to 0$).

The three types of equivalent loads used by Pecknold in modelling the effect of the rigid slab-column connection (Fig 1.37) act on an interior panel that is sufficiently far away from the boundary of the flat slab (Fig. 1.36). In all these cases, the slab structure has a two-fold symmetry, the columns being spaced at an equal distance in each direction of the plane ($l_x = a$; $l_y = e$). In the case of a system of horizontal loads, the only generalized interacting force which occurs in the analysis is the surface couple reaction M at the level of the slab-column nodes.

In order to establish the equation of the deflected surface of the plate under the three types of equivalent loads, Pecknold uses the expression in Fourier series given by Timoshenko-Lévy [7] for the transverse load:

$$p(x, y) = \frac{2M}{a^2 d} \sum_{m=1}^{\infty} Q_m \sin \frac{2m\pi x}{a}, \quad (-d \leqslant y \leqslant +d),$$

$$(m = 1, 2, 3, \ldots) \tag{1.67}$$

where Q_m is a function which depends on the loading distribution. Without presenting the solution $w(x, y)$ found, Pecknold passes directly to the determination of the effective plate width: the plate rotation φ_x at the center of the contact area is equated to the rotation φ_ξ of a beam subjected to a concentrated couple M. Both Stiglat and Pecknold do not include in the analysis the average value of the rotation φ_x along the width $2d$ of the area subjected to the couple M. The expression obtained by Pecknold for the effective plate width is of the form:

$$b_y = \frac{\dfrac{d}{a} \cdot \dfrac{a}{e}}{(1-\mu^2)\left(f_B + \dfrac{6}{\pi^3} \sum_{m=1}^{\infty} \dfrac{1}{m^3} Q_m A_m\right)} \cdot e, \tag{1.68}$$

1.2 State of the art. Main research trends

where

$$A_m = \frac{\frac{m\pi e}{a} \text{sh} \frac{2m\pi d}{a} - \text{sh} \frac{m\pi e}{a} \left[2\text{sh}\left(\frac{m\pi e}{a} - \frac{2m\pi d}{a}\right) \right.}{2\text{sh}^2 \frac{m\pi e}{a}} +$$

$$+ \frac{\left. \frac{2m\pi d}{a} \text{ch}\left(\frac{m\pi e}{a} - \frac{2m\pi d}{a}\right) \right]}{2\text{sh}^2 \frac{m\pi e}{a}}, \qquad (1.69)$$

and f_B is a function of c/a, denoting the "flexibility reduction for the beam". The expressions given for f_B and particularly those for the load function Q_m are involved. For illustration, we shall reproduce here only the expressions established by Pecknold for the "line loads" model (Fig. 1.37B), as these assume the most simple form:

$$f_B = \left(1 - \frac{c}{a}\right)\left(1 - \frac{2c}{a}\right); \quad Q_m = \frac{a}{2c} \sin \frac{2m\pi d}{a}. \qquad (1.70)$$

As a numerical application, Pecknold draws a comparison between the three procedures for evaluating the effective plate width b_y using a simple schematic diagram: the flat slab with columns of square cross-section ($c = d$), framing a network of square panels ($e = a$) (Fig. 1.38). The intermediate curve obtained for the "line loads" model (Fig. 1.37B) coincides with the curve obtained by the author using a different modelling procedure for the effect of grouping the surface couples in a simple row (Sub-sec. 4.2.2 — Table 4.8a and Fig. 4.32). The similitude of results is all the more interesting as Pecknold did not use the superposition of effects and, hence, neglected the influence of the adjacent columns in the row under consideration [68], [75] *.

The results of the study undertaken by Pecknold induced him to adopt the "rigid column" model (Fig. 1.37A), which leads to the highest values of the effective plate width b_y. Thus, for relatively slender columns ($2c = 2d = 0.140a$), he obtains the ratio $b_y/e = 1$, i.e. an effective plate width equal to the width of the panel (Fig. 1.38 — curve A). It is already known that the full panel width is used in evaluating the flexural rigidity of

---

* A more relevant comparison of numerical results was limited as Pecknold does not give tabulated values.

1 Introduction

the equivalent beam in the equivalent frame method, assuming vertical loads [154], [155], [160], [162]. In the case of horizontal forces, because of the manner in which the loads are transmitted

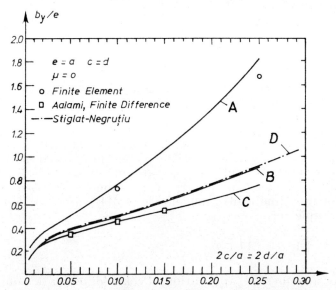

Figure 1.38 Numerical example illustrating Pecknold's procedure. Variation of effective plate width b_y as a function of procedure used for modelling the surface couple M effect. A. Rigid column. B. Line loads. C. Flexible column. D. Stiglat-Negruțiu.

to the slabs, only values $b_y < e$ can be considered. Using the procedure suggested herein (Sec. 4.2) for evaluating the effective plate width, the value $b_y/e = 1$ is obtained only for very large sizes of the slab-column contact area ($2c = 2d = 0.275a$) and this

Table

Slab structures of a two-fold symmetry

Comparative values of the effective plate width

$\xi = 0.50a$;

	$a/e = 0.50$		
$c/a = d/a$	$A - D$	N	$\Delta \%$
0.02	0.217	0.1780	+21.9%
0.04	0.271	0.2326	+16.5%
0.06	0.308	0.2834	+ 8.7%

1.2 State of the art. Main research trends

would correspond to larger capitals (Fig. 1.38 — curve D; Fig. 4.32).

1.2.4.5 *Allen-Darvall's procedure*. We shall conclude this survey of the procedures for modelling the effect of a surface couple on the plate and those used for evaluating *the effective plate width* by presenting the recent results due to Allen and Darvall [81]. The theoretical arguments underlying this procedure rely on previous research works [80] carried out by the first Australian author. Unfortunately, this paper was confined to a limited distribution. In the absence of these titles, we shall only make a comparative approach of the values b_y reported by Allen and Darvall in 1977 and those obtained by the author (Table 1.9).

The schematic diagram under consideration is identical with that used by Pecknold (Sub-sec. 1.2.4.4 — Fig. 1.36). However, Allen and Darvall approach only the first modelling procedure suggested by Pecknold (Fig. 1.37A), i.e. the "rigid column" model. It may be assumed that a concrete column is so rigid as to allow the plate to be plane within the column boundary at the plate-column junction, i.e. the column is axially infinitely rigid. The Australian authors observe that this assumption and the modelling procedure involved lead to the highest values of the effective plate width b_y. Besides, for the square interior panel, the values of b_y as obtained by these authors superpose over those calculated by Pecknold for this case (Fig. 1.38 — curve A). Similarly high are the values of b_y reported by Carpenter [62] and Brotchie [84]. As distinct from these approaches, the numerical data obtained by Tsuboi-Kawaguchi [59], and particularly those derived by Aalami with the finite difference treatment [72], coincide with those given by Pecknold for the third model under consideration (Fig. 1.38 — curve C). This would lead to the smallest values of b_y.

1.9

subjected to a set of horizontal forces
b_y/e, after Allen-Darvall and Negruțiu
$\mu = 0$

$a/e = 1.0$			$a/e = 2.0$		
$A - D$	N	$\Delta \%$	$A - D$	N	$\Delta \%$
0.429	0.3527	+21.6%	0.732	0.6266	+16.8%
0.532	0.4596	+15.75%	0.860	0.7902	+ 8.8%
0.602	0.5586	+ 7.8%	0.928	0.9332	− 0.56%

1 Introduction

As was shown above, the procedure given herein for evaluating the effective plate width leads to results that range between the two extreme zones. For the interior square ($e=a$) panel, these values coincide with those obtained by Pecknold for the second case (Fig. 1.38 — curves B and D). The adoption by Pecknold and Allen-Darvall of the "rigid column" assumption is justified neither by the interpretation of the numerical data obtained under this assumption (see Sub-sec. 1.2.4.4) nor by the qualitative analysis of the physical phenomenon. Nor do they consider in the analysis the effect of grouping the surface couples M in a simple row, which leads to a greater value φ_x of the slab rotation and hence to a smaller value of the parameter b_y.

Pecknold and Allen-Darvall had no knowledge of Negruțiu's studies [64], [68], [70], [73] concerning the distribution of the sectional stresses M_x, M_y and M_{xy} in the plate subjected to a surface couple M and to a set of surface couples, respectively. The characteristics of a local perturbation, specific to this type of load in the elastic range (see Secs. 3.3 and 4.1), show the strength reserves of the slabs in a structure subjected to a system of horizontal forces. As no such strength reserves due to the redistribution of stresses exist in the columns, the adoption in the structural analysis of smaller values for the bending stiffness of slabs and, hence, higher values for the relative column stiffness is fully justified.

1.2.5 Early approaches to the elastic analysis of slab structures

Before presenting the author's method of elastic analysis for slab structures (Chaps. 2 through 6), we shall discuss the only two contemporary works which use the surface couple model as a generalized interacting force. The first is Scholz's procedure [66] which relies on the results obtained by Stiglat. Whereas this procedure is confined to the study of a particular scheme, Pfaffinger's method of elastic analysis [71] is a generalized approach. Besides, the latter method is better with respect to the manner in which the analysis is performed and the level of mathematical treatment.

1.2.5.1 *Scholz' procedure.* The introduction of the surface couple as a generalized interacting force into the elastic analysis of slab structures was independently suggested (1969) by Scholz [66] and Negruțiu [64], [65]. This marked real progress in the study of slabs as structural members integrally built with the columns. As distinct from the general method of static and dynamic analysis given in the subsequent chapters, Scholz' procedure is of restricted applicability. This inconvenience is due to the particular category of slab structure considered, the specific type of loads and the simplifying assumptions adopted in the schematic diagram. Thus,

  a. Scholz considers a multistorey slab structure having at least one axis of symmetry, subjected to only one set of horizontal forces S_k;

  b. Because of the antisymmetry of this loading case, the zero moment points in the slabs of the floors are assumed to be located, for all the successive spans a_i, in the same vertical plane which is normal to the plane of the longitudinal frames in which the structure is divided (see Fig. 5.14);

  c. Under this assumption, Scholz suggests a sub-division of the structure into three possible schematic diagrams (Fig. 1.39): multistoried end frames (A and C) and multistoried interior frame (B). The static analysis, through the general force method, is carried out only for the interior frame which is assumed to be symmetric $a'_i = a''_i = a_i/2$ (Fig. 1.40);

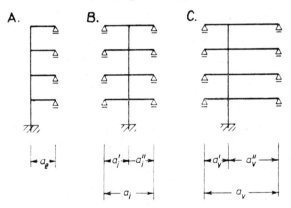

Figure 1.39 Schematic diagram for Scholz' procedure. Slab structures under horizontal loads.

A. End frame. B. Symmetric interior frame. C. Unsymmetric interior frame.

  d. Transversally, the structure is assumed to consist of an indefinite number of identical frames. Hence, for all the slabs constituting the floors, which are treated as equivalent beams, Scholz assumes only one mathematical model, i.e. the infinitely long plate with two parallel simply supported edges;

  e. In the chosen basic system, the columns are assumed to be continuous along the height of the structure; the only

1 Introduction

unknowns which occur are the couple reactions X_i acting at the slab-column nodes (Fig. 1.40). This assumption holds only in the case of a symmetrical schematic diagram. For any other type of scheme, including the multistoried end frame, one should have used a more complex basic system where the grouping of the generalized interacting forces would include both couple reactions and force reactions (See Chaps. 4 — 6).

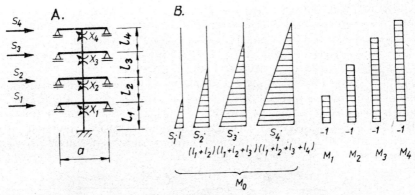

Figure 1.40 Scholz's procedure. Use of the general force method. A. Basic system. B. Auxiliary diagrams for moments in columns.

The only compatibility condition used by Scholz is the condition of zero relative rotation for the slab-column nodes. The coefficients δ_{ij} ($j \neq i$) of the unknowns X_i and the free terms Δ_{is} (see Sub-sec. 5.1.1) in the particular system of condition equations represent angular diaplacements of the columns. The rotation φ_x of the slab at the slab-column nodes occurs only in the expressions of type δ_{ii} along the main diagonal of the unit displacement matrix (5.1).

For illustration, Scholz considers a symmetric interior frame with four levels (Fig. 1.40), assuming that $l_1 = l_2 = \ldots = l$ and $S_1 = S_2 = \ldots = S$. The slabs at all levels are of equal thickness h and the column moment of inertia I_c is constant along the height $4l$ of the structure. For this particular scheme, the system of compatibility equations is given by:

$$(1 + \rho) X_1 + X_2 + X_3 + X_4 = 8\,Sl\,;$$
$$\vdots \qquad\qquad \vdots \qquad\qquad (1.71)$$
$$X_1 + 2X_2 + 3X_3 + (4 + \rho)X_4 = 15\,Sl$$

where $\rho = \dfrac{EI_c}{l}\,\varphi_x^{(M=1)}$ is a factor of the relative slab-column stiffness.

1.2 State of the art. Main research trends

Denoting by

$$\varphi_x^{(M=1)} = \left(\frac{M}{D} Z\right)_{M=1} \tag{1.72}$$

the slab rotation at $x = \xi = a/2$, due to a surface couple $M = 1$ acting at ξ, and assuming that $\mu = 0$, one finds that *:

$$\rho = \frac{12 I_c}{l h^3} Z. \tag{1.73}$$

Scholz gives several alternative solutions to the system of condition equations (1.71) as a function of parameter ρ (1.73). Once the values of the unknowns $X_1, X_2, \ldots, X_4$ are determined, we can plot the final diagram of the bending moments for the columns of the interior frame under study (Fig. 1.41). The distribution of the bending moments in the slabs of the structure is not given by the German author.**

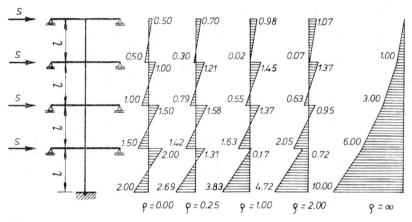

Figure 1.41 Numerical example illustrating Scholz' procedure. Final diagram for moments in columns depending on the relative stiffness factor ρ.

The values of the factor Z were derived by Scholz from solution $w(x, y)$, expressed in simple series (2.16), which was given by Stiglat for the infinitely long plate (Sub-secs. 1.2.4.3 and 2.2.1.2). By limiting the order of the last term summed in the series to $m = 60$, Scholz obtained less accurate values of the rotations φ_x than those reported in the sequel (Chaps. 3 — 6), which

---

* The factor Z is equivalent to the dimensionless coefficient k_φ^x used in this book to denote the unit angular displacements (Sub-secs. 2.3.1 and 2.3.2).
** The similar schematic diagram treated by Negruțiu with the general displacement method would correspond to $\rho = 0.185$ (Sub-sec. 5.2.2 — Fig. 5.18).

1 Introduction

are computed for a more severe accuracy condition (Sec. 3.4 — Table 3.24). Greater differences ($-$ 31.4 per cent $\leqslant \Delta \leqslant$ + 9.5 per cent) are observed between the values of the sectional bending moments M_x and M_y (Table 3.25). Scholz did not calculate the values of the corresponding torsional moments M_{xy}.

Mention should be made that all referrences to the accurate values obtained by Negruțiu are made to those values of the stresses that are calculated by Scholz on computer. But almost half of the amount of numerical data given by Scholz are "obtained by graphic interpolation". Besides, the value of Poisson's ratio adopted throughout his analysis was zero ($\mu = 0$).

Because of the restricted applicability, deriving from the excessive particularization of the schematic diagram, and the approximations adopted in the numerical computation, Scholz's procedure is of a moderate interest. However, it is the merit of the German researcher to rank among the few authors who used the surface couple model in studying the effect of the rigid connections characterizing slab structures.

1.2.5.2 *Pfaffinger's method*. Pfaffinger's latest contribution [71] in the field is invaluable. The effect of the couple reactions at the slab-column nodes is modelled here by introducing the surface couples into the analysis. In Pfaffinger-Thürlimann's procedure (Sub-sec. 1.2.3.2), the main approximation refers, as the authors themselves observe, to the influence of the column stiffness, which is neglected. This means that the effect of the rigid connection between the slabs and columns is disregarded in the analysis. In order to remove this inconvenience, Pfaffinger undertakes this time the study of this continuity effect, relying on the results obtained in his early approaches [48], [53].

Using an energy method based on the principle of minimum complementary energy

$$U^* = \frac{1}{2}\int \sigma_{ij}\varepsilon_{ij}\,dV = \min, \qquad (1.74)$$

he develops a method of analysis for slab structures which is of greater applicability than all the procedures presented so far. As Pfaffinger's study deals only with bodies of linear elastic behaviour, the strains ε_{ij} are linear functions of the stresses σ_{ij}. Hence, the complementary energy function U^* has in this linear case the same value as the strain energy U.

Pfaffinger's method of elastic analysis is characterized by the following:

a. The approach is confined to the most common case of slab structures which can be assembled by rectangular plates

1.2 State of the art. Main research trends

of constant thickness and by *columns with or without capitals*. The structure has at least one axis of symmetry (Fig. 1.42).

b. Two parallel edges of the slabs are assumed to be simply supported, whereas the remaining edges are supported by *edge-*

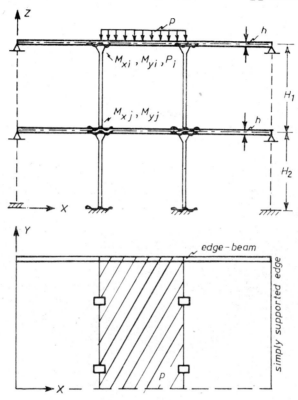

Figure 1.42 Schematic diagrams for Pfaffinger's method. Slab structures under vertical loads. Basic system.

beams of arbitrary flexural rigidity EI and torsional rigidity GT (Fig. 1.43).

c. The cross-section of all columns and capitals are rectangular (Figs. 1.42, 1.44 and 1.45). For small values of α, the classical linear distribution of the normal stresses σ_{zz} is also adopted at the region of the capitals ($0 \leqslant z \leqslant h_1$). In this manner, the state of stress along the entire length of a column could be defined as a function of z.

75

1 Introduction

d. The *generalized interacting forces* at the level of the slab-column nodes are of two types: the vertical load P, with a uniform stress distribution (Fig. 1.45A); and the two bending moments M_x and M_y^*, with corresponding linear stress distribution (Fig. 1.45B and C). These column reactions represent the unknowns of the system.

All the columns are assumed to be *rigidly connected with the slabs* and clamped at the foundations. After the assembly

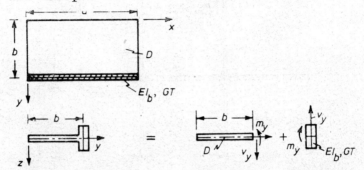

Figure 1.43 Pfaffinger's method. The edge-beam. Edge-beam conditions at $y = b$.

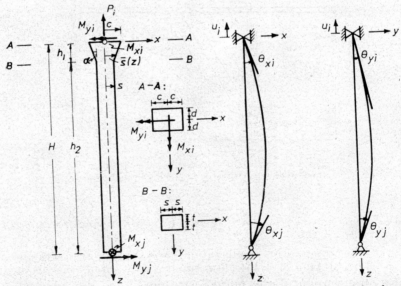

Figure 1.44 Pfaffinger's method. Generalized interacting forces at slab-column nodes. Sign convention.

---

* Pfaffinger does not refer to this type of generalized interacting force as a "surface couple", which was introduced by the author [64].

1.2 State of the art. Main research trends

of the total flexibility matrix, hinges and elastic foundations can be taken into account. The computational program allows also to treat the columns as *undeformable supports* or as *supports elastically deformable* along the z axis.

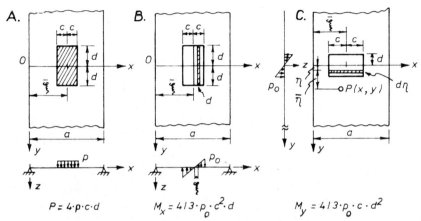

Figure 1.45 Mathematical modelling of the effect of generalized interacting forces (after Pfaffinger).
A. Force reaction P. B. Surface couple $\vec{M}_x \| O_y$. C. Surface couple $\vec{M}_x \| O_x$.

e. It is assumed that all the given loads $\mathbf{p}_l$ are applied on the plates only as transverse (vertical) loads. Pfaffinger's method *does not allow consideration of systems of horizontal loads*.

f. The principle of minimum complementary energy is used to establish the conditions of compatibility between columns and slabs, which determine the magnitude of column reactions $\mathbf{p}_k$.

The analytical calculations carried out by Pfaffinger develop in three steps. In the first step, mathematical models are built for the behaviour of slabs, edge beams and columns as structural members (Figs. 1.43, 1.44 and 1.45).

For the slab defined at b, Pfaffinger uses the solution $w(x, y)$, given by Eqn. (1.39), separated into a particular part w_0 and a homogeneous part w_h (see Sub-sec. 1.2.3.2). Solution w_0 for the infinitely long plate under the load $P = 4pcd$ (Fig. 1.45A) is, in this case too, that given by Girkmann [9] (see Sub-sec. 2.2.1.1. — formulae 2.11 — 2.13). For the same plate subjected to the surface couple $M_x = \dfrac{4}{3} p_0 c^2 d$ (Fig. 1.45B) and $M_y = \dfrac{4}{3} p_0 c d^2$ (Fig. 1.45 C), Pfaffinger uses solutions w_0 distinct from those employed in this work (see Sub-secs. 2.2.1.2 and 2.2.1.3). For instance, the particular solution established by Pfaffinger for the load $\vec{M}_x \| Oy$, which holds in the range $0 \leq y \leq d$, is expressed as

$$w_0(x, y) = \frac{3M_x}{2adc^3D} \sum_n \frac{1}{\alpha_n^6} \cos \alpha_n u(\sin \alpha_n c - \alpha_n c \cos \alpha_n c\{2 +$$
$$+ [\alpha_n y \operatorname{sh} \alpha_n y - (2 + \alpha_n d) \operatorname{ch} \alpha_n y]\} e^{-\alpha_n d} \sin \alpha_n x, \qquad (1.75)$$

where $\alpha_n = n\pi/a$.

The solution w_h of the fundamental homogeneous equation (1.1a) is identical with that used by Pfaffinger in a previous paper (Sub-sec. 1.2.3.2 — formula 1.40). The four constants of integration A_n, B_n, C_n and D_n can be determined from the total of four boundary conditions at $y = 0$ and $y = b$. For each term n of the superimposed solution $w(x, y)$, given by Eqn. (1.39), the following boundary equations are obtained:

$$\mathbf{a}_n = -[\mathbf{B}_n]^{-1} (\mathbf{L}_{k,n}\mathbf{p}_k + \mathbf{L}_{l,n}\mathbf{p}_l), \qquad (1.76)$$

in which $\mathbf{a}_n$ represents the constants of integration; $\mathbf{B}_n$ denotes the matrix of homogeneous boundary values; $\mathbf{L}_k$ is the load matrix corresponding to the general interacting forces $\mathbf{p}_k$; $\mathbf{L}_l$ denotes the load matrix corresponding to the given loads $\mathbf{p}_l$.

The edge-beams defined by the relative stiffness factors

$$K = \frac{EI_b}{aD}; \quad \gamma = \frac{GT}{aD} \qquad (1.77)$$

with respect to the cylindrical stiffness D of the plate (Fig. 1.43), allow the modelling of the boundary conditions at $y = 0$ and $y = b$. The two classical condition equations [7] are similar to those introduced by Rabe (Sub-sec. 1.2.3.1 — formulae 1.38). By varying K and γ, this equation allows one to take into account the most important kinds of elastically clamped and elastically supported edges.

Since the system of equations (1.76) allows the determination of the integration constants from the expression of w_h (1.40), the total deflection surface $w(x, y)$ given by Eqn. (1.39), which satisfies the boundary conditions, is determined. This elastic surface contains $\mathbf{p}_k$ as the only unknowns.

In the second step of the analytical calculations, Pfaffinger establishes the expressions for the complementary energy U^* for each slab and for each column. In this context, he defines the local flexibility matrix of plate $\mathbf{F}$ and of column $\mathbf{F}_c$, respectively.

1.2 State of the art. Main research trends

In the case of slabs, the elastic surface $w(x, y)$ of only one plate under the action of the local unknowns $\mathbf{p}_k$ and the given loads $\mathbf{p}_l$ can be written as

$$w(x, y) = \begin{Bmatrix} w_1(x, y, p_1 = 1) \\ \vdots \\ w_k(x, y, p_k = 1) \\ w_l(x, y, p_l = 1) \end{Bmatrix}^T \cdot \begin{Bmatrix} \mathbf{p}_k \\ \mathbf{p}_l \end{Bmatrix}. \tag{1.78}$$

For the total complementary energy U^* (1.74) of this plate, only the work done by the bending and torsional moments is considered:

$$U_p^* = U_1^* + U_2^* = \frac{1}{2} \begin{Bmatrix} \mathbf{p}_k \\ \mathbf{p}_l \end{Bmatrix}^T \left[\begin{array}{c|c} \mathbf{F} & \mathbf{L} \\ \hline [\mathbf{L}]^T & \mathbf{C} \end{array} \right] \begin{Bmatrix} \mathbf{p}_k \\ \mathbf{p}_l \end{Bmatrix}, \tag{1.79}$$

where U_1^*, U_2^* represent the complementary energy of the slab and the two edge beams, respectively; $\mathbf{L}$ is the load matrix which couples the unknowns with the given loads; $\mathbf{C}$ denotes the matrix of load flexibility. Thus, the matrix $\mathbf{C}$ will have no influence on the variation of U_p^* with respect to $\mathbf{p}_k$ and need not be determined. It is easy to see that $\{\boldsymbol{\delta}_k\} = \mathbf{F}\,\mathbf{p}_k$ and $\{\boldsymbol{\delta}_l\} = \mathbf{L}\,\mathbf{p}_l$ represent the generalized displacements under $\mathbf{p}_k$ and $\mathbf{p}_l$ at the points where $\mathbf{p}_k$ are applied.

In the case of columns (Fig. 1.44), the complementary energy for only one column is

$$U_c^* = \frac{1}{2} \{\mathbf{p}_k\}^T \mathbf{F}_c\, \mathbf{p}_k. \tag{1.80}$$

The flexibility matrices $\mathbf{F}$ (1.79) and $\mathbf{F}_c$ (1.80) are symmetric and positive definite.

In the last step of the analysis, Pfaffinger performs the required assembly operation. The total complementary energy U^* for the entire slab structure is obtained by summing up the expressions U_p^* given by (1.79) and U_c^* given by (1.80) for all plates and columns. Denoting by $\mathbf{P}_k$ and $\mathbf{P}_L$ the global vectors of column reactions and given loads, respectively, in the case of a structure with m slabs and n columns, the following result is obtained:

$$U^* = \frac{1}{2} \begin{Bmatrix} \mathbf{P}_k \\ \mathbf{P}_L \end{Bmatrix}^T \left(\sum_m [\overline{\mathbf{A}}]_m^t \left[\begin{array}{c|c} \mathbf{F} & \mathbf{L} \\ \hline [\mathbf{L}]^t & \mathbf{C} \end{array} \right]_m [\overline{\mathbf{A}}]_m \right) \begin{Bmatrix} \mathbf{P}_k \\ \mathbf{P}_L \end{Bmatrix} +$$

$$+ \frac{1}{2} \{\mathbf{P}_k\}^T \left(\sum_n [\mathbf{A}]_n^t [\mathbf{F}_c]_n [\mathbf{A}]_n \right) \mathbf{P}_k, \tag{1.81}$$

1   Introduction

where $\bar{\mathbf{A}}$ and $\mathbf{A}$ are the rectangular topological matrices corresponding to the slabs and columns, respectively. For instance, between the local unknowns $\mathbf{p}_k$ and the global unknowns $\mathbf{P}_k$, there exists the relationship

$$\mathbf{p}_k = [\mathbf{A}]_k \, \mathbf{P}_k. \tag{1.82}$$

The conditions of minimum U^* under the variables $\mathbf{P}_k$ are obtained from

$$\frac{\partial U^*}{\partial P_i} = 0 \tag{1.83}$$

which lead to the system of equations

$$\bar{\mathbf{F}}\mathbf{P}_k + \bar{\mathbf{L}}\mathbf{P}_L = 0, \tag{1.84}$$

which furnishes the column reactions $\mathbf{P}_k$ under the given loads $\mathbf{P}_L$.

The simplicity of Pfaffinger's method is only apparent. In the case of hinges between some columns and slabs or foundations certain generalized interacting forces P_i will be zero. These forces do not contribute to the complementary energy. The flexibility matrices $\mathbf{F}$, $\mathbf{F}_c$ and the load matrix $\mathbf{L}$ in (1.79), (1.80) and (1.81) should be modified in $\bar{\mathbf{F}}$, $\bar{\mathbf{L}}$ (1.84), respectively. Similar modifications that would make the analysis more involved, would be induced by the introduction of other modes of connection at the slab-column nodes. The main inconvenience of Pfaffinger's method derives from the fact that the basic system adopted (Fig. 1.42) does not allow the direct loading of the columns with given loads, e.g. with a system of horizontal forces. This fact can be also easily observed from the expression of the second term of U^* (1.81). Hence, the method of the Swiss researcher *does not allow the dynamic analysis of slab structures.*

For illustration, Pfaffinger undertakes the analysis of a symmetric, one-level slab structure with four interior columns (Fig. 1.46). The slab was investigated for a load q uniformly distributed throughout the surface of the floor. Poisson's ratio was taken as $\mu = 1/6$ and all Fourier series were evaluated with 50 terms. All the columns are rigidly clamped at the bottom and have square cross-sections. Calculations were made for different ratios s/c of the column and capital dimensions, as well as for two ratios E_c/E_s of the modulus of elasticity of the columns and the slab, respectively (Table 1.10). The results of the comparative approach for the load p uniformly distributed over only a portion of the floor are presented in the form of diagrams for the bending moments (Fig. 1.47). The following remarks should be made concerning Pfaffinger's numerical results. The negative

1.2 State of the art. Main research trends

moments m_x induced in the slab, in the region of the columns, pertaining to the slab structure by the load q, must be greater in absolute value when the effect of the rigid connection is neglected than it is when this additional restriction is introduced in the analysis. Pfaffinger's numerical data would derive in the opposite

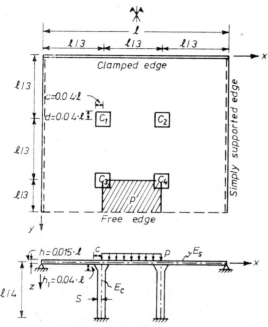

Figure 1.46 Numerical example illustrating Pfaffinger's method. Structures of simple symmetry. Partial uniformly distributed load p.

conclusion (Table 1.10 — Incompressible columns). Additionally, the order of magnitude of the rotation φ_x is unjustifiedly great when this interaction is neglected.

★

The conclusions of the foregoing critical survey of the present state of theoretical research in the field of slab structures could be summarized as follows :.

a) All works consulted assuming the hypotheses of the Theory of Elasticity proceed from recognized solutions of the fundamental equation of plates and neglect the effects due to the actual behaviour of reinforced concrete under loads.

b) With few exceptions (Lewe, Rabe, Pfaffinger), both classical and modern studies idealize the slab-column supports as point-like supports — even in the case of flat slabs with capitals.

1 Introduction

Table 1.10
Slab structures under the uniformly distributed load q

Values of the rotations φ_x and of the bending moments m_x in the column axes C_1 and C_3, after Pfaffinger

$c = d;\quad s = t;\quad \mu = 1/6$

E_c/E_s	s/c	Column C_1		Column C_3	
		$10^3 m_x/ql^2$	$10^5 \varphi_x \left/ \dfrac{ql^3}{D}\right.$	$10^3 m_x/ql^2$	$10^5 \varphi_k \left/ \dfrac{ql^3}{D}\right.$
Compressible columns					
10	1.00	-9.211 (-10.938)	-0.96	-11.949	-1.10
	0.50	-2.588 (-9.357)	6.95	-5.049 (-9.215)	9.99
1	1.00	-13.443	1.97	-14.920	2.45
	0.50	-11.952	0.66	-14.046	0.92
Incompressible columns					
10	1.00	-14.047	1.81	-15.247	2.19
	0.50	-13.890	-2.83	-15.272	-3.47
1	1.00	-14.066	2.40	-15.243	2.90
	0.50	-14.047	1.80	-15.247	2.18
Incompressible columns. No interaction					
—	—	-13.205	-21.50	-15.316	-27.40

Note. The values of m_x have been added in parantheses when the maximum bending moments occur at the edge of the column capitals and not at the central points.

c) In order to determine the state of stress in the slabs, most works resort exclusively to the zero transverse displacement condition of the type (1.7) or (1.42) at the slab-column nodes. If this approximation could be considered to be compatible with the point-like support assumption — assumed by Nádai, Marcus, Timoshenko, Bretthauer and others, it is not admissible with Lewe, Rabe and Pfaffinger-Thürlimann, who introduce the cross-sectional area of the columns in the analysis, ignoring however the supplementary condition of continuity for rotations.

d) Few are the authors who do not ignore the effect of the rigid slab-column connection, introducing the relative stiffness of the two members into the analysis. Such are Andersson (whose procedure cannot be considered as accurate due to other approximations), Bretthauer (however the couples which he introduces

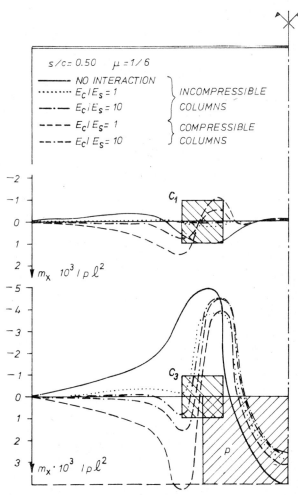

Figure 1.47 Numerical example illustrating Pfaffinger's method. Diagrams for sectional bending moments m_x along column rows C_1 $(y = l/3)$ and C_3 $(y = 2l/3)$.

as generalized interacting forces are concentrated and are not distributed on the cross-sectional area of the columns), Scholz (the validity of his procedure is however limited to particular

1 *Introduction*

types of multistoried symmetrical structures) and also Pfaffinger in his latest paper [71]. The last quoted author is the only one to give a complete solution that may be compared in this respect with the method which is the object of the present book. Mention should be made however that the Swiss researcher adopts a different mathematical model for the effect of the column rigidity, resorts to an energy principle and so chooses a different course in the structural analysis.

e) With reference to the design loads, Scholz's procedure is only valid for slab structures subjected to horizontal loads, whereas Pfaffinger's method — as well as all the already mentioned procedures — are limited exclusively to the cases of gravity loads.

f) The procedures like those due to Pecknold, Allen-Darvall and others are restricted to the evaluation of the effective plate width if the analysis of slab structures under horizontal loads is conducted by the equivalent frame method. Stiglat pursued a similar aim when be established the mathematical model of a surface couple applied to an infinitely long plate.

A comprehensive study of the effects of the rigid non-point-like slab (plate)-column (bar) connection is strictly necessary as it is pertinent for slab structures. Such a study was implemented by the author for three categories of plates and was the starting point of the present work. By incorporating in the analysis the surface couple as a generalized interacting force the author could develop a general method of elastic, static and dynamical analysis for slab structures subjected to arbitrary, gravity or seismic, vertical and horizontal loads.

1.3 Scope of the book

The examination of the main research trends in the field of slab structures, which was carried out in the preceding sections, was aimed at satisfying the questions that remained unanswered or at identifying further new questions (Chap. 1). The conclusions derived from the review of these approaches justify the author's study, the purpose of which is to establish a unified method of analysis for slab structures in the elastic range. The results thereby derived are original contributions in two distinct fields [75], [86]:

— in the Theory of Plates, new solutions were established to the Lagrange-Sophie Germain fundamental equation;

— in the research and design of slab structures, the author's method of static and dynamic analysis contributes to the theore-

1.3 Scope of the book

tical investigation and economical design of this category of structures.

From the range of problems which have been insufficiently studied so far, the effect of a surface couple M on plates ranks first in importance (Chaps. 2 and 3). Chapter 2 gives the theoretical developments and the new mathematical models of the mid-surface of plates deformed under the action of a surface couple. The formulae suggested for the numerical calculation of the elastic displacements, sectional stresses and boundary reactions are specified for three main categories of plates: the infinitely long plate; the rectangular plate with two simply supported parallel edges and the remaining edges free; and the rectangular plate simply supported on the entire boundary.

In order to satisfy the requirements of structural analysis (Chaps. 4, 5 and 6), computational algorithms were established both for loads pertaining to the surface couple M and to other conservative types of transverse loads: the locally distributed force P and the load q uniformly distributed throughout the plate surface. In the elastic analysis of slab structures, loads of type P or M may occur either as given loads or as generalized interacting forces.

Chapter 3 presents the conclusions of a numerical experiment — using tens of thousands of calculated values — concerning the behaviour of plates subjected to a surface couple M. The subsequent comparative approach identifies the geometrical and physico-mechanical parameters which influence the magnitude and distribution of the elastic displacements and stresses under the load M.

Chapter 4 gives preliminary elements to structural analysis by extending the results obtained for only one surface couple M and for only one locally distributed force P. Since the elastic plate is here treated as a structural member, problems specific to the general force method are first discussed (Sec. 4.1). The foregoing results allowed the consideration in the analysis of both the reaction forces of type P and the reaction couples of type M developed at the slab-column nodes. In calculations, these generalized interacting forces are introduced as locally distributed forces on the slab-column contact area. Given the geometrical and mechanical characteristics of slab structures which usually assume at least one axis of symmetry, it is desirable to group the unknowns. Genuine aspects of the distribution of the elastic displacements and stresses induced in plates by sets of unknowns are observed, using the superposition of effects. An attempt is made to extend this procedure for modelling the effect of sets of unknowns to the study of structures without symmetry. This effort was prompted

1 Introduction

by the lack of such approaches to these unsymmetrical structures.

Plates are also treated as structural members when the equivalent frame method is applied (Sec. 4.2). The results of the numerical experiment on the magnitude and distribution of rotations of plates subjected to only one surface couple M are extended to the study of the effect induced by sets of associated surface couples. A procedure is given for evaluating the effective slab width, which can be used in the analysis of slab structures subjected to horizontal forces.

The new method of static analysis for slab structures is described in Chap. 5. Owing to the form of the algorithms and the systematized treatment suggested in Chapts. 2—4, resort can be made to the general force method (Sec. 5.1). The general scheme of analysis is virtually applicable to any category of slab structures and any type of load. The schematic diagrams chosen to illustrate the manner in which the analysis must be conducted have at least one axis of symmetry. These schemes, which allow the grouping of unknowns, were comparatively investigated for two types of loads : Vertical loads acting across the slab ; and horizontal loads acting in the plane of the slab. Several examples of analysis are given to illustrate the behaviour of slab structures in the case of failure of supports. So far, the behaviour of slab structures under lateral loads and under the effect of the failure of supports has not been approached with the general force method.

The comparative studies highlight the main approximations resulting from the neglect of the rigid slab-column connection. These shortcomings were removed in the author's method by introducing the relative slab-column flexural stiffness in the analysis (Sec. 5.1).

Alternatively, the static analysis of slab structures subjected to a system of horizontal forces is made, in the case of multistoried symmetrical structures, resorting to the general displacement method. A new procedure is introduced for evaluating the bending stiffness of the equivalent beams in the elastic range (Sec. 5.2).

Chapter 6 gives a brief description of the manner in which the design method suggested for the static analysis can be extended to the dynamic analysis of slab structures. So far, such an approach has never been explored. The dynamic approach was made using the known free and forced vibration analysis of structures, namely the flexibility matrix method (Sec. 6.1). This dynamic method is illustrated through examples of analysis concerning natural and forced vibrations of mixed structures consisting of bars and plates (Sec. 6.2).

1.3 Scope of the book

Both in the static and dynamic treatments, the numerical calculations were carried out using only one simple relevant diagram of a slab structure. Owing to the simplicity of the diagram, the alternate structural versions under consideration could be comparatively examined. These versions were obtained by changing the design parameters: the in-plane sizes of slabs, the relative sizes of the column cross-section, the position of columns in the plane of the floor or the value of the physico-mechanical parameters.

The study of these alternate versions also showed the manner in which the slab behaviour is influenced by various parameters, such as the type of load (i.e. vertical, horizontal, locally distributed or distributed throughout the surface of the floor, failure of supports), the position and the contact area of the locally distributed loads or of the moving mass, the type of sets of associated generalized interacting forces, etc. The main design parameter which defines the behaviour of slab structures in the elastic range is the relative bending slabs-columns stiffness.

Using the auxiliary values tabulated in the Appendix, one can calculate the elastic displacements and the stresses induced in the square plates by the action of a surface couple M. The computational accuracy of these values, which are given in the form of dimensionless coefficients, satisfy the requirements set by design and research institutes.

2

New mathematical models for the deflected middle surface of plates

2.1 Preliminaries

2.1.1. Basic assumptions

The study of plates subjected to loads acting normal to their middle surface represents a complex problem of three-dimensional elasticity. In this most generalized form, the problem is at present of purely theoretical interest. In the applied research, the stresses σ_z, acting normal to the middle surface, as well as the influence of the tangential stresses τ_{xz} and τ_{yz} on the deflection of the plate, are often ignored. Hence, the study of the state of stress and deformation is considerably simplified, the solutions being only slightly different from those obtained by the exact calculation procedures.

Beside this simplified treatment, the Theory of Plates allows for the use of the basic assumptions underlying the Theory of Structures [7], [8], [9], [15], [16], [20], [21], [22], namely:
— the continuity, homogeneity and isotropy of the structural material;
— the perfectly elastic behaviour of the material within the limits of the given loads, the value of the modulus of longitudinal elasticity E (1.2) being the same for tensile and compressive stresses;
— the thickness h of the plate (Figs. 2.1 and 2.2) is not changed by deflection ($\sigma_z = 0$);
— the assumption of the straight normals (Fig. 2.1);
— the assumption of small displacements;
— the assumption of a proportional ratio between deformations and the corresponding stresses.

2. Models for the deflected middle surface of plates

Since the elastic displacements are small as compared to the thickness of the plate and can be expressed by the linear functions of stresses, the *principle of superposition of effects* holds in the Theory of Plates. The method proposed for modelling the effect of a surface couple is based on this principle (Sec. 2.2).

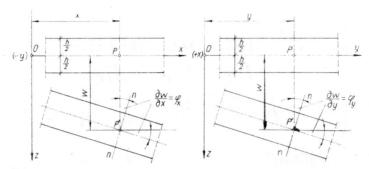

Figure 2.1 Elastic displacements of plates. Straight-normals hypothesis.

Throughout this book, a sign convention is adopted for the stresses M_x, M_y, M_{xy}, T_{xz} and T_{yz} (Fig. 2.2), which ensures that the torsional moments are equal ($M_{xy} = M_{yx}$) with regard to both value and sign. However, the sign convention adopted by Timoshenko [7], Girkmann [9] and Pfaffinger-Thürlimann [48],

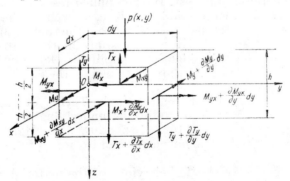

Figure 2.2 State of stress in plates under transverse loads. Sign convention.

leads to the equality $M_{xy} = -M_{yx}$ which is less suited for numerical applications. In our sign convention, the bending moments generate convexities in the direction of the gravity load $p(x, y)$, i.e. in the positive direction of the axis Oz. Since all the shearing forces are parallel to the axis Oz, for simplification, we have used the notation $T_{xz} = T_x$ and $T_{yz} = T_y$.

2.1 Preliminaries

2.1.2 General computation formulae

The fundamental equation of plates, which is known in the explicit form :

$$\frac{\partial^4 w}{\partial x^4} + 2\frac{\partial^4 w}{\partial x^2 \partial y^2} + \frac{\partial^4 w}{\partial y^4} = \frac{p(x,y)}{D} \qquad (2.0)$$

is equivalent to Eqn. (1.1), where the Laplace operator (Sub-sec. 1.2.2.1) is denoted by

$$\nabla^2 = \frac{\partial}{\partial x^2} + \frac{\partial}{\partial y^2}.$$

The cylindrical rigidity D of the plate is given by Eqn. (1.2).

Table 2.1

General computation formulae

Transverse elastic displacements (deflections)

$$w(x, y) = w(p, D, a, b, c, d, \xi, \eta, x, y) \qquad (2.1)$$

Angular elastic displacements (rotations)

$$\varphi_x = \frac{\partial w}{\partial x} \qquad (2.2) \qquad\qquad \varphi_y = \frac{\partial w}{\partial y} \qquad (2.3)$$

Bending moments

$$M_x = -D\left(\frac{\partial^2 w}{\partial x^2} + \mu \frac{\partial^2 w}{\partial y^2}\right) \qquad (2.4)$$

$$M_y = -D\left(\frac{\partial^2 w}{\partial y^2} + \mu \frac{\partial^2 w}{\partial x^2}\right) \qquad (2.5)$$

Torsional moments

$$M_{xy} = M_{yx} = -D(1-\mu)\frac{\partial^2 w}{\partial x \partial y} \qquad (2.6)$$

Shearing forces

$$T_x = -D\left(\frac{\partial^3 w}{\partial x^3} + \frac{\partial^3 w}{\partial x \partial y^2}\right) \qquad (2.7)$$

$$T_y = -D\left(\frac{\partial^3 w}{\partial y^3} + \frac{\partial^3 w}{\partial x^2 \partial y}\right) \qquad (2.8)$$

Boundary reactions

$$R_x = -D\left[\frac{\partial^3 w}{\partial x^3} + (2-\mu)\frac{\partial^3 w}{\partial x \partial y^2}\right] \qquad (2.9)$$

$$R_y = -D\left[\frac{\partial^3 w}{\partial y^3} + (2-\mu)\frac{\partial^3 w}{\partial x^2 \partial y}\right] \qquad (2.10)$$

2 *Models for the deflected middle surface of plates*

The function $w(x, y)$ — determined as a solution of the fourth-order differential equation (2.0), for a certain load and for given boundary conditions — represents the middle surface of the deflected plate. The elastic displacements w, φ_x, φ_y, the stresses M_x, M_y, M_{xy}, T_x, T_y and the boundary reactions R_x, R_y can be expressed using this function and its first-, second- and third-order partial derivatives (Table 2.1 — formulae 2.1—2.10).

2.1.3 Characteristics of the categories of plates investigated

Most authors dealing with slab structures neglect the existence of the rigid connection between columns and slabs. Hence, the few-existing theoretical studies on the effect of the surface couple on plates are incomplete (Sec. 1.2). In order to bridge this gap, the author has undertaken a parallel examination of three categories of plates subjected to a surface couple M (Secs. 2.2—2.3), namely [64], [75], [82]:

A. The *infinitely long plate*, simply supported along the two longitudinal edges (Figs. 2.3A, 2.5 and 2.6);

B. The *rectangular plate*, simply supported along two parallel edges, and free along the remaining edges (Figs. 2.3B, 2.8 and 2.9);

C. The *rectangular plate*, simply supported along the entire boundary (Figs. 2.3 C, 2.12 and 2.13).

The plates pertaining to category B, which reflect more faithfully the real systems, must be taken as a particular case of category A, which is in fact a theoretical model. The plate of category C is also a current case in structural design.

The main criteria which lead to the choice of the above three categories of plates are the following:

a) All the analysed categories represent *fundamental static schemes*. Using these schemes and their specific developments, we can achieve the mathematical modelling of other boundary conditions, i.e.: the *rigidly clamped edge*, the *free cantilever edge*, the *partially clamped edge* or the *elastically supported edge*. The first two of the above mentioned boundary conditions can be modelled by means of the author's results on the effect of a surface couple on plates.

b) A wide range of current structural systems can be assimilated with one of the three static schemes considered by the author. This simplification is possible since the design requirements

2.1 Preliminaries

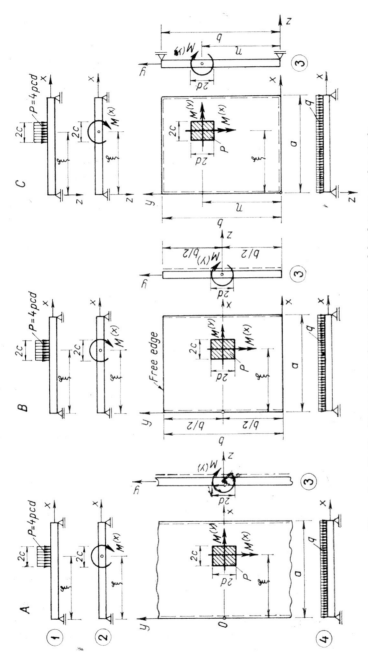

Figure 2.3 Static diagrams for plate categories and load types under study.

93

allow the assimilation of the edge beam with a simple support whenever its torsional stiffness is negligible [160], [162], etc.

c) For all the three categories of plates, we could establish solutions of the type (2.1) to the Lagrange-Sophie Germain equation (2.0). The resulting algorithms for elastic displacements, stresses and boundary reactions, functions of the current types of design loads (Sub-sec. 2.1.4), can be easily written in FORTRAN IV (Chap. 3).

An additional advantage of the infinitely long plate (scheme A) is that the time required by computer work is very short. This calculation rate allows us to obtain a significant amount of numerical data computed in conditions of *extreme accuracy*, as imposed by the programs. Taken as reference values, these data allowed a comparison between the results obtained by other authors in the sole case of the infinitely long plate (Sec. 3.4) and the author's data. Additionally, these accurate values enabled a comparison with the author's results pertaining to the finite boundary plate (scheme C) under the action of a surface couple, for which no data are available so far.

2.1.4 Characteristics of the types of design loads

The solutions of the Lagrange-Sophie Germain equation (2.0)

$$w(x, y) = w(P, D, a, b, c, d, \xi, \eta, x, y); \qquad (2.1\text{ a})$$

$$w(x, y) = w(M, D, a, b, c, d, \xi, \eta, x, y); \qquad (2.1\text{ b})$$

$$w(x, y) = w(q, D, a, b, x, y), \qquad (2.1\text{ c})$$

were obtained under this form for the following types of design loads:

1. the locally distributed load P on the loaded area $2c \times 2d$, or on the bar-plate area whenever P represents an interacting force (Fig. 2.3 — A1, B1, C1);

2. the surface couple M applied on the same contact area (Fig. 2.3 — A2, A3, B2, B3, C2, C3);

3. the load q uniformly distributed throughout the surface of the plate (Fig. 2.3— A4, B4, C4).

The functions $w(x, y)$ of type (2.1a), corresponding to the loading of the plate with a locally distributed force P (Sub-secs. 2.2.1.1, 2.2.2.1, 2.2.3.1), are used in the mathematical developments related to the modelling of the effect of the surface couple M. Additionally, these functions are directly used in structural analysis (Chaps. 4—6).

2.1 Preliminaries

The solutions $w(x, y)$ of type (2.1b), obtained in the case when a plate is subjected to a surface couple M, were established for two limit positions of the moment vector $\vec{M}$ in the plane of the plate:

— the moment vector $\vec{M}$ is parallel to the line of supports of the plate ($\vec{M} \parallel Oy$) (Sub-secs. 2.2.1.2, 2.2.2.2, 2.2.3.2), and

— the moment vector $\vec{M}$ is normal to the same line ($\vec{M} \parallel Ox$) (Sub-secs. 2.2.1.3, 2.2.2.3, 2.2.3.3).

Taken together, these solutions, corresponding to types M and P of loads, allow an accurate study of the effect of the rigid connection between bar and plate (column and slab or beam and shear-wall, respectively). Thus, the functions $w(x, y)$ of types (2.1a) and (2.1 b) are the main tool used in the *elastic analysis of slab structures* (Chaps. 5 and 6).

The load q uniformly distributed throughout the surface of the plate is a particular case of the locally distributed load $P = 4pcd$ (Sub-secs. 2.2.1.4, 2.2.2.4, 2.2.3.4). By introducing the function $w(x, y)$ of type (2.1 c), corresponding to the uniformly distributed load q for the three categories of plates, the author aims at a complete unified presentation of the algorithms used in writing programs for structural analysis *.

2.2 Equations of the middle surface of plates deflected by loads of type P, M and q

2.2.1 The infinitely long plate

2.2.1.1. *The locally distributed load P*. Let us consider an infinitely long plate, simply supported on its two longitudinal parallel edges and loaded by a locally distributed force $P = 4pcd$, applied at $(\xi, 0)$ (Fig. 2.4).

---

* Among the twelve mathematical models of the deflected mid-surface (Fig. 2.3) — presented in Sec. 2.2 — seven solutions were worked out by the author. Among the remaining solutions, four (A1, A4, C1 and C4) are known from classical literature [7], [9], [16], etc. The fifth solution (A2), corresponding to the loading of an infinitely long plate with a surface couple $\vec{M} \parallel Oy$, was given by Stiglat [60].

2 Models for the deflected middle surface of plates

The equation of the middle surface deflected under this load was established by Girkmann [9] in the form

$$w(x, y) = \frac{Pa^4}{2\pi^5 Dcd} \sum_m \frac{1}{m^5} \sin\frac{m\pi c}{a} \sin\frac{m\pi \xi}{a} \sin\frac{m\pi x}{a} Y_m$$

$$(m = 1, 2, 3, \ldots) \qquad (2.11)$$

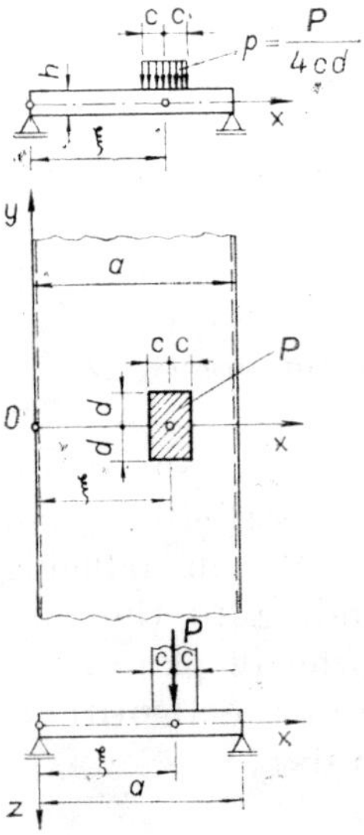

where,
— for $y \leq d$

$$Y_{m1} = 2 + \left[\frac{m\pi y}{a} \operatorname{sh}\frac{m\pi y}{a} - \left(2 + \frac{m\pi d}{a}\right)\operatorname{ch}\frac{m\pi y}{a}\right] e^{-\frac{m\pi d}{a}} \qquad (2.12)$$

— for $y \geq d$

$$Y_{m2} = \left[\left(2 + \frac{m\pi y}{a}\right) \operatorname{sh}\frac{m\pi d}{a} - \frac{m\pi d}{a}\operatorname{ch}\frac{m\pi d}{a}\right] e^{-\frac{m\pi y}{a}}. \qquad (2.13)$$

Figure 2.4 Infinitely long plate. Locally distributed load P at $(\xi, 0)$.

For the same boundary conditions and the same type of load, Timoshenko [7] gives an equivalent solution which holds only along the boundary and outside the loaded area, i.e. in the range $y \geq d$ ($y \leq -d$):

$$w(x, y) = \frac{Pa^4}{4\pi^5 Dcd} \sum_m \frac{1}{m^5} \sin\frac{m\pi c}{a} \sin\frac{m\pi \xi}{a} \sin\frac{m\pi x}{a} \times$$

$$\times \left\{\left[2 + \frac{m\pi}{a}(y-d)\right] e^{-\frac{m\pi}{a}(y-d)} - \left[2 + \frac{m\pi}{a}(y+d)\right] e^{-\frac{m\pi}{a}(y+d)}\right\}.$$

$$(m = 1, 2, 3, \ldots)$$

Since the extreme values of the stresses M_x, M_y, M_{xy} appear just within the loaded area of the plate ($-d \leq y \leq d$) Girkmann's formulae (2.11) — (2.13), which are valid throughout the plate, have been exclusively used in the book.

In order to simplify the formulae, in Eqns. (2.11)–(2.13) and all the subsequent mathematical developments, we introduce the notation

$$\alpha = \frac{m\pi}{a}; \quad (m = 1, 2, 3, \ldots). \qquad (2.14)$$

2.2 Plates deflected by loads of type P, M and q

Hence, we have

$$w(x, y) = \frac{Pa^4}{2\pi^5 Dcd} \sum_m \frac{1}{m^5} \sin \alpha c \sin \alpha \xi \sin \alpha x\, Y_m. \quad (2.11)$$

2.2.1.2 *The surface couple of vector $\vec{M} \parallel Oy$.*
In order to model the effect of a surface couple, the moment vector $\vec{M}$ of which is parallel to the line of supports (Oy axis), we have extended the procedure used by Girkmann [9] to establish the equation of the deflected mid-surface of an infinitely long plate subjected to a concentrated moment (Sub-sec. 1.2.4.1.).

The surface couple, whose moment is

$$M = P\varepsilon, \quad \text{where} \quad P = 4pcd, \quad (2.15)$$

acting at $(\xi, 0)$, is considered to be equivalent to the pair of locally distributed forces P and $-P$ (Fig. 2.5). The forces P act at $(\xi+\varepsilon, 0)$ and $(\xi, 0)$, the arm ε of the couple being arbitrarily small (Sub-sec. 1.2.4.3).

The equation of the mid-surface deflected under this load is obtained by substituting

$$P = \frac{M}{\varepsilon} \quad (2.15\,a)$$

in (2.11) and superposing the effects of the two forces

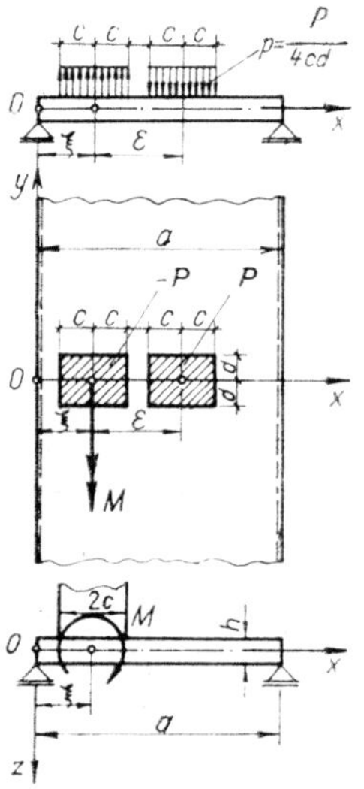

Figure 2.5 Infinitely long plate. Modelling of effect of surface couple M applied at $(\xi, 0)$, having moment vector $\vec{M}$ parallel to line of supports ($\vec{M} \parallel Oy$).

$$w(x, y) = \frac{Ma^4}{2\pi^5 Dcd\varepsilon} \sum_m \frac{1}{m^5} \sin \alpha c\, [\sin \alpha(\xi+\varepsilon) -$$

$$-\sin \alpha \xi]\sin \alpha x\, Y_m.$$

Passing to the limit, and noting that

$$\lim \frac{1}{\varepsilon} [\sin \alpha(\xi + \varepsilon) - \sin \alpha \xi] = \alpha \cos \alpha \xi,$$

we can write

$$w(x, y) = \frac{Ma^3}{2\pi^4 Dcd} \sum_m \frac{1}{m^4} \sin \alpha c \cos \alpha \xi \sin \alpha x\, Y_m. \quad (2.16)$$

$$(m = 1, 2, 3, \ldots)$$

2 Models for the deflected middle surface of plates

The expressions for the functions of y and that of α are given by (2.12), (2.13) and (2.14), respectively.

Equation (2.16) represents the deflected mid-surface of an infinitely long plate under the action of a surface couple acting at $(\xi, 0)$, whose moment vector $\vec{M}$ is parallel to the line of supports (Fig. 2.5). The solution (2.16) which was established by Stiglat [60] and used only for the range $y \leqslant d$, was extended by introducing the expression of Y_{m2} (2.13). This extension allowed the calculation of the elastic displacements and of the stresses produced by the load M throughout the plate [64].

2.2.1.3 The surface couple of vector $\vec{M} \parallel Ox$.

The effect of a surface couple whose moment vector is normal to the line of supports ($\vec{M} \parallel Ox$) is modelled in a similar manner, starting from solution (2.11) which is given for the infinitely long plate subjected to a locally distributed force P (Sub-sec. 2.2.1.1).

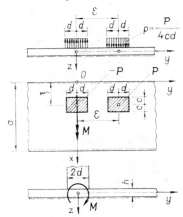

Figure 2.6 Infinitely long plate. Modelling of effect of surface couple M applied at $(\xi, 0)$, having moment vector normal to line of supports ($\vec{M} \parallel Ox$).

The surface couple of moment $M = P\varepsilon$, (Fig. 2.6), acting at the point $(\xi, 0)$ of the infinitely long plate, is considered to be equivalent with the pair of locally distributed forces P and $-P$. These forces act at (ξ, ε) and $(\xi, 0)$ respectively, the arm ε of the couple being arbitrarily small.

The equation of the mid-surface deflected under this load is obtained from (2.11) by superposing the effects of these two forces [82]. Putting

$$\alpha = \frac{m\pi}{a}, \qquad (2.14)$$

and recalling (2.15a)

— for $y \leqslant d$, (2.12) yields

$$w(x, y) = \frac{Ma^4}{2\pi^5 Dcd} \sum_m \frac{1}{m^5} \sin \alpha c \sin \alpha \xi \sin \alpha x \frac{e^{-\alpha d}}{\varepsilon} \times$$

$$\times \{\alpha y [\operatorname{sh} \alpha (y - \varepsilon) - \operatorname{sh} \alpha y] - \varepsilon \alpha \operatorname{sh} \alpha (y - \varepsilon) - (2 + \alpha d) \times$$

$$\times [\operatorname{ch} \alpha (y - \varepsilon) - \operatorname{ch} \alpha y]\};$$

2.2 Plates deflected by loads of type P, M and q

— for $y \geqslant d$, (2.13) yields

$$w(x, y) = \frac{Ma^4}{2\pi^5 D c d} \sum_m \frac{1}{m^5} \sin \alpha c \sin \alpha \xi \sin \alpha x \times$$

$$\times \left\{ \frac{e^{-\alpha(y-\varepsilon)} - e^{-\alpha y}}{\varepsilon} [(2 + \alpha y) \operatorname{sh} \alpha d - \alpha d \operatorname{ch} \alpha d] - \right.$$

$$\left. - e^{-\alpha(y-\varepsilon)} \cdot \alpha \operatorname{sh} \alpha d \right\}.$$

Passing to the limit, and noting that

$$\lim_{\varepsilon \to 0} \frac{1}{\varepsilon} [\operatorname{sh} \alpha(y - \varepsilon) - \operatorname{sh} \alpha y] = -\alpha \operatorname{ch} \alpha y;$$

$$\lim_{\varepsilon \to 0} \frac{1}{\varepsilon} [\operatorname{ch} \alpha(y - \varepsilon) - \operatorname{ch} \alpha y] = -\alpha \operatorname{sh} \alpha y,$$

$$\lim_{\varepsilon \to 0} \frac{1}{\varepsilon} [e^{-\alpha(y-\varepsilon)} - e^{-\alpha y}] = \alpha e^{-\alpha y},$$

we obtain:

— for $y \leqslant d$,

$$w(x, y) = \frac{Ma^3}{2\pi^4 D c d} \sum_m \frac{1}{m^4} \sin \alpha c \sin \alpha \xi \sin \alpha x \times$$

$$\times [(1 + \alpha d) \operatorname{sh} \alpha y - \alpha y \operatorname{ch} \alpha y] e^{-\alpha d}; \qquad (2.17)$$

$$(m = 1, 2, 3, \ldots)$$

— for $y \geqslant d$,

$$w(x, y) = \frac{Ma^3}{2\pi^4 D c d} \sum_m \frac{1}{m^4} \sin \alpha c \sin \alpha \xi \sin \alpha x \times$$

$$\times [(1 + \alpha y) \operatorname{sh} \alpha d - \alpha d \operatorname{ch} \alpha d] e^{-\alpha y}. \qquad (2.18)$$

$$(m = 1, 2, 3, \ldots)$$

Equations (2.17) and (2.18), established in [82], represent the deflected mid-surface of the infinitely long plate subjected to a surface couple at $(\xi, 0)$, whose moment vector $\vec{M}$ is normal to the line of supports (Fig. 2.6).

2 *Models for the deflected middle surface of plates*

For $y = d$, at the limit of their range of validity, both equations assume the same expression:

$$w(x, y) = \frac{Ma^3}{2\pi^4 Dcd} \sum_m \frac{1}{m^4} \sin\alpha c \sin\alpha\xi \sin\alpha x \times$$

$$\times [(1 + \alpha y)\operatorname{sh}\alpha d - \alpha d \operatorname{ch}\alpha d] e^{-\alpha d}. \tag{2.19}$$

$$(m = 1, 2, 3, \ldots)$$

Introducing in Eqn. (2.18), which is valid for $y \geq d$, the limit values $c \to 0$ and $d \to 0$ corresponding to a point loaded with the couple M_0, and since

$$\lim_{c \to 0} \frac{\sin\alpha c}{c} = \alpha; \quad \lim_{d \to 0} \frac{\operatorname{sh}\alpha d}{d} = \alpha; \quad \lim_{d \to 0} \operatorname{ch}\alpha d = 1,$$

we obtain

$$w(x, y) = \frac{M_0}{2\pi D} \sum_m \frac{1}{m} \sin\alpha\xi \sin\alpha x \cdot y e^{-\alpha y}. \tag{2.20}$$

$$(m = 1, 2, 3, \ldots)$$

Equation (2.20) represents the deflected mid-surface of an infinitely long plate subjected to the action of a *concentrated couple* at $(\xi, 0)$, whose vector moment $\vec{M_0}$ is parallel to the Ox axis. The same expression was obtained by Girkmann [9] by modelling the couple effect with the aid of a pair of *concentrated loads* P and $-P$, acting at (ξ, ε) and $(\xi, 0)$, respectively (Sub-sec. 1.2.4.1). Equation (2.20) cannot be obtained from Eqn. (2.17), which is valid for $y \leq d$, because if $d \to 0$, then $y \to d$ by negative values.

2.2.1.4 *The uniformly distributed load q.* The solution to the Lagrange-Sophie Germain equation (2.0) for the infinitely long plate under a load q uniformly distributed throughout the surface (Fig. 2.3—A4) is obtained by particularizing expression (2.11). To this solution, which holds in the case of the locally distributed load $P = 4pcd$, we associate in this particular case only the function Y_{m1} (2.12), which is valid in the range $y \leq d$. Observing that for the values of the geometrical parameters

$$c \to \frac{a}{2}; \quad d \to \infty; \quad \xi = \frac{a}{2},$$

2.2 Plates deflected by loads of type P, M and q

which correspond to this case,

$$p \to q \quad \text{and} \quad Y_{m1} \to 2 = \text{constant},$$

we obtain

$$w(x) = \frac{4qa^4}{\pi^5 D} \sum_m \frac{1}{m^5} \sin \alpha\, x, \qquad (m = 1, 2, 3, \ldots) \qquad (2.21)$$

where $\alpha = \dfrac{m\pi}{a}$ (2.14).

Equation (2.21) represents the middle surface of the infinitely long plate, deformed under the action of the uniformly distributed load q (Fig. 2.3—A.4).

As the function $w(x)$, given by (2.21), is exclusively dependent on x, its derivatives with respect to y are all zero. This means that we have a case of cylindrical bending. Consequently, the partial differential equation (2.0) is reduced to an ordinary equation of the form

$$\frac{d^4 w}{dx^4} = \frac{q(x)}{D} \qquad (2.0\text{a})$$

which is analogous to the fundamental equation for straight bars [9]. The stiffness EI of the bar of cross-section $1 \times h$ is substituted in (2.0a) by the cylindrical stiffness D (1.2) of the plates.

For $q(x) = q = $ constant, the integral of the equation (2.0a) has the following known finite form:

$$w(x) = \frac{qa^4}{24D}\left(\frac{x^4}{a^4} - 2\frac{x^3}{a^3} + \frac{x}{a}\right). \qquad (2.22)$$

Equation (2.22) can be considered to be the sum of the series (2.21). Indeed, since (2.22) is expanded in Fourier series as a function of x and with a period $L = 2a$, solution (2.21) is straightforward. However, mention should be made that the solution (2.22) holds only within the range $0 \leqslant x \leqslant a$, for which the integration constants were established. This explains the use of solution (2.21), which is more tedious, in writing the computation programs. However, this solution allows the study of loads applied in strips or in chess-board fashion, and also the extension of the author's method to the study of multi-spanned structures [75].

2.2.2 The rectangular plate with two parallel free edges

2.2.2.1 *The locally distributed load P.* The rectangular plate $a \times b$ is assumed to be simply supported on its two edges parallel to the Oy axis (Fig. 2.7). The equation of the mid-surface deflected under a locally distributed load $P = 4pcd$ is obtained from solution (2.11) — established for the case of the infinitely long plate subjected to the same type of load — and writing the conditions of the free edge in $y = \pm b/2$.

The general solution of the fundamental equation (2.0) is of the form

$$w(x, y) = w_1 + w_0, \qquad (2.23)$$

where

w_1 is the particular integral of the fundamental equation satisfying the boundary conditions of the problem,

and

w_0 denotes the integral of the fundamental homogeneous equation.

For w_1 we chose solution (2.11), knowing that when the boundary conditions are substituted in $y = \pm b/2$, we must introduce the expression Y_{m2} (2.13), which holds in the range $y \geqslant d$. Using the notation

$$S_m = \frac{P}{2\alpha^5 Dacd} \sin \alpha c \sin \alpha \xi, \qquad (2.24)$$

where α is defined by (2.14), the particular integral is

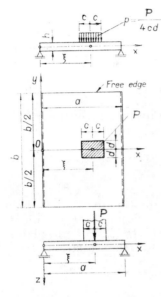

Figure 2.7 Rectangular plate with two parallel free edges. Locally distributed load P at $(\xi, 0)$.

$$w_1 = \sum_m S_m \sin \alpha x \, Y_m, \quad (m = 1, 2, 3, \ldots) \qquad (2.25)$$

the expressions (2.12) and (2.13) being valid for the function Y_m. The boundary coditions in $x = 0$ and $x = a$ are clearly satisfied.

For w_0, we choose a solution of the form

$$w_0 = \sum_m (a_m \operatorname{ch} \alpha y + \alpha y \, b_m \operatorname{sh} \alpha y + c_m \operatorname{sh} \alpha y + \alpha y d_m \operatorname{ch} \alpha y) \sin \alpha x,$$

$$(m = 1, 2, 3, \ldots)$$

where the known particular integrals of the biharmonic differential equation $\nabla^2 \nabla^2 F = 0$ appear.

2.2 Plates deflected by loads of type P, M and q

Owing to the geometrical and mechanical symmetry with respect to the axis Ox, the integration constants c_m and d_m become equal to zero. Taking (2.24) and (2.25) into consideration, the integral of the fundamental homogeneous equation becomes

$$w_0 = \sum_m S_m(A_m \operatorname{ch} \alpha y + \alpha y B_m \operatorname{sh} \alpha y) \sin \alpha x. \tag{2.26}$$

$$(m = 1, 2, 3, \ldots)$$

Denoting by

$$\Phi_m = A_m \operatorname{ch} \alpha y + \alpha y B_m \operatorname{sh} \alpha y, \tag{2.27}$$

the solution of the homogeneous fundamental equation will read

$$w_0 = \sum_m S_m \sin \alpha x \, \Phi_m \quad (m = 1,2,3, \ldots). \tag{2.26a}$$

With the above notation [82], the integration constants A_m and B_m become dimensionless. For the general solution $w(x, y)$ (2.23), where the expressions of w_1 (2.25) (with Y_{m_2} (2.13)) and w_0 (2.26) are inserted, these constants are determined by writing the boundary conditions corresponding to the free edges as

$$y = \pm \frac{b}{2} \rightarrow M_y = 0; \quad R_y = 0, \tag{2.28}$$

$$y = \pm \frac{b}{2} \begin{cases} \dfrac{\partial^2 w}{\partial y^2} + \mu \dfrac{\partial^2 w}{\partial x^2} = 0; & (2.28\text{a}) \\[2mm] \dfrac{\partial^3 w}{\partial y^3} + (2-\mu) \dfrac{\partial^3 w}{\partial x^2 \partial y} = 0, & (2.28\text{b}) \end{cases}$$

respectively.

The resulting system of transcendental equations is of the form

$$\begin{cases} \left[\left(-\dfrac{2\mu}{1-\mu} + \dfrac{\alpha b}{2}\right) \operatorname{sh} \alpha d - \alpha d \operatorname{ch} \alpha d\right] e^{-\frac{\alpha b}{2}} + \\[2mm] + A_m \operatorname{ch} \dfrac{\alpha b}{2} + B_m \left(\dfrac{2}{1-\mu} \operatorname{ch} \dfrac{\alpha b}{2} + \dfrac{\alpha b}{2} \operatorname{sh} \dfrac{\alpha b}{2}\right) = 0; \quad (2.29\text{a}) \\[2mm] \left[\left(\dfrac{3-\mu}{1-\mu} + \dfrac{\alpha b}{2}\right) \operatorname{sh} \alpha d - \alpha d \operatorname{ch} \alpha d\right] e^{-\frac{\alpha b}{2}} - \\[2mm] - A_m \operatorname{sh} \dfrac{\alpha b}{2} + B_m \left(\dfrac{1+\mu}{1-\mu} \operatorname{sh} \dfrac{\alpha b}{2} - \dfrac{\alpha b}{2} \operatorname{ch} \dfrac{\alpha b}{2}\right) = 0. \quad (2.29\text{b}) \end{cases}$$

2 *Models for the deflected middle surface of plates*

From (2.29a) and (2.29b), we derive

$$\begin{cases} A_m = \dfrac{1}{\operatorname{ch}\dfrac{\alpha b}{2}} \left\{ e^{-\dfrac{\alpha b}{2}} \left[\left(\dfrac{2\mu}{1-\mu} - \dfrac{\alpha b}{2} \right) \operatorname{sh} \alpha d + \alpha d \operatorname{ch} \alpha d \right] - \right. \\ \left. - \left(\dfrac{2}{1-\mu} \operatorname{ch} \dfrac{\alpha b}{2} + \dfrac{\alpha b}{2} \operatorname{sh} \dfrac{\alpha b}{2} \right) B_m \right\}, \hfill (2.30a) \\ B_m = -\dfrac{2e^{-\dfrac{\alpha b}{2}}}{\dfrac{3+\mu}{1-\mu} \operatorname{sh} \alpha b - \alpha b} \left\{ \left[\left(\dfrac{3-\mu}{1-\mu} + \dfrac{\alpha b}{2} \right) \operatorname{ch} \dfrac{\alpha b}{2} - \left(\dfrac{2\mu}{1-\mu} - \right. \right. \right. \\ \left. \left. \left. - \dfrac{\alpha b}{2} \right) \operatorname{sh} \dfrac{\alpha b}{2} \right] \operatorname{sh} \alpha d - \left(\operatorname{sh} \dfrac{\alpha b}{2} + \operatorname{ch} \dfrac{\alpha b}{2} \right) \alpha d \operatorname{ch} \alpha d \right\}. \hfill (2.30b) \end{cases}$$

Considering (2.23) and the expressions of the integrals w_1 (2.25) and w_0 (2.26 a) — where the integration constants A_m and B_m thus established are introduced — the solution of the fundamental equation (2.0) will be expressed as

$$w(x, y) = \sum_m S_m (Y_m + \Phi_m) \sin \alpha x. \qquad (m = 1,2,3,\ldots)$$

Introducing S_m (2.24) too, we obtain the design formula

$$w(x, y) = \frac{Pa^4}{2\pi^5 Dcd} \sum_m \frac{1}{m^5} \sin \alpha c \sin \alpha \xi \sin \alpha x (Y_m + \Phi_m),$$

$$(m = 1,2,3, \ldots) \quad (2.31)$$

where

$$\alpha = \frac{m\pi}{a} \qquad (2.14)$$

and the functions of y are given by Y_{m1} (2.12), Y_{m2} (2.13) and Φ_m (2.27).

Equation (2.31), established in [82], represents the mid-surface of a rectangular plate with two parallel free edges, deflected under a locally distributed load P applied in $(\xi, 0)$ (Fig. 2.7).

2.2.2.2 *The surface couple of vector $\vec{M} \parallel Oy$.* The modelling procedure used in the case of the infinitely long plates (Sub-sec. 2.2.1.2) is extended to the case of the rectangular plate with two parallel free edges [82]. In order to model the effect of a surface couple whose moment vector $\vec{M}$ is parallel to the line of supports (the Oy axis), the couple of moment $M = P\varepsilon$ acting at a point

2.2 Plates deflected by loads of type P, M and q

$(\xi, 0)$ of the rectangular plate is considered to be equivalent with the pair of locally distributed forces P and $-P$ (Fig. 2.8). These forces act at $(\xi + \varepsilon, 0)$ and $(\xi, 0)$, the arm ε of the couple tending towards an infinitesimal value.

The equation of the middle surface deflected under this load is obtained from w (2.31), using (2.15 a) and superposing the effects of the two forces:

$$w(x, y) = \frac{Ma^4}{2\pi^5 Dcd\varepsilon} \sum_m \frac{1}{m^5} \sin \alpha c \times$$

$$\times [\sin \alpha(\xi + \varepsilon) -$$

$$- \sin \alpha \xi] \sin \alpha x (Y_m + \Phi_m).$$

Passing to the limit for $\varepsilon \to 0$ leads to an expression similar to that (2.16) obtained for the case of the infinitely long plate, namely:

$$w(x, y) = \frac{Ma^3}{2\pi^4 Dcd} \sum_m \frac{1}{m^4} \times$$

$$\times \sin \alpha c \cos \alpha \xi \sin \alpha x (Y_m + \Phi_m),$$

$$(m = 1, 2, 3 \ldots) \quad (2.32)$$

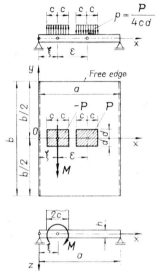

Figure 2.8 Rectangular plate with two parallel free edges. Modelling of effect of surface couple M applied at $(\xi, 0)$, having moment vector $\vec{M}$ parallel to line of supports $(\vec{M} \parallel Oy)$.

where α is given by (2.14) and the functions Y_m and Φ_m result from (2.12), (2.13) and (2.27), respectively, where (2.30 a) and (2.30 b) are inserted.

Equation (2.32), established in [82], represents the deflected mid-surface of a rectangular plate with two parallel free edges, subjected at $(\xi, 0)$ to a surface couple whose moment vector $\vec{M}$ is parallel to the line of supports (Fig. 2.8).

2.2.2.3 The surface couple of vector $\vec{M} \parallel Ox$.

The effect of a surface couple whose moment vector $\vec{M}$ is normal to the line of supports $(\vec{M} \parallel Ox)$ is established as in the case of the infinitely long plate (Sub-sec. 2.2.1.3). Here we start again from w (2.31), to which the functions of y (2.12), (2.13) and (2.27) are associated, corresponding to the rectangular plate with two parallel free edges, under the action of a locally distributed load P (Sub-sec. 2.2.2.1).

The surface couple whose moment $M = P\varepsilon$ acts at a point $(\xi, 0)$ of the rectangular plate, and is assumed to be equivalent

2 *Models for the deflected middle surface of plates*

to a pair of *locally distributed forces* P and $-P$ (Fig. 2.9). These forces are applied at (ξ, ε) and $(\xi, 0)$, the arm ε of the couple approaching zero.

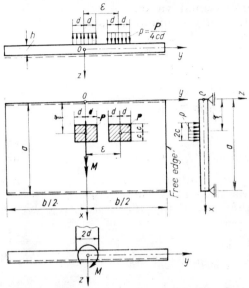

Figure 2.9 Rectangular plate with two parallel free edges. Modelling of effect of a surface couple M applied at $(\xi, 0)$, having moment vector normal to line of supports $(\vec{M} \parallel Ox)$.

The equation of the middle surface of the plate deformed under this load is obtained from w (2.31), with the aid of (2.15 a) and by superposing the effects of the two forces. Thus,

— for $0 \leqslant y \leqslant d$, according to (2.12),

$$\frac{1}{\varepsilon} Y_{m1} = \frac{1}{\varepsilon} [\alpha(y - \varepsilon) \operatorname{sh} \alpha(y - \varepsilon) - (2 + \alpha d) \operatorname{ch} \alpha(y - \varepsilon) -$$

$$- \alpha y \operatorname{sh} \alpha y + (2 + \alpha d) \operatorname{ch} \alpha y] e^{-\alpha d};$$

— for $d \leqslant y \leqslant \dfrac{b}{2}$, according to (2.13),

$$\frac{1}{\varepsilon} Y_{m2} = \frac{1}{\varepsilon} \{[2 + \alpha(y - \varepsilon)] \operatorname{sh} \alpha d - \alpha d \operatorname{ch} \alpha d\} e^{-\alpha(y-\varepsilon)} -$$

$$- \frac{1}{\varepsilon} [(2 + \alpha y) \operatorname{sh} \alpha d - \alpha d \operatorname{ch} \alpha d] e^{-\alpha y};$$

2.2 Plates deflected by loads of type P, M and q

— for $-\frac{b}{2} \leq y \leq \frac{b}{2}$, according to (2.27),

$$\frac{1}{\varepsilon}\Phi_m = \frac{1}{\varepsilon}\{A_m[\operatorname{ch}\alpha(y-\varepsilon) - \operatorname{ch}\alpha y] + \alpha y\, B_m[\operatorname{sh}\alpha(y-\varepsilon) - \operatorname{sh}\alpha y] - \alpha\varepsilon B_m \operatorname{sh}\alpha(y-\varepsilon)\}.$$

Passing to the limit for $\varepsilon \to 0$, we obtain expressions which are similar to those obtained in the case of the infinitely long plate (see (2.17) and (2.18)). Thus,

— for $0 \leq y \leq d$,

$$w(x,y) = \frac{Ma^3}{2\pi^4 Dcd}\sum_m \frac{1}{m^4}\sin\alpha c \sin\alpha\xi\sin\alpha x\{[(1+\alpha d)\operatorname{sh}\alpha y - \alpha y\operatorname{ch}\alpha y]e^{-\alpha d} - [(A_m + B_m)\operatorname{sh}\alpha y + \alpha y B_m\operatorname{ch}\alpha y]\}; \quad (2.33)$$

— for $d \leq y \leq \frac{b}{2}$,

$$w(x,y) = \frac{Ma^3}{2\pi^4 Dcd}\sum_m \frac{1}{m^4}\sin\alpha c \sin\alpha\xi\sin\alpha x\{[(1+\alpha y)\operatorname{sh}\alpha d - \alpha d\operatorname{ch}\alpha d]e^{-\alpha y} - [(A_m + B_m)\operatorname{sh}\alpha y + \alpha y B_m\operatorname{ch}\alpha y]\}, \quad (2.34)$$

$(m = 1, 2, 3, \ldots)$

where $\alpha = \frac{m\pi}{a}$ (2.14), and the integration constants A_m and B_m are given by (2.30 a) and (2.30 b).

Equations (2.33) and (2.34) — which are first published here — represent the deflected mid-surface of a rectangular plate with two parallel free edges, subjected in $(\xi, 0)$ to a surface couple whose moment vector $\vec{M}$ is normal to the line of supports (Fig. 2.9).

Solutions (2.31) and (2.32) associated to the loading case P (Fig. 2.7) and for $\vec{M} \| Oy$ (Fig. 2.8) can be used in modelling some symmetrical sets consisting of $2, 4, 6\ldots$ locally distributed loads P_i and $2, 4, 6\ldots$ surface couples M_i, respectively (Fig. 2.10). In this case too, P_i and M_i

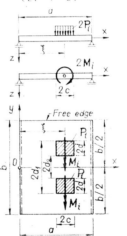

Figure 2.10 Rectangular plate with two parallel free edges. Schematic diagram for a symmetrical set of two locally distributed forces P_i and two surface couples M_i, having the moment vectors parallel to line of supports ($\vec{M}_i \| Oy$).

107

2 Models for the deflected middle surface of plates

type loads can be taken into account in the analysis both as given loads and as generalized interacting forces distributed on the slab-column contact area ($2c \times 2d$) [82].

2.2.2.4 *The uniformly distributed load q*. The equation of the deflected mid-surface of a rectangular plate with two free parallel edges, under a load q uniformly distributed throughout its surface (Fig. 2.3—B.4) can be obtained as a particular case of the general solution w (2.31), where the parameters c, d, ξ and P are replaced by

$$c = \frac{a}{2}; \quad d = \frac{b}{2}; \quad \xi = \frac{a}{2}; \quad P = 4pcd = qab.$$

We obtain

$$w(x, y) = \frac{2qa^4}{\pi^5 D} \sum_m \frac{1}{m^5} \sin \alpha x [(\Phi_m)_{d=\frac{b}{2}} + (Y_m)_{d=\frac{b}{2}}] \quad (2.35)$$

($m = 1, 2, 3, \ldots$).

In this particular case, the function Y_{m2} (2.13) holds only for $y = \frac{b}{2}$, ($y \geqslant d$), where the boundary conditions (2.28) were employed. In the remainder of the domain $a \times b$ we shall use the function Y_{m1} (2.12), which is valid for $y \leqslant \frac{b}{2} (y \leqslant d)$.

Solution (2.35) can be readily established using the procedure described in Sub-sec. 2.2.2.1. for the modelling of the effect of a locally distributed load P on a rectangular plate with two free edges.

Since the rectangular plate $a \times b$ is assumed to be simply supported along its two sides parallel to the axis Oy, the equation of the mid-surface deflected under the load q is established from the solution given for the infinitely long plate subjected to the same type of load

$$w(x, y) = \frac{4qa^4}{\pi^5 D} \sum_m \frac{1}{m^5} \sin \frac{m\pi}{a} x, \quad (2.21)$$

($m = 1, 3, 5, \ldots$)

and setting the free-edge conditions in $y = \pm \frac{b}{2}$ [82].

The general solution of the fundamental equation (2.0) is

$$w(x, y) = w_1 + w_0. \quad (2.23)$$

2.2 Plates deflected by loads of type P, M and q

As a particular integral, for w_1 we choose solution (2.21), which holds in the entire range $y \in R$. Denoting

$$\alpha = \frac{m\pi}{a} \quad (2.14) \quad \text{and} \quad Q_m = \frac{4q}{\alpha^5 D a}, \tag{2.36}$$

the particular integral becomes

$$w_1 = \sum_m Q_m \sin \alpha x \quad (m = 1, 3, 5, \ldots). \tag{2.37}$$

For the integral w_0 of the fundamental homogeneous equation, we choose the solution

$$w_0 = \sum_m Q_m(C_m \operatorname{ch} \alpha y + \alpha y\, D_m \operatorname{sh} \alpha y) \sin \alpha x \tag{2.38}$$

$$(m = 1, 2, 3, \ldots)$$

which is similar to solution (2.26).

The integration constants C_m and D_m are established by substituting in the general solution (2.23), where account is taken of (2.37) and (2.38), the boundary conditions for the free edges (2.28 a) and (2.28 b).

The resulting system of transcendental equations assumes the form

$$\begin{cases} (1-\mu)C_m \operatorname{ch} \alpha \frac{b}{2} + D_m \left[2\operatorname{ch} \alpha \frac{b}{2} + (1-\mu)\alpha \frac{b}{2} \operatorname{sh} \alpha \frac{b}{2} \right] - \mu = 0; \\ -(1-\mu)C_m \operatorname{sh} \alpha \frac{b}{2} + D_m \left[(1+\mu) \operatorname{sh} \alpha \frac{b}{2} - (1-\mu)\alpha \frac{b}{2} \operatorname{ch} \alpha \frac{b}{2} \right] = 0, \end{cases}$$

wherefrom we deduce the formulae for calculating the integration constants

$$\begin{cases} C_m = \frac{\mu}{1-\mu} \cdot \frac{(1+\mu) \operatorname{sh} \alpha \dfrac{b}{2} - (1-\mu)\alpha \dfrac{b}{2} \operatorname{ch} \alpha \dfrac{b}{2}}{(3+\mu) \operatorname{sh} \alpha \dfrac{b}{2} \operatorname{ch} \alpha \dfrac{b}{2} - (1-\mu)\alpha \dfrac{b}{2}} & (2.39\text{a}) \\ \\ D_m = \dfrac{\mu \operatorname{sh} \alpha \dfrac{b}{2}}{(3+\mu) \operatorname{sh} \alpha \dfrac{b}{2} \operatorname{ch} \alpha \dfrac{b}{2} - (1-\mu)\alpha \dfrac{b}{2}}. & (2.39\text{b}) \end{cases}$$

2 Models for the deflected middle surface of plates

Inserting the expressions of w_1 (2.37) and w_0 (2.38) in w (2.23), and taking into account (2.14) and (2.36), one finds that

$$w(x, y) = \frac{4qa^4}{\pi^5 D} \sum_m \frac{1}{m^5} (C_m \operatorname{ch} \alpha y + \alpha y \, D_m \operatorname{sh} \alpha y + 1) \sin \alpha x.$$

$$(m = 1, 2, 3, \ldots) \tag{2.40}$$

Equation (2.40), established in [82], represents the equation of the deflected mid-surface of a rectangular plate with two parallel free edges, subjected to a load q uniformly distributed throughout its surface (Fig. 2.3—B.4).

2.2.3 The rectangular plate simply supported along the entire boundary

A mathematical model of the effect of a surface couple M on finite boundary plates was established [64] as early as 1969. Basically, it is an extension of Stiglat's procedure [60] for the case of the infinitely long plate under the same type of load (Sub-sec. 1.2.4.3 — Fig. 1.33, Sub-sec. 2.2.1.2., — Fig. 2.5).

Later on, using a similar procedure, Pfaffinger reached another solution [71]. The model proposed by the Swiss researcher differs from the one suggested by the author inasmuch as it assumes another distribution rule for the component forces P of the couple M on the $2c \times 2d$ loaded area (Sub-sec. 1.2.5.2 — Fig. 1.45). No other models are known for the study of the behaviour of rectangular plates under the action of a surface couple.

A comparison of the numerical results obtained in this book and those given by the Swiss researcher allows us to state that the magnitude of the elastic deflections and of the stresses in the examined plate are unaffected by this difference between the two solutions for modelling the surface couple. Much larger differences appear when other design parameters, e.g. the contact area between the bar and the plate (column and slab) (Chaps. 3—6) are modified.

2.2.3.1 *The locally distributed load P.* Consider the rectangular plate $a \times b$ simply supported along its entire boundary and subjected to a locally distributed force $P = 4pcd$ acting at (ξ, η), i.e. at the centre of gravity of the loaded area (Fig. 2.11). The equation w (2.41), of the mid-surface of this finite plate, deflected under this load was established by Navier and this solution is among the first given in the theory of plates.

2.2 Plates deflected by loads of type P, M and q

In order to have under the summation sign only dimensionless terms, Navier's solution [7], [9], [16] is here written under the form

$$w(x, y) = \frac{4Pa^4}{\pi^6 Dcd} \sum_m \sum_n \frac{1}{m^5 n \left(1 + \dfrac{n^2}{m^2}\dfrac{a^2}{b^2}\right)^2} \sin\frac{m\pi c}{a} \times$$

$$\times \sin\frac{m\pi\xi}{a} \sin\frac{m\pi x}{a} \sin\frac{n\pi d}{b} \sin\frac{n\pi\eta}{b} \sin\frac{n\pi y}{b} \quad (2.41)$$

$(m = 1, 2, 3, \ldots; n = 1, 2, 3, \ldots).$

For simplification, we perform the following substitutions in Eqn. (2.41) and in all subsequent mathematical developments:

$$\alpha = \frac{m\pi}{a} \quad (m = 1, 2, 3, \ldots);$$

$$\beta = \frac{n\pi}{b} \quad (n = 1, 2, 3, \ldots). \quad (2.42)$$

Hence, we have

$$w(x, y) = \frac{4Pa^4}{\pi^6 Dcd} \times$$

$$\sum_m \sum_n \frac{1}{m^5 n \left(1 + \dfrac{a^2}{b^2}\dfrac{n^2}{m^2}\right)^2} \times$$

$\times \sin \alpha c \sin \alpha \xi \sin \alpha x \times$

$\times \sin \beta d \sin \beta \eta \sin \beta y$

$(m = 1, 2, 3, \ldots;$

$n = 1, 2, 3, \ldots). \quad (2.41\text{ a})$

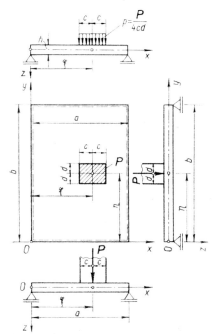

Figure 2.11 Rectangular plate simply supported on the entire boundary. Locally distributed load P at (ξ, η).

2.2.3.2 *The surface couple of vector $\vec{M} \parallel Oy$.* In order to model the effect of a surface couple whose moment vector $\vec{M}$ is parallel to the Oy axis, we consider the couple whose moment

$$M = P\varepsilon, \text{ where } P = 4pcd \quad (2.43)$$

111

2 Models for the deflected middle surface of plates

acts at the point (ξ, η) of the rectangular plate (Fig. 2.12). This couple is equivalent to the pair of locally distributed forces P and $-P$, applied at $(\xi + \varepsilon, \eta)$ and (ξ, η), respectively. The arm ε of the couple can be arbitrarily small.

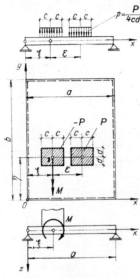

The equation of the mid-surface of the plate, deflected under this load, is obtained by substituting in (2.41) the expression for P given by

$$P = \frac{M}{\varepsilon} \qquad (2.43a)$$

and by superposing the effects of the two forces:

$$w(x, y) = \frac{4Ma^4}{\pi^6 Dcd\varepsilon} \times$$

$$\times \sum_m \sum_n \frac{1}{m^5 n \left(1 + \dfrac{a^2}{b^2}\dfrac{n^2}{m^2}\right)^2} \sin \alpha c \times$$

$$\times [\sin(\xi + \varepsilon) - \sin \alpha \xi] \times$$

$$\times \sin \alpha x \sin \beta d \sin \beta \eta \sin \beta y$$

Passing to the limit and seeing that

$$\lim_{\varepsilon \to 0} \frac{1}{\varepsilon}[\sin \alpha(\xi + \varepsilon) - \sin \alpha \xi] = \alpha \cos \alpha \xi,$$

Figure 2.12 Rectangular plate simply supported on the entire boundary. Modelling of effect of a surface couple M applied at (ξ, η), having the moment vector parallel to the Oy axis ($\vec{M} \parallel Oy$).

we obtain

$$w(x, y) = \frac{4Ma^3}{\pi^5 Dcd} \sum_m \sum_n \frac{1}{m^4 n \left(1 + \dfrac{a^2}{b^2}\dfrac{n^2}{m^2}\right)^2} \times$$

$$\times \sin \alpha c \cos \alpha \xi \sin \alpha x \sin \beta d \sin \beta \eta \sin \beta y \qquad (2.44)$$

$(m = 1, 2, 3, \ldots; n = 1, 2, 3, \ldots)$.

Equation (2.44), established in [64], represents the mid-surface of the rectangular plate deflected under the action of a surface couple applied at (ξ, η), the moment vector $\vec{M}$ being parallel to the Oy axis (Fig. 2.12).

Introducing in (2.44) the limit values $c \to 0$; $d \to 0$ corresponding to a point loaded by couple M_0 (Fig. 2.14 A), and

2.2 Plates deflected by loads of type P, M and q

noting that

$$\lim_{c \to 0} \frac{\sin \alpha c}{c} = \alpha ; \quad \lim_{d \to 0} \frac{\sin \beta d}{d} = \beta,$$

we obtain

$$w(x, y) = \frac{4 M_0 a^2}{\pi^3 D b} \sum_m \sum_n \frac{1}{m^3 \left(1 + \dfrac{a^2}{b^2} \dfrac{n^2}{m^2}\right)^2} \times$$

$$\times \cos \alpha \xi \sin \alpha x \sin \beta \eta \sin \beta y. \tag{2.45}$$

$$(m = 1, 2, 3, \ldots ; \; n = 1, 2, 3, \ldots)$$

Equation (2.45) represents the deflected mid-surface of a rectangular plate, simply supported along its entire boundary, subject to a *concentrated couple* acting at (ξ, η), whose moment vector $\vec{M}_0$ is parallel to the Oy axis (Fig. 2.14A).

2.2.3.3 The surface couple of vector $\vec{M} \parallel Ox$.

The effect of a surface couple whose moment vector $\vec{M}$ is normal to the Oy axis, i.e. parallel to the Ox axis (Fig. 2.13), is established by means of the same modelling procedure. The pair of locally distributed forces P and $-P$, which is equivalent to the couple M (2.43), acts at points $(\xi, \eta + \varepsilon)$ and (ξ, η) respectively, the arm ε of the couple approaching an infinitesimal value. Superposing the effects of the two forces, and noting that

$$\lim_{\varepsilon \to 0} \frac{1}{\varepsilon} [\sin \beta(\eta + \varepsilon) - \sin \beta \eta] =$$

$$= \beta \cos \beta \eta$$

and that

$$\alpha = \frac{m \pi}{a} ; \quad \beta = \frac{n \pi}{b}, \tag{2.42}$$

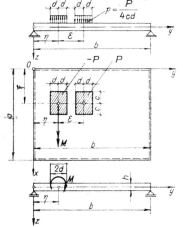

Figure 2.13 Rectangular plate simply supported on the entire boundary. Modelling of effect of a surface couple M applied at (ξ, η), having moment vector normal to the Oy axis $(\vec{M} \parallel Ox)$.

we obtain

$$w(x, y) = \frac{4Ma^4}{\pi^5 Dbcd} \sum_m \sum_n \frac{1}{m^5 \left(1 + \dfrac{a^2}{b^2} \dfrac{n^2}{m^2}\right)^2} \times$$

$$\times \sin \alpha c \sin \alpha \xi \sin \alpha x \sin \beta d \cos \beta \eta \sin \beta y.$$

$$(m = 1, 2, 3, \ldots ; \; n = 1, 2, 3, \ldots) \tag{2.46}$$

Equation (2.46), established in [64], represents the mid-surface of a rectangular plate, simply supported along its entire boundary, deflected under the action of a surface couple applied at (ξ, η), the moment vector $\vec{M}$ of which is parallel to the Ox axis (Fig. 2.13).

When the area subjected to the couple M tends towards zero, i.e. for the limit values (Fig. 2.14 B)

$$c \to 0, \quad d \to 0,$$

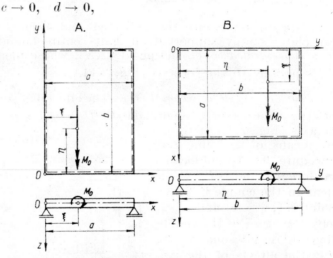

Figure 2.14 Rectangular plate simply supported on the entire boundary. A. Load induced by a concentrated couple M_0 applied at (ξ, η), having moment vector parallel to the Oy axis ($\vec{M}_0 \parallel Oy$). B. Concentrated couple M_0 applied at (ξ, η), having the moment vector normal to the Oy axis ($\vec{M}_0 \parallel Ox$).

solution (2.46) becomes

$$w(x, y) = \frac{4M_0 a^3}{\pi^3 Db^2} \sum_m \sum_n \frac{n}{m^4 \left(1 + \dfrac{a^2}{b^2} \dfrac{n^2}{m^2}\right)^2} \times$$

$$\times \sin \alpha \xi \sin \alpha x \cos \beta \eta \sin \beta y \tag{2.47}$$

$$(m = 1, 2, 3, \ldots ; \; n = 1, 2, 3, \ldots).$$

2.2 Plates deflected by loads of type P, M and q

Equation (2.47) represents the mid-surface of a rectangular plate, deflected under a *concentrated couple* applied at (ξ, η), the moment vector $\vec{M}_0$ of which is parallel to the Ox axis (Fig. 2.14B)

2.2.3.4 *The uniformly distributed load q.* The mid-surface of a rectangular plate $a \times b$, simply supported along its entire boundary, deflected under a load q uniformly distributed throughout its surface (Fig. 2.3 – C.4) is given by the function

$$w(x, y) = \frac{16qa^4}{\pi^6 D} \sum_m \sum_n \frac{1}{m^5 n \left(1 + \frac{a^2}{b^2} \frac{n^2}{m^2}\right)^2} \sin \alpha x \sin \beta y,$$

$$(m = 1, 3, 5, \ldots; \ n = 1, 3, 5, \ldots), \tag{2.48}$$

where

$$\alpha = \frac{m\pi}{a}; \quad \beta = \frac{n\pi}{b}. \tag{2.42}$$

Solution (2.48) of the fundamental equation (2.0) is obtained by particularization of the Navier solution (2.41) for a locally distributed load $P = 4pcd$.

Similarly to solution (2.21), corresponding to the infinitely long plate under a uniformly distributed load q, solution (2.48) allows an extension of the range of applicability of the structural analysis method presented in this book. The computation programs allow the consideration of the loads distributed in strips or in chess-board fashion and the approach of finite boundary plates that are continuous over several spans [75].

2.3 Algorithms for computing elastic displacements, sectional stresses and boundary reactions

2.3.1 Systematization of numerical computations

The elastic displacements w, φ_x, φ_y, the stresses M_x, M_y, M_{xy}, T_x, T_y and the boundary reactions R_x, R_y are calculated starting from the functions $w(x, y)$, which are clearly determined for certain boundary conditions and certain types of loads (Sub-sec. 2.2.1—2.2.3). The expressions resulting from the use of the general formulae (2.1)—(2.10) can sometimes assume very sophisticated forms (see Sub-sec. 2.3.2).

In order to simplify the numerical computation involved in the analysis of slab structures, we give below a systematization

2 Models for the deflected middle surface of plates

suited for data processing. It is a simplified system of writing the algorithms for ten dimensionless coefficients [68], [70], [75]

$$k_w,\ k_\varphi^x,\ k_\varphi^y,\ k_M^x,\ k_M^y,\ k_M^{xy},\ k_T^x,\ k_T^y,\ k_R^x,\ k_R^y \qquad (2.49)$$

which completely define the states of stress and deformation for a given case (Table 2.2).

Table 2.2

Practical computation formulae

Symbol	Type of load			Unit of measurement	Reference number
	P [N]	M [Nm]	q [N/m²]		
Transverse displacement					
$w =$	$\dfrac{Pa^2}{D} k_w$	$\dfrac{Ma}{D} k_w$	$\dfrac{qa^4}{D} k_w$	[m]	(2.51)
Rotations					
$\varphi_x =$	$\dfrac{Pa}{D} k_\varphi^x$	$\dfrac{M}{D} k_\varphi^x$	$\dfrac{qa^3}{D} k_\varphi^x$	[rad]	(2.52)
$\varphi_y =$	$\dfrac{Pa}{D} k_\varphi^y$	$\dfrac{M}{D} k_\varphi^y$	$\dfrac{qa^3}{D} k_\varphi^y$	[rad]	(2.53)
Moments					
$M_x =$	$P k_M^x$	$\dfrac{M}{a} k_M^x$	$qa^2 k_M^x$	[Nm/m]	(2.54)
$M_y =$	$P k_M^y$	$\dfrac{M}{a} k_M^y$	$qa^2 k_M^y$	[Nm/m]	(2.55)
$M_{xy} =$	$P k_M^{xy}$	$\dfrac{M}{a} k_M^{xy}$	$qa^2 k_M^{xy}$	[Nm/m]	(2.56)
Shearing forces					
$T_x =$	$\dfrac{P}{a} k_T^x$	$\dfrac{M}{a^2} k_T^x$	$qa\, k_T^x$	[N/m]	(2.57)
$T_y =$	$\dfrac{P}{a} k_T^y$	$\dfrac{M}{a^2} k_T^y$	$qa\, k_T^y$	[N/m]	(2.58)
Boundary reactions					
$R_x =$	$\dfrac{P}{a} k_R^x$	$\dfrac{M}{a^2} k_R^x$	$qa\, k_R^x$	[N/m]	(2.59)
$R_y =$	$\dfrac{P}{a} k_R^y$	$\dfrac{M}{a^2} k_R^x$	$qa\, k_R^x$	[N/m]	(2.60)

2.3 Computation algorithms

Indeed, the expressions of these dimensionless coefficients allow us to determine at any arbitrarily chosen point (x, y) in the plane of the plate the following static magnitudes, depending on the category of the plate and the type of load, whether P, M or q:

1. k_w — transverse elastic displacement (deflection);
2. k_φ^x — angular elastic displacement (rotation) in the direction of the Ox axis;
3. k_φ^y — angular elastic displacement (rotation) in the direction of the Oy axis;
4. k_M^x — bending moment in the direction of the Ox axis;
5. k_M^y — bending moment in the direction of the Oy axis;
6. k_M^{xy} — torsional moment;
7. k_T^x — shear force in the direction of the Ox axis;
8. k_T^y — shear force in the direction of the Oy axis;
9. k_R^x — boundary reaction in the direction of the Ox axis;
10. k_R^y — boundary reaction in the direction of the Oy axis.

Usually, the symbol of each dimensionless coefficient representing a certain static quantity is defined by three indices

$$k_r^{st} = k(a, b, c, d, \xi, \eta, \mu, x, y), \tag{2.49a}$$

where

r defines the nature of the elastic displacement (deflection or rotation), and the nature of the stresses (bending or torsional moment, shear force, and boundary reaction) respectively;

s defines the in-plane axis (Ox or Oy) in the direction of which the static quantity defined by k_r is calculated;

t defines the type of load (P, M or q) causing the static quantity to be determined.

This simplified system of notation can be extended also to the study of plates subjected to sets of distributed loads P or to sets of associated surface couples M (Chaps. 4, 5 and 6). In order to discriminate from the coefficients established for only one load of type P or M, which are denoted by k, the dimensionless coefficients determined for sets of loads P or M have been denoted by K:

$$K_w, K_\varphi^x, K_\varphi^y, K_M^x, K_M^y, K_M^{xy}, K_T^x, K_T^y, K_R^x, K_R^y. \tag{2.50}$$

In structural analysis, where superposition of effects occurs under different types of loads (P, M, q), each static quantity associated is defined by a notation bearing three indices:

$$K_r^{st} \tag{2.50a}$$

where the meanings of r, s and t are those established in (2.49a).

2 *Models for the deflected middle surface of plates*

The practical computation formulae (2.51)—(2.60) have a general range of applicability. Indeed, their structure remains unchanged :
- for any of the categories of plates under consideration, i.e. for any boundary conditions,
- for any direction of the moment vector $\vec{M}$ of the surface couple in the plane of the plate, and
- for any distribution (throughout the plate, on strips or in chess-board fashion) of the design load q.

Using the formulae given in Table 2.2, we can simultaneously introduce into calculation the two forms under which the load of type P or M may appear in the study of slab structures (Chaps. 5 and 6), namely :
- *given loads*, either forces of type P_i or couples M_i that are locally distributed on the contact areas $2c_i \times 2d_i$, or
- *generalized interacting forces* X_j, either reaction-forces or reaction-couples that are locally distributed on the contact areas $2c_j \times 2d_j$.

The simplified system of writing the computational algorithms allows the separation of the lengthy calculations required for their numerical computation (summation of series of trigonometrical, hyperbolic and exponential functions), which can be carried out only on computers. Once the value of the dimensionless coefficients k_r^{st} is known, the remaining calculations required for :
- the study of the behaviour of plates under different types of loads (Chaps. 3 and 4),

and
- the elastic, static and dynamic, analysis of slab structures (Chaps. 5 and 6).

can be carried out without using large-sized computers.

2.3.2 Expressions for dimensionless coefficients

The functions $w(x, y)$ and their partial derivatives of first-, second- and third-orders allow us to establish the expressions of the ten static magnitudes which define the elastic displacements w, φ_x, φ_y, the stresses M_x, M_y, M_{xy}, T_x, T_y and the boundary reactions R_x, R_y (Table 2.1).

The expressions of the dimensionless coefficients k_r^{st} (2.49) required by the computer programs can be easily obtained by applying the practical computational formulae (2.51)—(2.60) given in Table 2.2. The results of these calculations are given below for each category of plates and type of load [68], [70], [75], [82].

2.3 Computation algorithms

2.3.2.1 The infinitely long plate. The equations of the deflected middle surface are as follows:

$\dfrac{\text{Load } P}{(\text{Fig. 2.4})}$: $w(x, y) = \dfrac{Pa^4}{2\pi^5 Dcd} \sum\limits_m \mathcal{A}_m \sin \alpha x\, Y_m$

$\hspace{6cm}(m = 1, 2, 3, \ldots) \hspace{1cm}$ (2.11a)

$\dfrac{\text{Load } \vec{M} \parallel Oy}{(\text{Fig. 2.5})}$: $w(x, y) = \dfrac{Ma^3}{2\pi^4 Dcd} \sum\limits_m \mathcal{B}_m \sin \alpha x\, Y_m$

$\hspace{6cm}(m = 1, 2, 3, \ldots) \hspace{1cm}$ (2.16a)

$\dfrac{\text{Load } \vec{M} \parallel Ox}{(\text{Fig. 2.6})}$:

— for $y \leqslant d$, $w(x, y) = \dfrac{Ma^3}{2\pi^4 Dcd} \sum\limits_m m\mathcal{A}_m \sin \alpha x [(1 +$

$+ \alpha d)\,\text{sh}\,\alpha y - \alpha y\,\text{ch}\,\alpha y]\,e^{-\alpha d} \hspace{0.5cm} (m = 1, 2, 3, \ldots) \hspace{0.5cm}$ (2.17a)

— for $y \geqslant d$, $w(x, y) = \dfrac{Ma^3}{2\pi^4 Dcd} \sum\limits_m m\mathcal{A}_m \sin \alpha x [(1 +$

$+ \alpha y)\,\text{sh}\,\alpha d - \alpha d\,\text{ch}\,\alpha d]\,e^{-\alpha y} \hspace{0.5cm} (m = 1, 2, 3, \ldots) \hspace{0.5cm}$ (2.18a)

$\dfrac{\text{Load } q}{(\text{Fig. 2.3—A.4})}$: $w(x) = \dfrac{4qa^4}{\pi^5 D} \sum\limits_m \dfrac{1}{m^5} \sin \alpha x$

$\hspace{6cm}(m = 1, 3, 5, \ldots) \hspace{1cm}$ (2.21)

$\hspace{2cm} w(x) = \dfrac{qa^4}{24D}\left(\dfrac{x^4}{a^4} - 2\dfrac{x^3}{a^3} + \dfrac{x}{a}\right) \hspace{1cm}$ (2.22)

Notation: $\mathcal{A}_m = \dfrac{1}{m^5} \sin \alpha c \sin \alpha \xi$

$\hspace{2cm} \mathcal{B}_m = \dfrac{1}{m^4} \sin \alpha c \cos \alpha \xi$

— for $y \leqslant d$, $Y_{m1} = 2 + [\alpha y\,\text{sh}\,\alpha y - (2 + \alpha d)\,\text{ch}\,\alpha y]\,e^{-\alpha d}$ (2.12a)

— for $y \geqslant d$, $Y_{m2} = [(2 + \alpha y)\,\text{sh}\,\alpha d - \alpha d\,\text{ch}\,\alpha d]\,e^{-\alpha y}$ (2.13a)

$\alpha = \dfrac{m\pi}{a}; \hspace{3cm} m = 1, 2, 3, \ldots \hspace{1cm}$ (2.14)

The expressions for the dimensionless coefficients that are required in the study of the infinitely long plate are given in Tables 2.3—2.6.

2 Models for the deflected middle surface of plates

Table 2.3
The infinitely long plate — Load P
Expressions of the dimensionless coefficients

$\boxed{y \leq d}$ $\quad k_w = \dfrac{a^2}{2\pi^5 cd} \sum_m \mathcal{A}_m \sin \alpha x \, Y_{m_1}$ \hfill (2.61)

$k_\varphi^x = \dfrac{a^2}{2\pi^4 cd} \sum_m m \mathcal{A}_m \cos \alpha x \, Y_{m_1}$ \hfill (2.62)

$k_\varphi^y = \dfrac{a^2}{2\pi^4 cd} \sum_m m \mathcal{A}_m \sin \alpha x [\alpha y \operatorname{ch} \alpha y - (1 + \alpha d) \operatorname{sh} \alpha y] e^{-\alpha d}$ \hfill (2.63)

$k_M^x = \dfrac{a^2}{2\pi^3 cd} \sum_m m^2 \mathcal{A}_m \sin \alpha x [Y_{m_1} + \mu(\alpha d \operatorname{ch} \alpha y - \alpha y \operatorname{sh} \alpha y) e^{-\alpha d}]$ \hfill (2.64)

$k_M^y = \dfrac{a^2}{2\pi^3 cd} \sum_m m^2 \mathcal{A}_m \sin \alpha x [(\alpha d \operatorname{ch} \alpha y - \alpha y \operatorname{sh} \alpha y) e^{-\alpha d} + \mu Y_{m_1}]$ \hfill (2.65)

$k_M^{xy} = \dfrac{(1-\mu)a^2}{2\pi^3 cd} \sum_m m^2 \mathcal{A}_m \cos \alpha x [(1 + \alpha d) \operatorname{sh} \alpha y - \alpha y \operatorname{ch} \alpha y] e^{-\alpha d}$ \hfill (2.66)

$k_T^x = \dfrac{a^2}{\pi^2 cd} \sum_m m^3 \mathcal{A}_m \cos \alpha x (1 - e^{-\alpha d} \operatorname{ch} \alpha y)$ \hfill (2.67)

$k_T^y = -\dfrac{a^2}{\pi^2 cd} \sum_m m^3 \mathcal{A}_m \sin \alpha x \, e^{-\alpha d} \operatorname{sh} \alpha y$ \hfill (2.68)

$k_R^x = \dfrac{a^2}{2\pi^2 cd} \sum_m m^3 \mathcal{A}_m \cos \alpha x [Y_{m_1} + (2 - \mu)(\alpha d \operatorname{ch} \alpha y - \alpha y \operatorname{sh} \alpha y) e^{-\alpha d}]$ \hfill (2.69)

$\boxed{y \geq d}$ $\quad k_w = \dfrac{a^2}{2\pi^5 cd} \sum_m \mathcal{A}_m \sin \alpha x \, Y_{m_2}$ \hfill (2.61′)

$k_\varphi^x = \dfrac{a^2}{2\pi^4 cd} \sum_m m \mathcal{A}_m \cos \alpha x \, Y_{m_2}$ \hfill (2.62′)

$k_\varphi^y = \dfrac{a^2}{2\pi^4 cd} \sum_m m \mathcal{A}_m \sin \alpha x [\alpha d \operatorname{ch} \alpha d - (1 + \alpha y) \operatorname{sh} \alpha d] e^{-\alpha y}$ \hfill (2.63′)

$k_M^x = \dfrac{a^2}{2\pi^3 cd} \sum_m m^2 \mathcal{A}_m \sin \alpha x [Y_{m_2} + \mu(\alpha d \operatorname{ch} \alpha d - \alpha y \operatorname{sh} \alpha d) e^{-\alpha y}]$ \hfill (2.64′)

$k_M^y = \dfrac{a^2}{2\pi^3 cd} \sum_m m^2 \mathcal{A}_m \sin \alpha x [(\alpha d \operatorname{ch} \alpha d - \alpha y \operatorname{sh} \alpha d) e^{-\alpha y} + \mu Y_{m_2}]$ \hfill (2.65′)

2.3 Computation algorithms

Table 2.3 (continued)

$$k_M^{xy} = \frac{(1-\mu)a^2}{2\pi^3 cd} \sum_m m^2 \mathcal{A}_m \cos \alpha x [(1+\alpha y) \operatorname{sh} \alpha d - \alpha d \operatorname{ch} \alpha d] e^{-\alpha y} \qquad (2.66')$$

$$k_T^x = \frac{a^2}{\pi^2 cd} \sum_m m^3 \mathcal{A}_m \cos \alpha x \cdot e^{-\alpha y} \operatorname{sh} \alpha d \qquad (2.67')$$

$$k_T^y = -\frac{a^2}{\pi^2 cd} \sum_m m^3 \mathcal{A}_m \sin \alpha x \cdot e^{-\alpha y} \operatorname{sh} \alpha d \qquad (2.68')$$

$$k_R^x = \frac{a^2}{2\pi^2 cd} \sum_m m^3 \mathcal{A}_m \cos \alpha x [Y_{m_2} + (2-\mu)(\alpha d \operatorname{ch} \alpha d - \alpha y \operatorname{sh} \alpha d) e^{-\alpha y}] \qquad (2.69')$$

Table 2.4

The infinitely long plate — Load $\vec{M} \parallel Oy$
Expressions of the dimensionless coefficients

$$\boxed{y \leqslant d} \quad k_w = \frac{a^2}{2\pi^4 cd} \sum_m \mathcal{B}_m \sin \alpha x \, Y_{m_1} \qquad (2.71)$$

$$k_\varphi^x = \frac{a^2}{2\pi^3 cd} \sum_m m \mathcal{B}_m \cos \alpha x \, Y_{m_1} \qquad (2.72)$$

$$k_\varphi^y = \frac{a^2}{2\pi^3 cd} \sum_m m \mathcal{B}_m \sin \alpha x [\alpha y \operatorname{ch} \alpha y - (1+\alpha d) \operatorname{sh} \alpha y] e^{-\alpha d} \qquad (2.73)$$

$$k_M^x = \frac{a^2}{2\pi^2 cd} \sum_m m^2 \mathcal{B}_m \sin \alpha x [Y_{m_1} + \mu(\alpha d \operatorname{ch} \alpha y - \alpha y \operatorname{sh} \alpha y) e^{-\alpha d}] \qquad (2.74)$$

$$k_M^y = \frac{a^2}{2\pi^2 cd} \sum_m m^2 \mathcal{B}_m \sin \alpha x [(\alpha d \operatorname{ch} \alpha y - \alpha y \operatorname{sh} \alpha y) e^{-\alpha d} + \mu Y_{m_1}] \qquad (2.75)$$

$$k_M^{xy} = \frac{(1-\mu)a^2}{2\pi^2 cd} \sum_m m^2 \mathcal{B}_m \cos \alpha x [(1+\alpha d) \operatorname{sh} \alpha y - \alpha y \operatorname{ch} \alpha y] e^{-\alpha d} \qquad (2.76)$$

$$k_T^x = \frac{a^2}{\pi cd} \sum_m m^3 \mathcal{B}_m \cos \alpha x (1 - e^{-\alpha d} \cdot \operatorname{ch} \alpha y) \qquad (2.77)$$

$$k_T^y = -\frac{a^2}{\pi cd} \sum_m m^3 \mathcal{B}_m \sin \alpha x \cdot e^{-\alpha d} \operatorname{sh} \alpha y \qquad (2.78)$$

$$k_R^x = \frac{a^2}{2\pi cd} \sum_m m^3 \mathcal{B}_m \cos \alpha x [Y_{m_1} + (2-\mu)(\alpha d \operatorname{ch} \alpha y - \alpha y \operatorname{sh} \alpha y) e^{-\alpha d}] \qquad (2.79)$$

2 Models for the deflected middle surface of plates

Table 2.4 (continued)

$$\boxed{y \geqslant d} \quad k_w = \frac{a^2}{2\pi^4 cd} \sum_m \mathscr{B}_m \sin \alpha x \, Y_{m_2} \tag{2.71'}$$

$$k_\varphi^x = \frac{a^2}{2\pi^3 cd} \sum_m m \mathscr{B}_m \cos \alpha x \, Y_{m_2} \tag{2.72'}$$

$$k_\varphi^y = \frac{a^2}{2\pi^3 cd} \sum_m m \mathscr{B}_m \sin \alpha x [\alpha d \, \text{ch} \, \alpha d - (1 + \alpha y) \, \text{sh} \, \alpha d] e^{-\alpha y} \tag{2.73'}$$

$$k_M^x = \frac{a^2}{2\pi^2 cd} \sum_m m^2 \mathscr{B}_m \sin \alpha x \, [Y_{m_2} + \mu(\alpha d \, \text{ch} \, \alpha d - \alpha y \, \text{sh} \, \alpha d) e^{-\alpha y}] \tag{2.74'}$$

$$k_M^y = \frac{a^2}{2\pi^2 cd} \sum_m m^2 \mathscr{B}_m \sin \alpha x [(\alpha d \, \text{ch} \, \alpha d - \alpha y \, \text{sh} \, \alpha d) e^{-\alpha y} + \mu Y_{m_2}] \tag{2.75'}$$

$$k_M^{xy} = \frac{(1-\mu)a^2}{2\pi^2 cd} \sum_m m^2 \mathscr{B}_m \cos \alpha x [(1 + \alpha y) \, \text{sh} \, \alpha d - \alpha d \, \text{ch} \, \alpha d] e^{-\alpha y} \tag{2.76'}$$

$$k_T^x = \frac{a^2}{\pi cd} \sum_m m^3 \mathscr{B}_m \cos \alpha x \cdot e^{-\alpha y} \, \text{sh} \, \alpha d \tag{2.77'}$$

$$k_T^y = -\frac{a^2}{\pi cd} \sum_m m^3 \mathscr{B}_m \sin \alpha x \cdot e^{-\alpha y} \, \text{sh} \, \alpha d \tag{2.78'}$$

$$k_R^x = \frac{a^2}{2\pi cd} \sum_m m^3 \mathscr{B}_m \cos \alpha x [Y_{m_2} + (2-\mu)(\alpha d \, \text{ch} \, \alpha d - \alpha y \, \text{sh} \, \alpha d) e^{-\alpha y}] \tag{2.79'}$$

Table 2.5

The infinitely long plate — Load $\vec{M} \parallel Ox$
Expressions of the dimensionless coefficients

$$\boxed{y \leqslant d} \quad k_w = \frac{a^2}{2\pi^4 cd} \sum_m m \mathscr{A}_m \sin \alpha x [(1 + \alpha d) \, \text{sh} \, \alpha y - \alpha y \, \text{ch} \, \alpha y] e^{-\alpha d} \tag{2.81}$$

$$k_\varphi^x = \frac{a^2}{2\pi^3 cd} \sum_m m^2 \mathscr{A}_m \cos \alpha x [(1 + \alpha d) \text{sh} \, \alpha y - \alpha y \, \text{ch} \, \alpha y] e^{-\alpha d} \tag{2.82}$$

$$k_\varphi^y = \frac{a^2}{2\pi^3 cd} \sum_m m^2 \mathscr{A}_m \sin \alpha x (\alpha d \, \text{ch} \, \alpha y - \alpha y \, \text{sh} \, \alpha y) e^{-\alpha d} \tag{2.83}$$

$$k_M^x = \frac{a^2}{2\pi^2 cd} \sum_m m^3 \mathscr{A}_m \sin \alpha x \{(1 + \alpha d) \, \text{sh} \, \alpha y - \alpha y \, \text{ch} \, \alpha y + \\ + \mu[(1 - \alpha d) \, \text{sh} \, \alpha y + \alpha y \, \text{ch} \, \alpha y]\} \, e^{-\alpha d} \tag{2.84}$$

2.3 Computation algorithms

Table 2.5 (continued)

$$k_M^y = \frac{a^2}{2\pi^2 cd} \sum_m m^3 \mathcal{A}_m \sin \alpha x \{(1 - \alpha d) \operatorname{sh} \alpha y + \alpha y \operatorname{ch} \alpha y +$$
$$+ \mu[(1 + \alpha d)\operatorname{sh} \alpha y - \alpha y \operatorname{ch} \alpha y]\} e^{-\alpha d} \quad (2.85)$$

$$k_M^{xy} = \frac{(1 - \mu)a^2}{2\pi^2 cd} \sum_m m^3 \mathcal{A}_m \cos \alpha x (\alpha y \operatorname{sh} \alpha y - \alpha d \operatorname{ch} \alpha y) \, e^{-\alpha d} \quad (2.86)$$

$$k_T^x = \frac{a^2}{\pi cd} \sum_m m^4 \mathcal{A}_m \cos \alpha x \cdot e^{-\alpha d} \cdot \operatorname{sh} \alpha y \quad (2.87)$$

$$k_T^y = \frac{a^2}{\pi cd} \sum_m m^4 \mathcal{A}_m \sin \alpha x \cdot e^{-\alpha d} \cdot \operatorname{ch} \alpha y \quad (2.88)$$

$$k_R^x = \frac{a^2}{2\pi cd} \sum_m m^4 \mathcal{A}_m \cos \alpha x \{(1 + \alpha d)\operatorname{sh} \alpha y - \alpha y \operatorname{ch} \alpha y +$$
$$+ (2 - \mu)[(1 - \alpha d)\operatorname{sh} \alpha y + \alpha y \operatorname{ch} \alpha y]\} e^{-\alpha d} \quad (2.89)$$

$\boxed{y \geq d}$
$$k_w = \frac{a^2}{2\pi^4 cd} \sum_m m \mathcal{A}_m \sin \alpha x [(1 + \alpha y)\operatorname{sh} \alpha d - \alpha d \operatorname{ch} \alpha d] e^{-\alpha y} \quad (2.81')$$

$$k_\varphi^x = \frac{a^2}{2\pi^3 cd} \sum_m m^2 \mathcal{A}_m \cos \alpha x [(1 + \alpha y)\operatorname{sh} \alpha d - \alpha d \operatorname{ch} \alpha d] e^{-\alpha y} \quad (2.82')$$

$$k_\varphi^y = \frac{a^2}{2\pi^3 cd} \sum_m m^2 \mathcal{A}_m \sin \alpha x (\alpha d \operatorname{ch} \alpha d - \alpha y \operatorname{sh} \alpha d) \, e^{-\alpha y} \quad (2.83')$$

$$k_M^x = \frac{a^2}{2\pi^2 cd} \sum_m m^3 \mathcal{A}_m \sin \alpha x \{(1 + \alpha y)\operatorname{sh} \alpha d - \alpha d \operatorname{ch} \alpha d +$$
$$+ \mu[(1 - \alpha y)\operatorname{sh} \alpha d + \alpha d \operatorname{ch} \alpha d]\} e^{-\alpha y} \quad (2.84')$$

$$k_M^y = \frac{a^2}{2\pi^2 cd} \sum_m m^3 \mathcal{A}_m \sin \alpha x \{(1 - \alpha y)\operatorname{sh} \alpha d + \alpha d \operatorname{ch} \alpha d +$$
$$+ \mu[(1 + \alpha y) \operatorname{sh} \alpha d - \alpha d \operatorname{ch} \alpha d]\} e^{-\alpha y} \quad (2.85')$$

$$k_M^{xy} = \frac{(1 - \mu)a^2}{2\pi^2 cd} \sum_m m^3 \mathcal{A}_m \cos \alpha x (\alpha y \operatorname{sh} \alpha d - \alpha d \operatorname{ch} \alpha d) e^{-\alpha y} \quad (2.86')$$

$$k_T^x = \frac{a^2}{\pi cd} \sum_m m^4 \mathcal{A}_m \cos \alpha x \cdot e^{-\alpha y} \cdot \operatorname{sh} \alpha d \quad (2.87')$$

$$k_T^y = -\frac{a^2}{\pi cd} \sum_m m^4 \mathcal{A}_m \sin \alpha x \, e^{-\alpha y} \cdot \operatorname{sh} \alpha d \quad (2.88')$$

$$k_R^x = \frac{a^2}{2\pi cd} \sum_m m^4 \mathcal{A}_m \cos \alpha x \{(1 + \alpha y)\operatorname{sh} \alpha d - \alpha d \operatorname{ch} \alpha d +$$
$$+ (2 - \mu)[(1 - \alpha y)\operatorname{sh} \alpha d + \alpha d \operatorname{ch} \alpha d]\} e^{-\alpha y} \quad (2.89')$$

2 Models for the deflected middle surface of plates

Table 2.6
The infinitely long plate — Load q
Expressions of the dimensionless coefficients

$k_w = \dfrac{4}{\pi^5} \sum\limits_m \dfrac{1}{m^5} \sin \alpha x$	(2.91)	$k_w = \dfrac{1}{24}\left(\dfrac{x^4}{a^4} - 2\dfrac{x^3}{a^3} + \dfrac{x}{a}\right)$	(2.91')
$k_\varphi^x = \dfrac{4}{\pi^4} \sum\limits_m \dfrac{1}{m^4} \cos \alpha x$	(2.92)	$k_\varphi^x = \dfrac{1}{24}\left(4\dfrac{x^3}{a^3} - 6\dfrac{x^2}{a^2} + 1\right)$	(2.92')
$k_\varphi^y = 0$	(2.93)	$k_\varphi^y = 0$	(2.93')
$k_M^x = \dfrac{4}{\pi^3} \sum\limits_m \dfrac{1}{m^3} \sin \alpha x$	(2.94)	$k_M^x = \dfrac{1}{2}\left(\dfrac{x}{a} - \dfrac{x^2}{a^2}\right)$	(2.94')
$k_M^y = \mu\, k_M^x$	(2.95)	$k_M^y = \mu\, k_M^x$	(2.95')
$k_M^{xy} = 0$	(2.96)	$k_M^{xy} = 0$	(2.96')
$k_T^x = \dfrac{4}{\pi^2} \sum\limits_m \dfrac{1}{m^2} \cos \alpha x$	(2.97)	$k_T^x = \dfrac{1}{2} - \dfrac{x}{a}$	(2.97')
$k_T^y = 0$	(2.98)	$k_T^y = 0$	(2.98')
$k_R^x = \dfrac{4}{\pi^2} \sum\limits_m \dfrac{1}{m^2} \cos \alpha x$	(2.99)	$k_R^x = \dfrac{1}{2} - \dfrac{x}{a}$	(2.99')

2.3.2.2 *The rectangular plate with two parallel free edges.* The equations of the deflected middle surface are as follows:

$\dfrac{\text{Load } P}{\text{(Fig. 2.7)}}$: $w(x, y) = \dfrac{Pa^4}{2\pi^5 Dcd} \sum\limits_m \mathcal{A}_m \sin \alpha x (Y_m + \Phi_m)$

$(m = 1, 2, 3, \ldots)$ \hfill (2.31 a)

$\dfrac{\text{Load } \vec{M} \parallel Oy}{\text{(Fig. 2.8)}}$: $w(x, y) = \dfrac{Ma^3}{2\pi^4 Dcd} \sum\limits_m \mathcal{B}_m \sin \alpha x (Y_m + \Phi_m)$

$(m = 1, 2, 3, \ldots)$ \hfill (2.32 a)

$\dfrac{\text{Load } \vec{M} \parallel Ox}{\text{(Fig. 2.9)}}$:

— for $0 \leqslant y \leqslant d$, $w(x, y) = \dfrac{Ma^3}{2\pi^4 Dcd} \sum\limits_m m\, \mathcal{A}_m \sin \alpha x \times$

$\times \{[(1 + \alpha d)\,\text{sh}\,\alpha y - \alpha y\,\text{ch}\,\alpha y]e^{-\alpha d} -$
$- [(A_m + B_m)\text{sh}\,\alpha y + \alpha y B_m \text{ch}\,\alpha y]\}$

$(m = 1, 2, 3, \ldots)$ \hfill (2.33 a)

2.3 Computation algorithms

— for $d \leqslant y \leqslant \dfrac{b}{2}$, $w(x, y) = \dfrac{Ma^3}{2\pi^4 Dcd} \sum\limits_{m} m \mathcal{A}_m \sin \alpha x \times$

$\times \{[(1 + \alpha y) \operatorname{sh} \alpha d - \alpha d \operatorname{ch} \alpha d] e^{-\alpha y} -$

$- [(A_m + B_m) \operatorname{sh} \alpha y + \alpha y B_m \operatorname{ch} \alpha y]\}$

$(m = 1, 2, 3, \ldots)$ \hfill (2.34a)

$\dfrac{\text{Load } q}{\text{(Fig. 2.3—B.4)}}$: $w(x, y) = \dfrac{4qa^4}{\pi^5 D} \sum\limits_{m} \dfrac{1}{m^5} \sin \alpha x (C_m \operatorname{ch} \alpha y +$

$+ \alpha y D_m \operatorname{sh} \alpha y + 1)$ $(m = 1, 2, 3, \ldots)$ \hfill (2.40)

Notation: $\mathcal{A}_m = \dfrac{1}{m^5} \sin \alpha c \sin \alpha \xi$; $\mathcal{B}_m = \dfrac{1}{m^4} \sin \alpha c \cos \alpha \xi$

— for $0 \leqslant y \leqslant d$, $\quad Y_{m1} = 2 + [\alpha y \operatorname{sh} \alpha y - (2 + \alpha d) \operatorname{ch} \alpha y] e^{-\alpha d}$

\hfill (2.12a)

— for $d \leqslant y \leqslant \dfrac{b}{2}$, $\quad Y_{m2} = [(2 + \alpha y) \operatorname{sh} \alpha d - \alpha d \operatorname{ch} \alpha d] e^{-\alpha y}$

\hfill (2.13a)

$\Phi_m = A_m \operatorname{ch} \alpha y + \alpha y B_m \operatorname{sh} \alpha y$ \hfill (2.27)

$A_m = \dfrac{1}{\operatorname{ch} \dfrac{\alpha b}{2}} \left\{ e^{-\frac{\alpha b}{2}} \left[\left(\dfrac{2\mu}{1 - \mu} - \dfrac{\alpha b}{2} \right) \operatorname{sh} \alpha d + \alpha d \operatorname{ch} \alpha d \right] - \right.$

$\left. - \left(\dfrac{2}{1 - \mu} \operatorname{ch} \dfrac{\alpha b}{2} + \dfrac{\alpha b}{2} \operatorname{sh} \dfrac{\alpha b}{2} \right) B_m \right\}$ \hfill (2.30a)

$B_m = - \dfrac{2e^{-\frac{\alpha b}{2}}}{\dfrac{3 + \mu}{1 - \mu} \operatorname{sh} \alpha b - \alpha b} \times$

$\times \left\{ \left[\left(\dfrac{3 - \mu}{1 - \mu} + \dfrac{\alpha b}{2} \right) \operatorname{ch} \dfrac{\alpha b}{2} - \left(\dfrac{2\mu}{1 - \mu} - \dfrac{\alpha b}{2} \right) \operatorname{sh} \dfrac{\alpha b}{2} \right] \operatorname{sh} \alpha d - \right.$

$\left. - \left(\operatorname{sh} \dfrac{\alpha b}{2} + \operatorname{ch} \dfrac{\alpha b}{2} \right) \alpha d \operatorname{ch} \alpha d \right\}$ \hfill (2.30b)

$C_m = \dfrac{\mu}{1 - \mu} \dfrac{(1 + \mu) \operatorname{sh} \dfrac{\alpha b}{2} - (1 - \mu) \dfrac{\alpha b}{2} \operatorname{ch} \dfrac{\alpha b}{2}}{(3 + \mu) \operatorname{sh} \dfrac{\alpha b}{2} \operatorname{ch} \dfrac{\alpha b}{2} - (1 - \mu) \dfrac{\alpha b}{2}}$ \hfill (2.39a)

2 Models for the deflected middle surface of plates

$$D_m = \frac{\mu \, \text{sh} \, \frac{\alpha b}{2}}{(3+\mu) \, \text{sh} \, \frac{\alpha b}{2} \, \text{ch} \, \frac{\alpha b}{2} - (1-\mu) \frac{\alpha b}{2}} \qquad (2.39\text{ b})$$

$$\alpha = \frac{m\pi}{a}; \quad m = 1, 2, 3, \ldots \qquad (2.14)$$

The expresions for the dimensionless coefficients that are required in the study of the rectangular plate with two parallel free edges are as given in Tables 2.7—2.10.

Table 2.7

The rectangular plate with two free parallel edges — Load P
Expressions of the dimensionless coefficients

$$\boxed{y \leq d} \quad k_w = \frac{a^2}{2\pi^5 cd} \sum_m \mathcal{A}_m \sin \alpha x (Y_{m_1} + \Phi_m) \qquad (2.101)$$

$$k_\varphi^x = \frac{a^2}{2\pi^4 cd} \sum_m m \mathcal{A}_m \cos \alpha x (Y_{m_1} + \Phi_m) \qquad (2.102)$$

$$k_\varphi^y = \frac{a^2}{2\pi^4 cd} \sum_m m \mathcal{A}_m \sin \alpha x \{[\alpha y \, \text{ch} \, \alpha y - (1+\alpha d) \, \text{sh} \, \alpha y] e^{-\alpha d} +$$
$$+ (A_m + B_m) \text{sh} \, \alpha y + \alpha y B_m \, \text{ch} \, \alpha y\} \qquad (2.103)$$

$$k_M^x = \frac{a^2}{2\pi^3 cd} \sum_m m^2 \mathcal{A}_m \sin \alpha x \{Y_{m_1} + \Phi_m + \mu[(\alpha d \, \text{ch} \, \alpha y - \alpha y \, \text{sh} \, \alpha y) e^{-\alpha d} -$$
$$- (A_m + 2B_m) \text{ch} \, \alpha y - \alpha y B_m \, \text{sh} \, \alpha y]\} \qquad (2.104)$$

$$k_M^y = \frac{a^2}{2\pi^3 cd} \sum_m m^2 \mathcal{A}_m \sin \alpha x [(\alpha d \, \text{ch} \, \alpha y - \alpha y \, \text{sh} \, \alpha y) e^{-\alpha d} -$$
$$- (A_m + 2B_m) \text{ch} \, \alpha y - \alpha y B_m \, \text{sh} \, \alpha y + \mu(Y_{m_1} + \Phi_m)] \qquad (2.105)$$

$$k_M^{xy} = \frac{(1-\mu)a^2}{2\pi^3 cd} \sum_m m^2 \mathcal{A}_m \cos \alpha x \{[(1+\alpha d) \text{sh} \, \alpha y - \alpha y \, \text{ch} \, \alpha y] e^{-\alpha d} -$$
$$- (A_m + B_m) \text{sh} \, \alpha y - \alpha y B_m \text{ch} \, \alpha y\} \qquad (2.106)$$

$$k_T^x = \frac{a^2}{2\pi^2 cd} \sum_m m^3 \mathcal{A}_m \cos \alpha x [Y_{m_1} + \Phi_m + (\alpha d \, \text{ch} \, \alpha y - \alpha y \, \text{sh} \, \alpha y) e^{-\alpha d} -$$
$$- (A_m + 2B_m) \text{ch} \, \alpha y - \alpha y B_m \, \text{sh} \, \alpha y] \qquad (2.107)$$

2.3 Computation algorithms

Table 2.7 (continued)

$$k_T^y = -\frac{a^2}{\pi^2 cd} \sum_m m^3 \mathcal{A}_m \sin \alpha x (e^{-\alpha d} + B_m) \operatorname{sh} \alpha y \qquad (2.108)$$

$$k_R^x = \frac{a^2}{2\pi^2 cd} \sum_m m^3 \mathcal{A}_m \cos \alpha x \{ Y_{m_1} + \Phi_m + (2-\mu)[(\alpha d \operatorname{ch} \alpha y - \alpha y \operatorname{sh} \alpha y) e^{-\alpha d} -$$

$$- (A_m + 2B_m) \operatorname{ch} \alpha y - \alpha y B_m \operatorname{sh} \alpha y]\} \qquad (2.109)$$

$\boxed{y \geqslant d}\ k_w = \dfrac{a^2}{2\pi^5 cd} \sum_m \mathcal{A}_m \sin \alpha x (Y_{m_2} + \Phi_m) \qquad (2.101')$

$$k_\varphi^x = \frac{a^2}{2\pi^4 cd} \sum_m m \mathcal{A}_m \cos \alpha x (Y_{m_2} + \Phi_m) \qquad (2.102')$$

$$k_\varphi^y = \frac{a^2}{2\pi^4 cd} \sum_m m \mathcal{A}_m \sin \alpha x \{ [\alpha d \operatorname{ch} \alpha d - (1 + \alpha y) \operatorname{sh} \alpha d] e^{-\alpha y} +$$

$$+ (A_m + B_m) \operatorname{sh} \alpha y + \alpha y B_m \operatorname{ch} \alpha y \} \qquad (2.103')$$

$$k_M^x = \frac{a^2}{2\pi^3 cd} \sum_m m^2 \mathcal{A}_m \sin \alpha x \{ Y_{m_2} + \Phi_m + \mu[(\alpha d \operatorname{ch} \alpha d - \alpha y \operatorname{sh} \alpha d) e^{-\alpha y} -$$

$$- (A_m + 2B_m) \operatorname{ch} \alpha y - \alpha y B_m \operatorname{sh} \alpha y]\} \qquad (2.104')$$

$$k_M^y = \frac{a^2}{2\pi^3 cd} \sum_m m^2 \mathcal{A}_m \sin \alpha x [(\alpha d \operatorname{ch} \alpha d - \alpha y \operatorname{sh} \alpha d) e^{-\alpha y} -$$

$$- (A_m + 2B_m) \operatorname{ch} \alpha y - \alpha y B_m \operatorname{sh} \alpha y + \mu(Y_{m_2} + \Phi_m)] \qquad (2.105')$$

$$k_M^{xy} = \frac{(1-\mu)a^2}{2\pi^3 cd} \sum_m m^2 \mathcal{A}_m \cos \alpha x \{ [(1 + \alpha y) \operatorname{sh} \alpha d - \alpha d \operatorname{ch} \alpha d] e^{-\alpha y} -$$

$$- (A_m + B_m) \operatorname{sh} \alpha y - \alpha y B_m \operatorname{ch} \alpha y \} \qquad (2.106')$$

$$k_T^x = \frac{a^2}{2\pi^2 cd} \sum_m m^3 \mathcal{A}_m \cos \alpha x [Y_{m_2} + \Phi_m + (\alpha d \operatorname{ch} \alpha d - \alpha y \operatorname{sh} \alpha d) e^{-\alpha y} -$$

$$- (A_m + 2B_m) \operatorname{ch} \alpha y - \alpha y B_m \operatorname{sh} \alpha y] \qquad (2.107')$$

$$k_T^y = -\frac{a^2}{\pi^2 cd} \sum_m m^3 \mathcal{A}_m \sin \alpha x (e^{-\alpha y} \cdot \operatorname{sh} \alpha d + B_m \operatorname{sh} \alpha y) \qquad (2.108')$$

$$k_R^x = \frac{a^2}{2\pi^2 cd} \sum_m m^3 \mathcal{A}_m \cos \alpha x \{ Y_{m_2} + \Phi_m + (2-\mu)[(\alpha d \operatorname{ch} \alpha d -$$

$$- \alpha y \operatorname{sh} \alpha d) e^{-\alpha y} - (A_m + 2B_m) \operatorname{ch} \alpha y - \alpha y B_m \operatorname{sh} \alpha y]\} \qquad (2.109')$$

2 Models for the deflected middle surface of plates

Table 2.8

The rectangular plate with two free parallel edges — Load $\overrightarrow{M} \parallel Oy$
Expressions of the dimensionless coefficients

$\boxed{y \leqslant d}$ $\quad k_w = \dfrac{a^2}{2\pi^4 cd} \sum_m \mathscr{B}_m \sin \alpha x (Y_{m_1} + \Phi_m)$ \hfill (2.111)

$k_\varphi^x = \dfrac{a^2}{2\pi^3 cd} \sum_m m \mathscr{B}_m \cos \alpha x (Y_{m_1} + \Phi_m)$ \hfill (2.112)

$k_\varphi^y = \dfrac{a^2}{2\pi^3 cd} \sum_m m \mathscr{B}_m \sin \alpha x \{ [\alpha y \operatorname{ch} \alpha y - (1 + \alpha d) \operatorname{sh} \alpha y] e^{-\alpha d} +$
$\qquad + (A_m + B_m) \operatorname{sh} \alpha y + \alpha y B_m \operatorname{ch} \alpha y \}$ \hfill (2.113)

$k_M^x = \dfrac{a^2}{2\pi^2 cd} \sum_m m^2 \mathscr{B}_m \sin \alpha x \{ Y_{m_1} + \Phi_m + \mu [(\alpha d \operatorname{ch} \alpha y - \alpha y \operatorname{sh} \alpha y) e^{-\alpha d} -$
$\qquad - (A_m + 2B_m) \operatorname{ch} \alpha y - \alpha y B_m \operatorname{sh} \alpha y] \}$ \hfill (2.114)

$k_M^y = \dfrac{a^2}{2\pi^2 cd} \sum_m m^2 \mathscr{B}_m \sin \alpha x [(\alpha d \operatorname{ch} \alpha y - \alpha y \operatorname{sh} \alpha y) e^{-\alpha d} -$
$\qquad - (A_m + 2B_m) \operatorname{ch} \alpha y - \alpha y B_m \operatorname{sh} \alpha y + \mu (Y_{m_1} + \Phi_m)]$ \hfill (2.115)

$k_M^{xy} = \dfrac{(1 - \mu) a^2}{2\pi^2 cd} \sum_m m^2 \mathscr{B}_m \cos \alpha x \{ [(1 + \alpha d) \operatorname{sh} \alpha y - \alpha y \operatorname{ch} \alpha y] e^{-\alpha d} -$
$\qquad - (A_m + B_m) \operatorname{sh} \alpha y - \alpha y B_m \operatorname{ch} \alpha y \}$ \hfill (2.116)

$k_T^x = \dfrac{a^2}{2\pi cd} \sum_m m^3 \mathscr{B}_m \cos \alpha x [Y_{m_1} + \Phi_m + (\alpha d \operatorname{ch} \alpha y - \alpha y \operatorname{sh} \alpha y) e^{-\alpha d} -$
$\qquad - (A_m + 2B_m) \operatorname{ch} \alpha y - \alpha y B_m \operatorname{sh} \alpha y]$ \hfill (2.117)

$k_T^y = - \dfrac{a^2}{\pi cd} \sum_m m^3 \mathscr{B}_m \sin \alpha x (e^{-\alpha d} + B_m) \operatorname{sh} \alpha y$ \hfill (2.118)

$k_R^x = \dfrac{a^2}{2\pi cd} \sum_m m^3 \mathscr{B}_m \cos \alpha x \{ Y_{m_1} + \Phi_m + (2 - \mu) [(\alpha d \operatorname{ch} \alpha y - \alpha y \operatorname{sh} \alpha y) e^{-\alpha d} -$
$\qquad - (A_m + 2B_m) \operatorname{ch} \alpha y - \alpha y B_m \operatorname{sh} \alpha y] \}$ \hfill (2.119)

$\boxed{y \geqslant d}$ $\quad k_w = \dfrac{a^2}{2\pi^4 cd} \sum_m \mathscr{B}_m \sin \alpha x (Y_{m_2} + \Phi_m)$ \hfill (2.111')

$k_\varphi^x = \dfrac{a^2}{2\pi^3 cd} \sum_m m \mathscr{B}_m \cos \alpha x (Y_{m_2} + \Phi_m)$ \hfill (2.112')

2.3 *Computation algorithms*

Table 2.8 (continued)

$$k_\Phi^y = \frac{a^2}{2\pi^3 cd} \sum_m m\mathcal{B}_m \sin \alpha x \{[\alpha d \operatorname{ch} \alpha d - (1 + \alpha y)\operatorname{sh} \alpha d]e^{-\alpha y} +$$

$$+ (A_m + B_m)\operatorname{sh} \alpha y + \alpha y B_m \operatorname{ch} \alpha y\} \qquad (2.113')$$

$$k_M^x = \frac{a^2}{2\pi^2 cd} \sum_m m^2 \mathcal{B}_m \sin \alpha x \{Y_{m_2} + \Phi_m + \mu[(\alpha d \operatorname{ch} \alpha d - \alpha y \operatorname{sh} \alpha d)e^{-\alpha y} -$$

$$- (A_m + 2B_m)\operatorname{ch} \alpha y - \alpha y B_m \operatorname{sh} \alpha y]\} \qquad (2.114')$$

$$k_M^y = \frac{a^2}{2\pi^2 cd} \sum_m m^2 \mathcal{B}_m \sin \alpha x [(\alpha d \operatorname{ch} \alpha d - \alpha y \operatorname{sh} \alpha d)e^{-\alpha y} -$$

$$- (A_m + 2B_m)\operatorname{ch} \alpha y - \alpha y B_m \operatorname{sh} \alpha y + \mu(Y_{m_2} + \Phi_m)] \qquad (2.115')$$

$$k_M^{xy} = \frac{(1-\mu)a^2}{2\pi^2 cd} \sum_m m^2 \mathcal{B}_m \cos \alpha x \{[(1 + \alpha y)\operatorname{sh} \alpha d - \alpha d \operatorname{ch} \alpha d]e^{-\alpha y} -$$

$$- (A_m + B_m)\operatorname{sh} \alpha y - \alpha y B_m \operatorname{ch} \alpha y\} \qquad (2.116')$$

$$k_T^x = \frac{a^2}{2\pi cd} \sum_m m^3 \mathcal{B}_m \cos \alpha x [Y_{m_2} + \Phi_m + (\alpha d \operatorname{ch} \alpha d - \alpha y \operatorname{sh} \alpha d)e^{-\alpha y} -$$

$$- (A_m + 2B_m)\operatorname{ch} \alpha y - \alpha y B_m \operatorname{sh} \alpha y] \qquad (2.117')$$

$$k_T^y = -\frac{a^2}{\pi^2 cd} \sum_m m^3 \mathcal{B}_m \sin \alpha x (e^{-\alpha y} \operatorname{sh} \alpha d + B_m \operatorname{sh} \alpha y) \qquad (2.118')$$

$$k_R^x = \frac{a^2}{2\pi^2 cd} \sum_m m^2 \mathcal{B}_m \cos \alpha x \{Y_{m_2} + \Phi_m + (2-\mu)[\alpha d \operatorname{ch} \alpha d - \alpha y \operatorname{sh} \alpha d]e^{-\alpha y} -$$

$$- (A_m + 2B_m)\operatorname{ch} \alpha y - \alpha y B_m \operatorname{sh} \alpha y]\} \qquad (2.119')$$

Table 2.9

The rectangular plate with two free parallel edges — Load $\vec{M} \parallel Ox$
Expressions of the dimensionless coefficients

$$\boxed{y \leqslant d} \quad k_w = \frac{a^2}{2\pi^4 cd} \sum_m m \mathcal{A}_m \sin \alpha x \{[(1 + \alpha d)\operatorname{sh} \alpha y - \alpha y \operatorname{ch} \alpha y]e^{-\alpha d} -$$

$$- [(A_m + B_m)\operatorname{sh} \alpha y + \alpha y B_m \operatorname{ch} \alpha y]\} \qquad (2.121)$$

$$k_\Phi^x = \frac{a^2}{2\pi^3 cd} \sum_m m^2 \mathcal{A}_m \cos \alpha x \{[(1 + \alpha d)\operatorname{sh} \alpha y - \alpha y \operatorname{ch} \alpha y]e^{-\alpha d} -$$

$$- [(A_m + B_m)\operatorname{sh} \alpha y + \alpha y B_m \operatorname{ch} \alpha y]\} \qquad (2.122)$$

2. Models for the deflected middle surface of plates

Table 2.9 (continued)

$$k_\varphi^y = \frac{a^2}{2\pi^3 cd} \sum_m m^2 \mathfrak{A}_m \sin \alpha x \{(\alpha d \operatorname{ch} \alpha y - \alpha y \operatorname{sh} \alpha y) e^{-\alpha d} -$$
$$- [(A_m + 2B_m) \operatorname{ch} \alpha y + \alpha y B_m \operatorname{sh} \alpha y]\} \tag{2.123}$$

$$k_M^x = \frac{a^2}{2\pi^2 cd} \sum_m m^3 \mathfrak{A}_m \sin \alpha x \{[(1 + \alpha d) \operatorname{sh} \alpha y - \alpha y \operatorname{ch} \alpha y] e^{-\alpha d} -$$
$$- (A_m + B_m) \operatorname{sh} \alpha y + \alpha y B_m \operatorname{ch} \alpha y] + \mu[(1 - \alpha d) \operatorname{sh} \alpha y + \alpha y \operatorname{ch} \alpha y] e^{-\alpha d} +$$
$$+ \mu[(A_m + 3B_m) \operatorname{sh} \alpha y + \alpha y B_m \operatorname{ch} \alpha y]\} \tag{2.124}$$

$$k_M^y = \frac{a^2}{2\pi^2 cd} \sum_m m^3 \mathfrak{A}_m \sin \alpha x \{[(1 - \alpha d) \operatorname{sh} \alpha y + \alpha y \operatorname{ch} \alpha y] e^{-\alpha d} +$$
$$+ (A_m + 3B_m) \operatorname{sh} \alpha y + \alpha y B_m \operatorname{ch} \alpha y + \mu[(1 + \alpha d) \operatorname{sh} \alpha y - \alpha y \operatorname{ch} \alpha y] e^{-\alpha d} -$$
$$- \mu[(A_m + B_m) \operatorname{sh} \alpha y + \alpha y B_m \operatorname{ch} \alpha y]\} \tag{2.125}$$

$$k_M^{xy} = \frac{(1-\mu)a^2}{2\pi^2 cd} \sum_m m^3 \mathfrak{A}_m \cos \alpha x [(\alpha y \operatorname{sh} \alpha y - \alpha d \operatorname{ch} \alpha y) e^{-\alpha d} +$$
$$+ (A_m + 2B_m) \operatorname{ch} \alpha y + \alpha y B_m \operatorname{sh} \alpha y] \tag{2.126}$$

$$k_T^x = \frac{a^2}{\pi cd} \sum_m m^4 \mathfrak{A}_m \cos \alpha x (e^{-\alpha d} + B_m) \operatorname{sh} \alpha y \tag{2.127}$$

$$k_T^y = \frac{a^2}{\pi cd} \sum_m m^4 \mathfrak{A}_m \sin \alpha x (e^{-\alpha d} + B_m) \operatorname{ch} \alpha y \tag{2.128}$$

$$k_R^x = \frac{a^2}{2\pi cd} \sum_m m^4 \mathfrak{A} \cos \alpha x \{[(1 + \alpha d) \operatorname{sh} \alpha y - \alpha y \operatorname{ch} \alpha y] e^{-\alpha d} -$$
$$- [(A_m + B_m) \operatorname{sh} \alpha y + \alpha y B_m \operatorname{ch} \alpha y] + (2 - \mu)[(1 - \alpha d) \operatorname{sh} \alpha y +$$
$$+ \alpha y \operatorname{ch} \alpha y] e^{-\alpha d} + (2 - \mu)[(A_m + 3B_m) \operatorname{sh} \alpha y + \alpha y B_m \operatorname{ch} \alpha y]\} \tag{2.129}$$

$\boxed{y \geqslant d}$
$$k_w = \frac{a^2}{2\pi^4 cd} \sum_m m \mathfrak{A}_m \sin \alpha x \{[(1 + \alpha y) \operatorname{sh} \alpha d - \alpha d \operatorname{ch} \alpha d] e^{-\alpha y} -$$
$$- [(A_m + B_m) \operatorname{sh} \alpha y + \alpha y B_m \operatorname{ch} \alpha y]\} \tag{2.121'}$$

$$k_\varphi^x = \frac{a^2}{2\pi^3 cd} \sum_m m^2 \mathfrak{A}_m \cos \alpha x \{[(1 + \alpha y) \operatorname{sh} \alpha d - \alpha d \operatorname{ch} \alpha d] e^{-\alpha y} -$$
$$- [(A_m + B_m) \operatorname{sh} \alpha y + \alpha y B_m \operatorname{ch} \alpha y]\} \tag{2.122'}$$

$$k_\varphi^y = \frac{a^2}{2\pi^3 cd} \sum_m m^2 \mathfrak{A}_m \sin \alpha x \{(\alpha d \operatorname{ch} \alpha d - \alpha y \operatorname{sh} \alpha d) e^{-\alpha y} -$$
$$- [(A_m + 2B_m) \operatorname{ch} \alpha y + \alpha y B_m \operatorname{sh} \alpha y]\} \tag{2.123'}$$

2.3 Computation algorithms

Table 2.9 (continued)

$$k_M^x = \frac{a^2}{2\pi^2 cd} \sum_m m^3 \mathcal{A}_m \sin \alpha x \{[(1 + \alpha y)\operatorname{sh} \alpha d - \alpha d \operatorname{ch} \alpha d]e^{-\alpha y} -$$

$$- [(A_m + B_m)\operatorname{sh} \alpha y + \alpha y B_m \operatorname{ch} \alpha y] + \mu[(1 - \alpha y)\operatorname{sh} \alpha d + \alpha d \operatorname{ch} \alpha d]e^{-\alpha y} +$$

$$+ \mu[(A_m + 3B_m)\operatorname{sh} \alpha y + \alpha y B_m \operatorname{ch} \alpha y]\} \qquad (2.124')$$

$$k_M^y = \frac{a^2}{2\pi^2 cd} \sum_m m^3 \mathcal{A}_m \sin \alpha x \{[(1 - \alpha y)\operatorname{sh} \alpha d + \alpha d \operatorname{ch} \alpha d]e^{-\alpha y} +$$

$$+ (A_m + 3B_m)\operatorname{sh} \alpha y + \alpha y B_m \operatorname{ch} \alpha y + \mu[(1 + \alpha y)\operatorname{sh} \alpha d - \alpha d \operatorname{ch} \alpha d]e^{-\alpha y} -$$

$$- \mu[(A_m + B_m)\operatorname{sh} \alpha y + \alpha y B_m \operatorname{ch} \alpha y]\} \qquad (2.125')$$

$$k_M^{xy} = \frac{(1-\mu)a^2}{2\pi^2 cd} \sum_m m^3 \mathcal{A}_m \cos \alpha x [(\alpha y \operatorname{sh} \alpha d - \alpha d \operatorname{ch} \alpha d)e^{-\alpha y} +$$

$$+ (A_m + 2B_m)\operatorname{ch} \alpha y + \alpha y B_m \operatorname{sh} \alpha y]\} \qquad (1.126')$$

$$k_T^x = \frac{a^2}{\pi cd} \sum_m m^4 \mathcal{A}_m \cos \alpha x (e^{-\alpha y}\operatorname{sh} \alpha d + B_m \operatorname{sh} \alpha y) \qquad (1.127')$$

$$k_T^y = -\frac{a^2}{\pi cd} \sum_m m^4 \mathcal{A}_m \sin \alpha x (e^{-\alpha y}\operatorname{sh} \alpha d - B_m \operatorname{ch} \alpha y) \qquad (1.128')$$

$$k_R^x = \frac{a^2}{2\pi cd} \sum_m m^4 \mathcal{A}_m \cos \alpha x \{[(1 + \alpha y)\operatorname{sh} \alpha d - \alpha d \operatorname{ch} \alpha d]e^{-\alpha y} -$$

$$- [(A_m + B_m)\operatorname{sh} \alpha y + \alpha y B_m \operatorname{ch} \alpha y] + (2 - \mu)[(1 - \alpha y)\operatorname{sh} \alpha d +$$

$$+ \alpha d \operatorname{ch} \alpha d]e^{-\alpha y} + (2 - \mu)[(A_m + 3B_m)\operatorname{sh} \alpha y + \alpha y B_m \operatorname{ch} \alpha y]\} \qquad (1.129')$$

Table 2.10

The rectangular plate with two free parallel edges — Load q
Expressions of the dimensionless coefficients

$$k_w = \frac{4}{\pi^5} \sum_m \frac{1}{m^5} \sin \alpha x (C_m \operatorname{ch} \alpha y + \alpha y D_m \operatorname{sh} \alpha y + 1) \qquad (2.131)$$

$$k_\varphi^x = \frac{4}{\pi^4} \sum_m \frac{1}{m^4} \cos \alpha x (C_m \operatorname{ch} \alpha y + \alpha y D_m \operatorname{sh} \alpha y + 1) \qquad (2.132)$$

$$k_\varphi^y = \frac{4}{\pi^4} \sum_m \frac{1}{m^4} \sin \alpha x [(C_m + D_m)\operatorname{sh} \alpha y + \alpha y D_m \operatorname{ch} \alpha y] \qquad (2.133)$$

2 Models for the deflected middle surface of plates

Table 2.10 (continued)

$$k_M^x = \frac{4}{\pi^3} \sum_m \frac{1}{m^3} \sin \alpha x \{C_m \operatorname{ch} \alpha y + \alpha y D_m \operatorname{sh} \alpha y + 1 -$$
$$- \mu[(C_m + 2D_m)\operatorname{ch} \alpha y + \alpha y D_m \operatorname{sh} \alpha y]\} \qquad (2.134)$$

$$k_M^y = \frac{4}{\pi^3} \sum_m \frac{1}{m^3} \sin \alpha x \{-[(C_m + 2D_m)\operatorname{ch} \alpha y + \alpha y D_m \operatorname{sh} \alpha y] +$$
$$+ \mu(C_m \operatorname{ch} \alpha y + \alpha y D_m \operatorname{sh} \alpha y + 1)\} \qquad (2.135)$$

$$k_M^{xy} = -\frac{4(1-\mu)}{\pi^3} \sum_m \cos \alpha x [(C_m + D_m)\operatorname{sh} \alpha y + \alpha y D_m \operatorname{ch} \alpha y] \qquad (2.136)$$

$$k_T^x = \frac{4}{\pi^2} \sum_m \frac{1}{m^2} \cos \alpha x (1 - 2D_m \operatorname{ch} \alpha y) \qquad (2.137)$$

$$k_T^y = -\frac{8}{\pi^2} \sum_m \frac{1}{m^2} \sin \alpha x D_m \operatorname{sh} \alpha y \qquad (2.138)$$

$$k_R^x = \frac{4}{\pi^2} \sum_m \frac{1}{m^2} \cos \alpha x \{C_m \operatorname{ch} \alpha y + \alpha y D_m \operatorname{sh} \alpha y + 1 -$$
$$- (2 - \mu)[(C_m + 2D_m)\operatorname{ch} \alpha y + \alpha y D_m \operatorname{sh} \alpha y]\} \qquad (2.139)$$

2.3.2.3 The rectangular plate simply supported along the entire boundary. The equations of the deflected middle surface are as follows

$$\frac{\text{Load } P}{(\text{Fig. 2.11})} : w(x,y) = \frac{4Pa^4}{\pi^6 Dcd} \sum_m \sum_n \mathcal{A}_{mn} \sin \alpha x \sin \beta y$$

$$(m = 1, 2, 3, \ldots; \; n = 1, 2, 3, \ldots) \qquad (2.41a)$$

$$\frac{\text{Load } \vec{M} \parallel Oy}{(\text{Fig. 2.12})} : w(x,y) = \frac{4Ma^3}{\pi^5 Dcd} \sum_m \sum_n \mathcal{B}_{mn} \sin \alpha x \sin \beta y$$

$$(m = 1, 2, 3, \ldots; \; n = 1, 2, 3, \ldots) \qquad (2.44a)$$

$$\frac{\text{Load } \vec{M} \parallel Ox}{(\text{Fig. 2.13})} : w(x,y) = \frac{4Ma^4}{\pi^5 Dbcd} \sum_m \sum_n \mathcal{C}_{mn} \sin \alpha x \sin \beta y$$

$$(m = 1, 2, 3, \ldots; \; n = 1, 2, 3, \ldots) \qquad (2.46a)$$

$$\frac{\text{Load } q}{(\text{Fig. 2.3-C.4})} : w(x,y) = \frac{16qa^4}{\pi^6 D} \sum_m \sum_n \mathcal{D}_{mn} \sin \alpha x \sin \beta y$$

$$(m = 1, 3, 5, \ldots; \; n = 1, 3, 5, \ldots) \qquad (2.48a)$$

2.3 Computation algorithms

Notation:

$$\mathcal{A}_{mn} = \frac{1}{m^5 n \left(1 + \dfrac{a^2}{b^2} \dfrac{n^2}{m^2}\right)^2} \sin \alpha c \sin \alpha \xi \sin \beta d \sin \beta \eta$$

$$\mathcal{B}_{mn} = \frac{1}{m^4 n \left(1 + \dfrac{a^2}{b^2} \dfrac{n^2}{m^2}\right)^2} \sin \alpha c \cos \alpha \xi \sin \beta d \sin \beta \eta$$

$$\mathcal{C}_{mn} = \frac{1}{m^5 \left(1 + \dfrac{a^2}{b^2} \dfrac{n^2}{m^2}\right)^2} \sin \alpha c \sin \alpha \xi \sin \beta d \cos \beta \eta$$

$$\mathcal{D}_{mn} = \frac{1}{m^5 n \left(1 + \dfrac{a^2}{b^2} \dfrac{n^2}{m^2}\right)^2}$$

$$\alpha = \frac{m\pi}{a} \quad (m = 1, 2, 3, \ldots); \quad \beta = \frac{n\pi}{b} \quad (n = 1, 2, 3, \ldots)$$

(2.42)

The expressions for the dimensionless coefficients required in the study of the rectangular plate simply supported along the entire boundary are given in Tables 2.11–2.14.

Table 2.11
The rectangular plate — Load P
Expressions of the dimensionless coefficients

$$k_w = \frac{4a^2}{\pi^6 cd} \sum_m \sum_n \mathcal{A}_{mn} \sin \alpha x \sin \beta y \qquad (2.141)$$

$$k_\varphi^x = \frac{4a^2}{\pi^5 cd} \sum_m \sum_n m \mathcal{A}_{mn} \cos \alpha x \sin \beta y \qquad (2.142)$$

$$k_\varphi^y = \frac{4a^2}{\pi^5 cd} \sum_m \sum_n \frac{a}{b} n \mathcal{A}_{mn} \sin \alpha x \cos \beta y \qquad (2.143)$$

$$k_M^x = \frac{4a^2}{\pi^4 cd} \sum_m \sum_n \left(m^2 + \mu \frac{a^2}{b^2} n^2\right) \mathcal{A}_{mn} \sin \alpha x \sin \beta y \qquad (2.144)$$

$$k_M^y = \frac{4a^2}{\pi^4 cd} \sum_m \sum_n \left(\frac{a^2}{b^2} n^2 + \mu m^2\right) \mathcal{A}_{mn} \sin \alpha x \sin \beta y \qquad (2.145)$$

$$k_M^{xy} = -\frac{4(1-\mu)a^3}{\pi^4 bcd} \sum_m \sum_n mn \mathcal{A}_{mn} \cos \alpha x \cos \beta y \qquad (2.146)$$

Table 2.11 (continued)

$$k_T^x = \frac{4a^2}{\pi^3 cd} \sum_m \sum_n \left(m^3 + \frac{a^2}{b^2} mn^2\right) \mathcal{A}_{mn} \cos \alpha x \sin \beta y \qquad (2.147)$$

$$k_T^y = \frac{4a^3}{\pi^3 bcd} \sum_m \sum_n \left(\frac{a^2}{b^2} n^3 + m^2 n\right) \mathcal{A}_{mn} \sin \alpha x \cos \beta y \qquad (2.148)$$

$$k_R^x = \frac{4a^2}{\pi^3 cd} \sum_m \sum_n \left[m^3 + (2-\mu)\frac{a^2}{b^2} mn^2\right] \mathcal{A}_{mn} \cos \alpha x \sin \beta y \qquad (2.149)$$

$$k_R^y = \frac{4a^3}{\pi^3 bcd} \sum_m \sum_n \left[\frac{a^2}{b^2} n^3 + (2-\mu)m^2 n\right] \mathcal{A}_{mn} \sin \alpha x \cos \beta y \qquad (2.150)$$

Table 2.12

The rectangular plate — Load $\vec{M} \parallel Oy$
Expressions of the dimensionless coefficients

$$k_w = \frac{4a^2}{\pi^5 cd} \sum_m \sum_n \mathcal{B}_{mn} \sin \alpha x \sin \beta y \qquad (2.151)$$

$$k_\varphi^x = \frac{4a^2}{\pi^4 cd} \sum_m \sum_n m \mathcal{B}_{mn} \cos \alpha x \sin \beta y \qquad (2.152)$$

$$k_\varphi^y = \frac{4a^2}{\pi^4 cd} \sum_m \sum_n \frac{a}{b} n \mathcal{B}_{mn} \sin \alpha x \cos \beta y \qquad (2.153)$$

$$k_M^x = \frac{4a^2}{\pi^3 cd} \sum_m \sum_n \left(m^2 + \mu \frac{a^2}{b^2} n^2\right) \mathcal{B}_{mn} \sin \alpha x \sin \beta y \qquad (2.154)$$

$$k_M^y = \frac{4a^2}{\pi^3 cd} \sum_m \sum_n \left(\frac{a^2}{b^2} n^2 + \mu m^2\right) \mathcal{B}_{mn} \sin \alpha x \sin \beta y \qquad (2.155)$$

$$k_M^{xy} = -\frac{4(1-\mu)a^3}{\pi^3 bcd} \sum_m \sum_n mn \mathcal{B}_{mn} \cos \alpha x \cos \beta y \qquad (2.156)$$

$$k_T^x = \frac{4a^2}{\pi^2 cd} \sum_m \sum_n \left(m^3 + \frac{a^2}{b^2} mn^2\right) \mathcal{B}_{mn} \cos \alpha x \sin \beta y \qquad (2.157)$$

$$k_T^y = \frac{4a^3}{\pi^2 bcd} \sum_m \sum_n \left(\frac{a^2}{b^2} n^3 + m^2 n\right) \mathcal{B}_{mn} \sin \alpha x \cos \beta y \qquad (2.158)$$

$$k_R^x = \frac{4a^2}{\pi^2 cd} \sum_m \sum_n \left[m^3 + (2-\mu)\frac{a^2}{b^2} mn^2\right] \mathcal{B}_{mn} \cos \alpha x \sin \beta y \qquad (2.159)$$

$$k_R^y = \frac{4a^2}{\pi^2 cd} \sum_m \sum_n \frac{a}{b} \left[\frac{a^2}{b^2} n^3 + (2-\mu)m^2 n\right] \mathcal{B}_{mn} \sin \alpha x \cos \beta y \qquad (2.160)$$

2.3 Computation algorithms

Table 2.13
The rectangular plate — Load $\vec{M} \| Ox$
Expressions of the dimensionless coefficients

$$k_w = \frac{4a^3}{\pi^5 bcd} \sum_m \sum_n \mathcal{C}_{mn} \sin \alpha x \sin \beta y \tag{2.161}$$

$$k_\varphi^x = \frac{4a^3}{\pi^4 bcd} \sum_m \sum_n m \mathcal{C}_{mn} \cos \alpha x \sin \beta y \tag{2.162}$$

$$k_\varphi^y = \frac{4a^3}{\pi^4 bcd} \sum_m \sum_n \frac{a}{b} n \mathcal{C}_{mn} \sin \alpha x \cos \beta y \tag{2.163}$$

$$k_M^x = \frac{4a^3}{\pi^3 bcd} \sum_m \sum_n \left(m^2 + \mu \frac{a^2}{b^2} n^2 \right) \mathcal{C}_{mn} \sin \alpha x \sin \beta y \tag{2.164}$$

$$k_M^y = \frac{4a^3}{\pi^3 bcd} \sum_m \sum_n \left(\frac{a^2}{b^2} n^2 + \mu m^2 \right) \mathcal{C}_{mn} \sin \alpha x \sin \beta y \tag{2.165}$$

$$k_M^{xy} = -\frac{4(1-\mu)a^4}{\pi^3 b^2 cd} \sum_m \sum_n mn \mathcal{C}_{mn} \cos \alpha x \cos \beta y \tag{2.166}$$

$$k_T^x = \frac{4a^3}{\pi^2 bcd} \sum_m \sum_n \left(m^3 + \frac{a^2}{b^2} mn^2 \right) \mathcal{C}_{mn} \cos \alpha x \sin \beta y \tag{2.167}$$

$$k_T^y = \frac{4a^4}{\pi^2 b^2 cd} \sum_m \sum_n \left(\frac{a^2}{b^2} n^3 + m^2 n \right) \mathcal{C}_{mn} \sin \alpha x \cos \beta y \tag{2.168}$$

$$k_R^x = \frac{4a^3}{\pi^2 bcd} \sum_m \sum_n \left[m^3 + (2-\mu) \frac{a^2}{b^2} mn^2 \right] \mathcal{C}_{mn} \cos \alpha x \sin \beta y \tag{2.169}$$

$$k_R^y = \frac{4a^4}{\pi^3 b^2 cd} \sum_m \sum_n \left[\frac{a^2}{b^2} n^3 + (2-\mu) m^2 n \right] \mathcal{C}_{mn} \sin \alpha x \cos \beta y \tag{2.170}$$

Table 2.14
The rectangular plate — Load q
Expressions of the dimensionless coefficients

$$k_w = \frac{16}{\pi^6} \sum_m \sum_n \mathcal{D}_{mn} \sin \alpha x \sin \beta y \tag{2.171}$$

$$k_\varphi^x = \frac{16}{\pi^5} \sum_m \sum_n m \mathcal{D}_{mn} \cos \alpha x \sin \beta y \tag{2.172}$$

$$k_\varphi^y = \frac{16}{\pi^5} \sum_m \sum_n \frac{a}{b} n \mathcal{D}_{mn} \sin \alpha x \cos \beta y \tag{2.173}$$

2 Models for the deflected middle surface of plates

Table 2.14 (continued)

$$k_M^x = \frac{16}{\pi^4} \sum_m \sum_n \left(m^2 + \mu \frac{a^2}{b^2} n^2\right) \mathfrak{D}_{mn} \sin \alpha x \sin \beta y \qquad (2.174)$$

$$k_M^y = \frac{16}{\pi^4} \sum_m \sum_n \left(\frac{a^2}{b^2} n^2 + \mu m^2\right) \mathfrak{D}_{mn} \sin \alpha x \sin \beta y \qquad (2.175)$$

$$k_M^{xy} = -\frac{16(1-\mu)a}{\pi^4 b} \sum_m \sum_n mn \mathfrak{D}_{mn} \cos \alpha x \cos \beta y \qquad (2.176)$$

$$k_T^x = \frac{16}{\pi^3} \sum_m \sum_n \left(m^3 + \frac{a^2}{b^2} mn^2\right) \mathfrak{D}_{mn} \cos \alpha x \sin \beta y \qquad (2.177)$$

$$k_T^y = \frac{16a}{\pi^3 b} \sum_m \sum_n \left(\frac{a^2}{b^2} n^3 + m^2 n\right) \mathfrak{D}_{mn} \sin \alpha x \cos \beta y \qquad (2.178)$$

$$k_R^x = \frac{16}{\pi^3} \sum_m \sum_n \left[m^3 + (2-\mu)\frac{a^2}{b^2} mn^2\right] \mathfrak{D}_{mn} \cos \alpha x \sin \beta y \qquad (2.179)$$

$$k_R^y = \frac{16a}{\pi^3 b} \sum_m \sum_n \left[\frac{a^2}{b^2} n^3 + (2-\mu)m^2 n\right] \mathfrak{D}_{mn} \sin \alpha x \cos \beta y \qquad (2.180)$$

3

Behaviour of plates subjected to a surface couple

3.1 Aim of the numerical study. Computation parameters and grid

In order to study the effect of a surface couple on plates, new solutions to the Lagrange-Sophie Germain equation (2.0) were obtained (Sub-secs. 2.2.1.2, 2.2.1.3, 2.2.2.2, 2.2.2.3, 2.2.3.2 and 2.2.3.3). From these six solutions, we derived the formulae allowing the numerical determination of the elastic displacements, stresses and boundary reactions at any point (x, y) in the plane of the plate under the load M (Sub-sec. 2.3.2 — Tables 2.4, 2.5, 2.8, 2.9, 2.12 and 2.13).

An exhaustive numerical experiment on the behaviour of plates subjected to the action of a surface couple is mainly justified by the lack of such approaches in the literature on the Theory of Plates (see Sec. 1.2). Additionally, by introducing the surface couple M as a generalized interacting force in the elastic analysis of slab structures, account can be taken of the effect of the rigid bar-plate (column-slab) connection.

The importance of the comparative studies given herein derives from the conclusions pertaining to: the accurate modelling of the physical phenomenon, the similar behaviour of the categories of plates compared, the characteristics of a local perturbation specific to the effects induced by a surface couple, the relatively high values of the stresses M_x, M_y and M_{xy} about the loaded area (Secs. 3.2, 3.3 and Chap. 4) and the consequences induced by incorporating the rigid connection effect in structural analysis (Chaps. 5 and 6).

★

The numerical values of the dimensionless coefficients k_r^{st} (2.49 a) were computed using programs written in FORTRAN IV language on the basis of the algorithms presented in Tables 2.3—2.14.

3 Behaviour of plates subjected to a surface couple

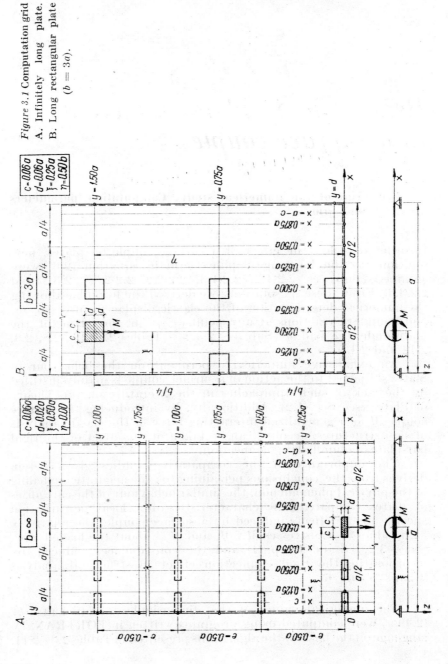

Figure 3.1 Computation grid
A. Infinitely long plate.
B. Long rectangular plate ($b = 3a$).

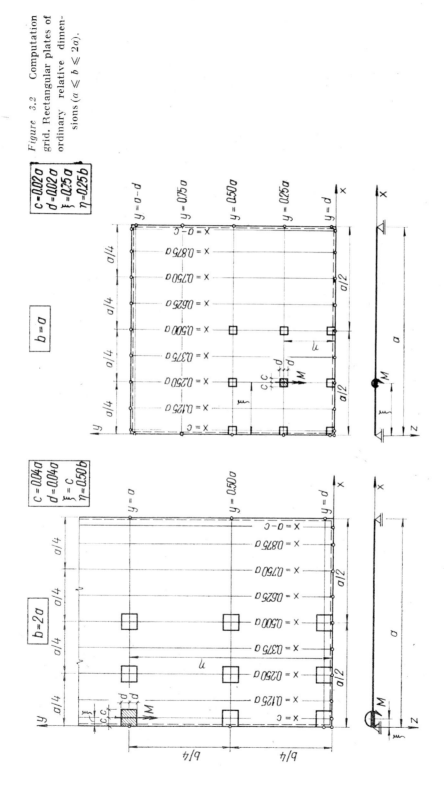

Figure 3.2 Computation grid. Rectangular plates of ordinary relative dimensions ($a \leqslant b \leqslant 2a$).

3 Behaviour of plates subjected to a surface couple

In the numerical studies, account was taken of a large number of alternatives where the values of the geometrical and physical parameters a, b, c, d, ξ, η and μ, defined in Sub-sec. 1.2.1, were chosen. In the comparative studies carried out on the infinitely long plate we must also introduce the equal distance e between the column rows, measured lengthwise the plate (Fig. 3.1A).

For the computation grid (Table 3.1), the density was taken constant across the plate (module $a/8$) for the infinitely long (Fig. 3.1A) and finite boundary plates (Figs. 3.1B and 3.2). In order to introduce the edge columns in calculation, in the grid we inserted the points $x = c$ and $x = a - c$ transverse to the plate, and $y = d$ and $y = b - d$ in the longitudinal direction of the plate, respectively (Figs. 3.1 and 3.2).

In the case of the infinitely long plate, where no restrictions appeared with respect to the amount of computation, the same module ($a/8$) of the grid was retained in the longitudinal direction. For this category of plates, the grid was extended to the vertical planes $y = \pm 2a$ (Table 3.1).

Table 3.1

Computation grid. Co-ordinates of the basic nodes

Abscissae x	Ordinates y		
$a \leqslant b \leqslant \infty$	$a \leqslant b \leqslant 3a$	$b = \infty$	
0	0	0	
c	d	$\pm d$	
$0.125a$		$\pm 0.125a$	$\pm 1.125a$
$0.250a$	$0.250b$	$\pm 0.250a$	$\pm 1.250a$
$0.375a$		$\pm 0.375a$	$\pm 1.375a$
$0.500a$	$0.500b$	$\pm 0.500a$	$\pm 1.500a$
$0.625a$		$\pm 0.625a$	$\pm 1.625a$
$0.750a$	$0.750b$	$\pm 0.750a$	$\pm 1.750a$
$0.875a$		$\pm 0.875a$	$\pm 1.875a$
$a - c$	$b - d$		
a	b	$\pm a$	$\pm 2a$

For points farther away from the area subjected to the surface couple M (the plane $y=0$) all the values of the basic dimensionless coefficients k_w, k_φ^x, k_φ^y, k_M^x, k_M^y and k_M^{xy} can be assumed to be negligible. However, the ordinates $y = \pm 2a$, which frame the computation grid in the case of the infinitely long strip, could correspond to the edges of a finite boundary plate, where $b = 4a$. This limit case by far exceeds the sizes of

3.1 Aim of the numerical study

the very long rectangular plate ($b = 3a$) considered in our comparative studies (Secs. 3.2 and 3.3 and Chap. 4). In the case of finite boundary plates, we had to lower the density of the grid along the lengthwise direction of the plate (module $b/4$) because the computation rate is here much slower than that required for infinitely long plates (Table 3.1).

For the comparative study of the elastic displacements (dimensionless coefficients k_w, k_φ^x, and k_φ^y), the density of the basic grid is satisfactory. In order to investigate the stress distribution (dimensionless coefficients k_M^x, k_M^y and k_M^{xy}), the points of abscissae $x = \xi - c$ and $x = \xi + c$, which delimit the area subjected to the couple M, were inserted across the grid. The extreme values of the bending moments appear exactly on the boundary of this contact area (Sec. 3.3 and Chap. 4).

The programs implemented here allow an unrestrictive introduction of any of the current values of the design parameters. In defining the computation grid and in the choice of the values of the relative geometrical parameters b/a, c/a, d/a, ξ/a and η/a, account was taken of the requirements of an exhaustive comparative study. The practical interest of the chosen alternatives was also considered.

The only physical parameter appearing in the expression of the dimensionless coefficients is Poisson's ratio μ. In addition to the values $\mu = 1/3$ for steel and $\mu = 1/6$ for concrete, the value $\mu = 0$ was also considered in calculations. This permits a comparison of the author's results with the results reported by other researchers who, for simplification, assume that Poisson's ratio is equal to zero.

In compliance with the aims pursued, the following sections give a presentation, in the form of tables and diagrams, of the relevant results and conclusions stemming from processing the 95,000 static quantities thus computed [68], [70], [75].

3.2 Factors influencing the magnitude and distribution of elastic displacements of plates subjected to a surface couple

The author's comparative studies highlight the influence of the geometrical and physical parameters on the variation of the sectional stresses and elastic displacements in plates subjected to a load M. The behaviour of plates subjected to a surface couple was examined for two categories of plates, considered as basic static schemes

3 Behaviour of plates subjected to a surface couple

(Sub-sec. 2.1.3):
— the infinitely long plate ($b = \infty$),
and
— the rectangular plate, simply supported along the entire boundary ($a \leqslant b \leqslant 3a$).

Generally, the moment vector M was assumed to be parallel to the line of supports, namely parallel to the longer side b, in the case of rectangular plates.

3.2.1 Magnitude and distribution of elastic displacements along the plate width

A comparison between the manner in which the infinitely long plate ($b = \infty$) and the very long rectangular plate ($b = 3a$) deform under a surface couple $\vec{M} \parallel Oy$, shows that the deflected mid-surfaces coincide (Fig. 3.3). Since the transverse axes of symmetry of the two categories of plates under consideration are defined by the vertical planes $y = \eta = 0$ (for $b = \pm \infty$) and $y = \eta = 1.5a$ (for $b = 3a$) respectively, the comparison was carried out for three characteristic positions of the surface couple M along these two axes:

— the edge position, adjacent to the support of the plate ($\xi = c$);
— the intermediate position, at the quarter of the span ($\xi = 0.25a$);
— the central position, at mid-span ($\xi = 0.50a$).

The characteristic locations will be frequently used in the comparative approach to the behaviour of plates subjected to locally distributed loads of the type M and P.

We can easily observe that the two deflected mid-surfaces can be fully superposed in the transverse plane $y = \eta$ for any of the three characteristic positions of M (Fig. 3.3 — diagram $y = 0$, $y = 1.5a$). This coincidence is retained irrespective of the sizes ($2c \times 2d$) of the area loaded by couple M for the central position $\xi = 0.5a$ (Table 3.2 — columns $b = \infty$ and $b = 3a$) and for the edge position $\xi = c$ (Table 3.10 — column $10^5 \, k_w$).

The idea of a similar (sometimes identical) behaviour of the two categories of plates under the load M only reinforces the results obtained by other authors, Timoshenko in particular [7], concerning the behaviour of these plates under loads different from M (e.g. load q).

3.2 Magnitude and distribution of elastic displacements

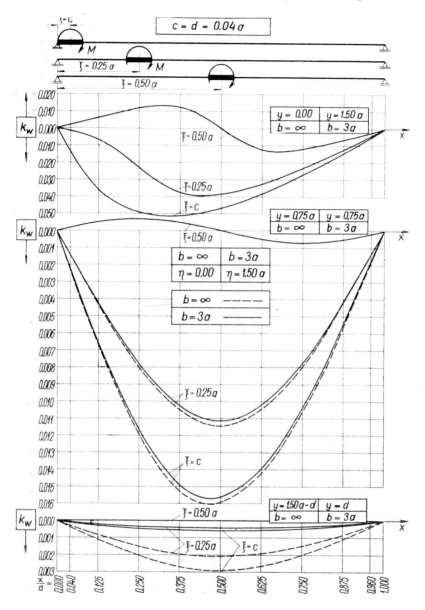

Figure 3.3 Infinitely long plate and long rectangular plate ($b = 3a$) subjected to a surface couple $\vec{M} \parallel Oy$. Deflected mid-surface. Comparative displacement values w (dimensionless coefficients k_w) in transverse planes $y = $ constant.

143

Table 3.2

Plates subjected to a surface couple $\vec{M} \parallel Oy$ at ($\xi = 0.50a$; $\eta = 0.50b$) — Deflected mid-surface

Comparative values of coefficient $10^5 \, k_w$

x	$c = d = 0.02\,a$				$c = d = 0.04a$				$c = d = 0.06a$			
	$b = \infty$	$b = 3a$	$b = 2a$	$b = a$	$b = \infty$	$b = 3a$	$b = 2a$	$b = a$	$b = \infty$	$b = 3a$	$b = 2a$	$b = a$
c	−110	−110	−110	−101	−218	−218	−218	−208	−322	−322	−322	−310
$0.125\,a$	−662	−662	−662	−635	−658	−658	−658	−633	−651	−651	−651	−627
$0.250a$	−1157	−1157	−1157	−1116	−1147	−1147	−1146	−1110	−1130	−1130	−1130	−1096
$0.375a$	−1235	−1235	−1235	−1198	−1211	−1211	−1211	−1182	−1170	−1170	−1170	−1148
$0.500a$	0	0	0	0	0	0	0	0	0	0	0	0
$0.625a$	1235	1235	1235	1198	1211	1211	1211	1182	1170	1170	1170	1148
$0.750a$	1157	1157	1157	1116	1147	1147	1146	1110	1130	1130	1130	1096
$0.875a$	662	662	662	635	658	658	658	633	651	651	651	627
$a - c$	110	110	110	101	218	218	218	208	322	322	322	310
Plane	$y = 0$	$y = 0$	$y = 0.50b$	$y = 0.50b$	$y = 0$	$y = 0$	$y = 0.50b$	$y = 0.50b$	$y = 0$	$y = 0$	$y = 0.50b$	$y = 0.50b$

3.2 Magnitude and distribution of elastic displacements

The similar behaviour of the infinitely long plate and the very long rectangular plate to be followed throughout the book (see Secs. 3.2, 3.3 and Chaps. 4, 5 and 6) is one on the essential conclusions of the numerical study above. This observation is all the more important as the functions $w(x, y)$ (2.16) and (2.44), which define the deflected mid-surfaces of the two categories of plates, are essentially distinct : simple Fourier series of hyperbolic and exponential functions in the infinitely-long plate case (Table 2.4) and double series of trigonometric functions in the finite-boundary plate case (Table 2.12). This similar behaviour under a new-type of load — the surface couple — is an implicit test of the accuracy of the obtained solutions $w(x, y)$ (see Sec. 2.2).

Further interpretation of the comparative data leads to the conclusion that the physical phenomenon is accurately modelled. Thus, we notice :

— the absence of the inflection points along the curves k_w corresponding to the edge position $\xi = c$ of the couple M, for which the bending moments M_x do not change sign throughout the width of the plate (see Sec. 3.3);

— the coincidence of the inflection point which appears for the positions $\xi = 0.25a$ and $\xi = 0.50a$ in the centroid (ξ, η) of the area loaded by the couple, with the point where the moment M_x is zero (Sec. 3.3 and Chap. 4);

— the influence of the transverse supports on the curves k_w in the case of the rectangular plate $b = 3a$, i.e. the separation of the two mid-surfaces into the vertical planes $y = 0.75a$ and $y = 1.5a - d$, which are sufficiently far from the loaded area;

— the tendency of the deflected mid-surfaces, which are non-symmetrical in the plane $y = \eta$, to become symmetrical with respect to the longitudinal axis $x = 0.5a$ of the plate. This tendency is even stronger as the distance from the loaded area increases. This phenomenon, which we shall call "tendency towards symmetrization", is observed in the study of elastic displacements, and is evident from the diagrams of the bending moments M_x and M_y (Sec. 3.3).

A comparative approach of the finite boundary plates ($a \leqslant b \leqslant 3a$) under the action of a surface couple $\overline{M} \parallel Oy$ shows that the same conclusions may be derived regarding the similar behaviour of these plates and the correct modelling of the physical phenomenon (Fig. 3.4). The higher rigidity of the square plate $b = a$ as compared to that of the very long rectangular plate $b = 3a$ results from the curves k_w, corresponding to the positions $\xi = c$ and $\xi = 0.25a$ ($\eta = 0.50b$) of the surface couple. For $\xi = 0.50a$, the curves nearly coincide in the entire range $2a \leqslant b \leqslant \infty$,

3 *Behaviour of plates subjected to a surface couple*

very small differences appearing only in the range $a \leqslant b \leqslant 2a$ for this position of the couple (Table 3.2). The values given in Table 3.2 show the extent to which the distribution of transverse displacements is influenced by the size of the area subjected to the surface couple (parameters c and d).

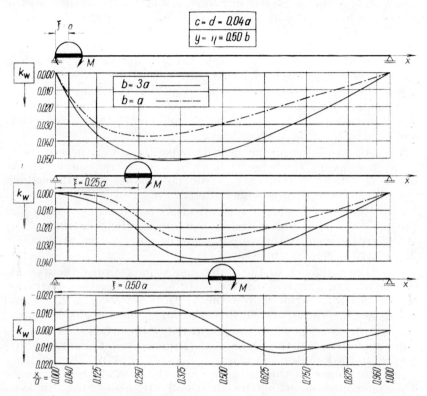

Figure 3.4 Rectangular plates ($a \leqslant b \leqslant 3a$) subjected to a surface couple $\vec{M} \| Oy$. Deflected mid-surface. Comparative values of displacement w in transverse plane of symmetry $y = \eta = 0.50b$.

Finally, mention should be made that the stiffness of the plates against the action of a couple $\vec{M} \| Oy$ is highest for the central position $\xi = 0.50a$ (Figs. 3.3 and 3.4), whereas in the case of plates subjected to a locally distributed load P, the lowest values of the ordinates k_w correspond to the edge position $\xi = c$ of this load (Sec. 4.1.2).

In structural analysis, account is taken of the transverse displacements w and of the plate rotations φ_x and φ_y produced

Table 3.3

The infinitely long plate subjected to a surface couple $\vec{M} \| Oy$ at $(\xi, 0)$

Design values of transverse displacement w and rotations φ_x

at $(x = \xi; y = 0)$

The dimensionless coefficient $10^5 \, k_w$

$c = d$	$\xi = c$	m	$\xi = 2c$	m	$\xi = 0.060a$	m	$\xi = 0.075a$	m	$\xi = 0.125a$	m	$\xi = 0.250a$	m	$\xi = 0.500a$	m
0.01a	590	50	986	30	1887	60	2092	40	2493	20	2319	40	0	20
0.02a	952	50	1503	60	1880	50	2088	40	2483	40	2319	50	0	20
0.03a	1234	70	1869	70	1869	70	2080	50	2481	20	2317	40	0	20
0.04a	1463	60	2127	50	1853	60	2068	60	2471	40	2314	50	0	20
0.06a	1809	60	2426	40	1809	60	2033	60	2450	20	2305	40	0	20

The dimensionless coefficient $10^5 \, k_\varphi^x$

$c = d$	$\xi = c$	m	$\xi = 2c$	m	$\xi = 0.060a$	m	$\xi = 0.075a$	m	$\xi = 0.125a$	m	$\xi = 0.250a$	m	$\xi = 0.500a$	m
0.01a	57163	150	51314	130	42223	200	40467	200	36526	200	31657	200	28878	200
0.02a	46109	130	40103	190	36332	150	35032	200	31063	200	26170	200	23409	200
0.03a	39670	170	33678	140	33678	140	31868	140	27876	170	22962	170	20198	170
0.04a	35107	80	29158	80	31533	130	29673	180	25618	180	20692	80	17914	130
0.06a	28683	170	22808	120	28683	170	26697	90	22484	170	17483	170	14702	170

The dimensionless coefficient $k_\varphi^y = 0$ ($m = 20$)

Note. The order of the last summed term in the simple series is m.

3 Behaviour of plates subjected to a surface couple

by a surface couple M. These values are calculated at the loading points $(x = \xi; y = 0)$ in the case of the infinitely long plate (Table 3.3), and $(x = \xi; y = \eta)$ in the case of the rectangular plate (Table 3.4).

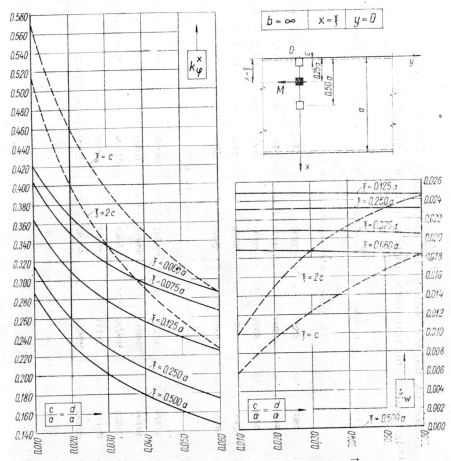

Figure 3.5 Infinitely long plate subjected to a surface couple $\vec{M} \parallel Oy$. Computational values of linear displacements w and rotations $\varphi_x(x = \xi; y = 0)$.

Next, a comparative examination (Table 3.3 and Fig. 3.5) in the case of the infinitely long plate shows how the position ξ of the surface couple bears on the distribution of the elastic displacements, and also the variation of the coefficients k_w and k_φ^x, function of parameters c and d, in the entire range of practical interest, namely $0.01a \leqslant c, d \leqslant 0.06a$ (slab structures without capitals).

3.2 Magnitude and distribution of elastic displacements

In the case of finite boundaries plates ($a \leqslant b \leqslant 3a$), if an average size of the loaded area ($c = d = 0.04a$) is constant, the influence of the parameter b/a is paramount for the symmetrical (Table 3.4) and unsymmetrical positions of the surface couple (Table 3.5).

The particular case of the square plate ($b = a$) was studied in more detail, the variation of the transverse displacements w and of the rotations φ_x being given as a function of the position (ξ, η) and also of the dimensions ($2c \times 2d$) of the area subjected to the couple M (Table 3.6 and Fig. 3.6).

Table 3.4

Rectangular plates subjected to a surface couple $\vec{M} \| Oy$ at (ξ, η)
Comparative values of transverse displacements w and rotations φ_x and φ_y at ($x = \xi$; $y = \eta$)

c = d = 0.040a

b/a	x/a	y/a	$10^5 k_w$	$10^5 k_\varphi^x$	k_φ^y	$m = n$
			$\xi = 0.50\,a \quad \eta = 0.50\,b$			
1.0		0.500	0	17789	0	43
1.5		0.750	0	17898	0	57
2.0	0.500	1.000	0	17910	0	69
3.0		1.500	0	17924	0	92
∞^*		0.000	0 —	— 17914	0 —	20 130
			$\xi = 0.25\,a \quad \eta = 0.50\,b$			
1.0		0.500	1473	18135	0	43
1.5		0.750	2058	19880	0	57
2.0	0.250	1.000	2245	20464	0	69
3.0		1.500	2309	20679	0	92
∞^*		0.000	2314 —	— 20692	0 —	50 80
			$\xi = c \quad \eta = 0.50\,b$			
1.0		0.500	1247	29745	0	43
1.5		0.750	1400	33510	0	57
2.0	0.040	1.000	1447	34676	0	69
3.0		1.500	1461	35081	0	92
∞^*		0.000	1463 —	— 35107	0 —	60 80

* Simple series.

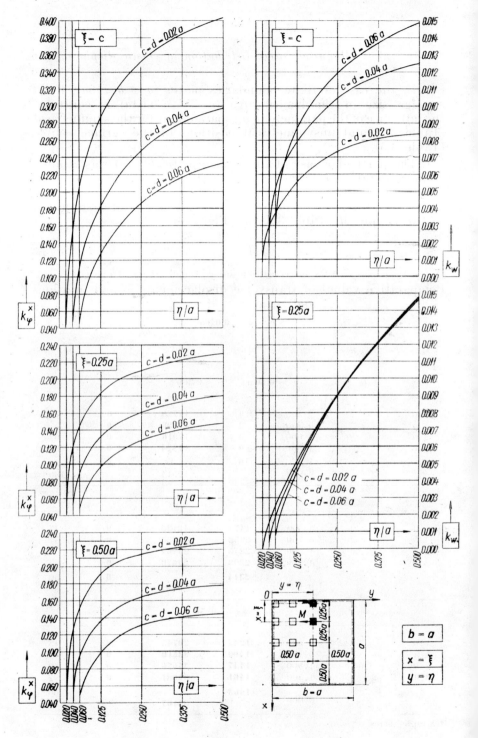

Figure 3.6 Square plate subjected to a surface couple $\vec{M} \parallel Oy$. Computational values of linear displacements w and rotations φ_x ($x = \xi$; $y = \eta$).

3.2 Magnitude and distribution of elastic displacements

Table 3.5

Rectangular plates ($a \leqslant b \leqslant 3a$) subjected to a surface couple $\vec{M} \| Oy$ at (ξ, η)

Comparative values of transverse displacements w and rotations φ_x at ($\mathbf{x} = \xi$; $\mathbf{y} = \eta$)

$c = d = 0.040\,a$; $\xi = 0.25\,a$; $\eta = 0.25\,b$

b/a	x/a	y/a	$10^5\,k_w$	$10^5\,k_\varphi^x$	$m = n$
1.0		0.250	907	16119	43
1.5	0.250	0.375	1498	18093	57
2.0		0.500	1859	19250	69
3.0		0.750	2183	20284	92

Table 3.6

The square plate subjected to a surface couple $\vec{M} \| Oy$ at (ξ, η)

Design values of transverse displacements w and rotations φ_x at ($\mathbf{x} = \xi$; $\mathbf{y} = \eta$)

$c = d$	The dimensionless coefficient $10^5\,k_w$			The dimensionless coefficient $10^5\,k_\varphi^x$		
	$\xi = c$	$\xi = 0.25a$	$\xi = 0.50a$	$\xi = c$	$\xi = 0.25a$	$\xi = 0.50a$
$\eta = d$						
$0.02a$	87	9	0	4286	4330	4347
$0.04a$	194	38	0	4660	5211	5275
$0.06a$	282	82	0	4495	4824	4955
$\eta = 0.25b$						
$0.02a$	736	910	0	35219	21044	21447
$0.04a$	1045	907	0	24749	16119	16534
$0.06a$	1194	902	0	18626	12798	13191
$\eta = 0.50b$						
$0.02a$	840	1477	0	40368	23043	22751
$0.04a$	1247	1473	0	29745	18135	17789
$0.06a$	1487	1466	0	23383	14800	14448

Note. The order of the last summed term in the double series is $m = n = 42$.   The order of the last summed term in the double series is $m = n = 43$.

3 Behaviour of plates subjected to a surface couple

3.2.2 Magnitude and distribution of elastic displacements along the plate length

An extension of the comparative study of the magnitude and distribution of elastic displacements of the entire surface of plates reinforces earlier conclusions concerning:
— *the accurate modelling* of the physical phenomenon by the new solutions established in Secs. 2.2 and 2.3;
— *the similar behaviour* of the very long rectangular plate and the infinitely long plate;
— the *"tendency toward symmetrization"* of the deflected mid-surface, especially in the transverse planes $y = $ constant, which are sufficiently far away from the area subjected to the surface couple $\vec{M} \| Oy$ (Figs. 3.7—3.11).

One of the particular aspects to be considered in the design activity is the large extent of the region (along the plate length) where the elastic surfaces deflected under the load M are very close to one another or are almost coincident (Fig. 3.8).

The influence of the position of the surface couple along the plate length, which can be made evident only in the case of finite boundary plates, has scarcely been approached in the literature. The first approach in this respect, which became possible owing to the solutions of the type (2.32), (2.33), (2.34), (2.44) and (2.46) [64], [82], consists of the comparative study of a square plate deflected by a surface couple $\vec{M} \| Oy$, at two different positions : $\eta = 0.25a$ and $\eta = 0.50a$ (Fig. 3.9). For the same abscissa $\xi = 0.25a$, the parameter η bears heavily on the magnitude of the transverse displacements. Mention should be made that the tendency toward symmetrization is evident even in this limit case ($b = a$).

The density of the computation grid, i.e. the great number of computed values for the dimensionless coefficient k_w, permits a three-dimensional representation of the mid-surfaces deflected under the load M. The surfaces $w(x, y)$ corresponding to the infinitely long plate (Fig. 3.10) and to the square plate (Fig. 3.11) were represented assuming that the surface couple $\vec{M} \| Oy$ has two characteristic positions : the edge ($\xi = c$) and the central position ($\xi = 0.50a$). For both categories of plates, the ordinate of the centroid (ξ, η) of the area loaded by the couple M, corresponds to the transverse axis of symmetry of the plate : $\eta = 0$ for $b = \pm \infty$ and $\eta = 0.50b$ for $b = a$, respectively.

For both categories of plates under consideration, and assuming that the surface couple M has an antisymmetrical position $\xi = 0.50\ a$, the longitudinal axis of symmetry $x = 0.50a$ remains

3.2 Magnitude and distribution of elastic displacements

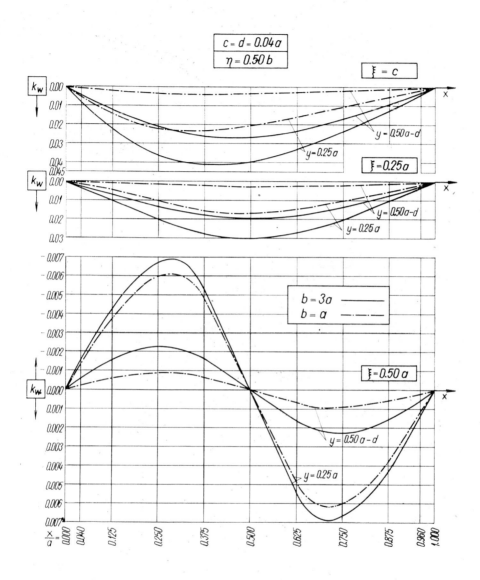

Figure 3.7 Finite-boundary rectangular plates ($a \leqslant b \leqslant 3a$) subjected to a surface couple $\vec{M} \| Oy$. Deflected mid-surface. Comparative values of displacements w in transverse planes in the vicinity of symmetry plane $y = \eta = 0.50\ b$.

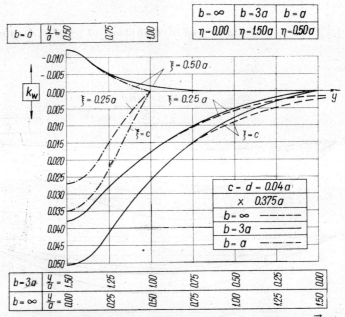

Figure 3.8 Plates $a \leqslant b \leqslant \infty$ subjected to a surface couple $\vec{M} \| Oy$. Deflected mid-surface. Comparative values of displacement w in longitudinal plane $x = 0.375\,a$.

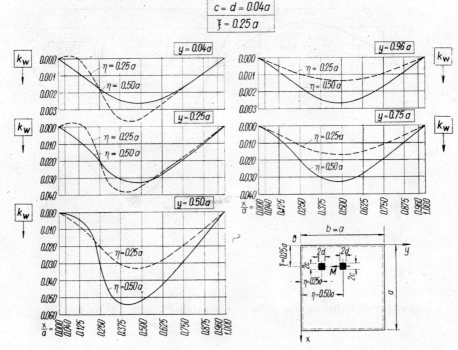

Figure 3.9 Square plate subjected to a surface couple $\vec{M} \| Oy$. Deflected mid-surface. Variation of displacements w with position η of couple and with distance y from loaded surface.

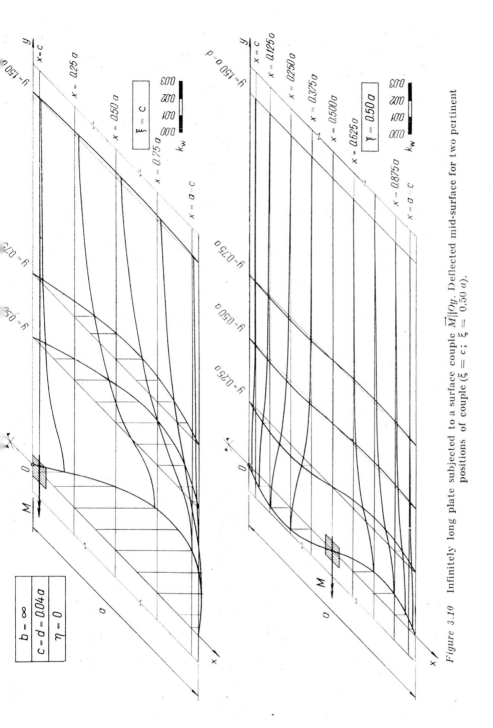

Figure 3.10 Infinitely long plate subjected to a surface couple $\vec{M}\|Oy$. Deflected mid-surface for two pertinent positions of couple ($\xi = c$; $\xi = 0.50\,a$).

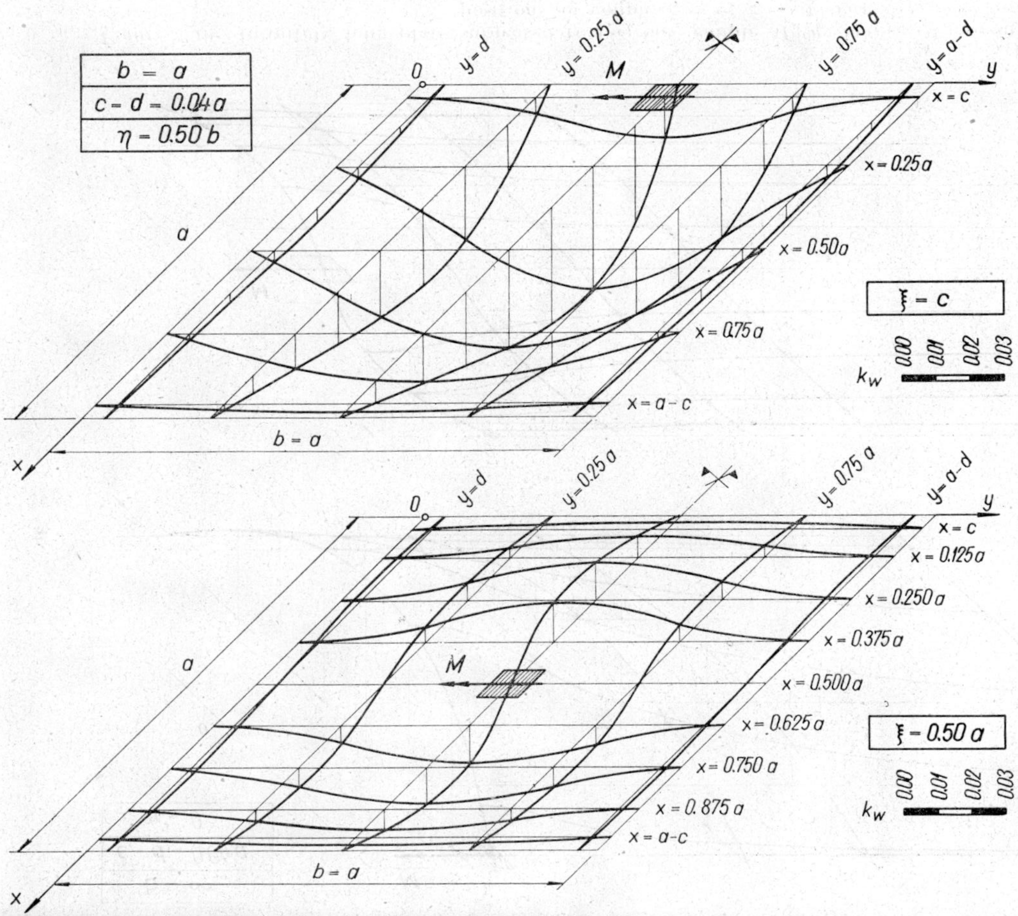

Figure 3.11 Square plate subjected to a surface couple $\vec{M} \| Oy$. Deflected mid-surface for two pertinent positions of couple ($\xi = c$; $\eta = 0.50\,b$) and ($\xi = 0.50\,a$; $\eta = 0.50\,b$).

3.3 Magnitude and distribution of stresses

undeformed ($k_w = 0$). The three-dimensional representation of the elastic surfaces $w(x, y)$ given by (2.16) and (2.44) reinforces in a relevant manner our former conclusions regarding the accurate modelling of the physical phenomenon.

The relatively high values of the dimensionless coefficient k_w in the vertical plane $y = \eta$ decrease substantially along the length of the plate. However, the perturbation characteristics of the load M will be made evident in the subsequent study of the distribution of stresses in plates (Sec. 3.3).

3.3 Factors influencing the magnitude and distribution of stresses in plates subjected to a surface couple

Using the systematic developments presented in Sub-sec. 2.3.1, the study of the variation of stresses throughout the plate is reduced to the study of the distribution of the values of the dimensionless coefficients k_M^x, k_M^y, k_M^{xy} (Table 2.2) in the characteristic points of the computation grid [70], [75].

Irrespective of the type of load (P, M or q), the simple series of hyperbolic and exponential functions in the expressions of the dimensionless coefficients which define the state of stress in the infinitely long plate (Tables 2.3—2.6) can be easily performed on computers. However, this is not the case of the finite boundary plates, where computations are carried out at much slower rate as the dimensionless coefficients are expressed in double trigonometric series. This distinction is strict when we compute the stresses M_x, M_y and M_{xy} in the all-edge simply supported rectangular plate under load M (Tables 2.11—2.14). Since the convergence of the corresponding expressions used in the computation of shear forces and boundary reactions for this category of plates is very slow, a comparative study similar to the approach made in this sub-section is out of question. The expressions for the dimensionless coefficients k_T^x and k_T^y, and especially of k_R^x and k_R^y (presented in Sub-sec. 2.3.2), can be used to solve individual design problems.

In order to reduce the amount of numerically processed data, the comparative study on the value and distribution of the bending moments induced in loading the plate by a surface couple M, was carried out only in the case of a moment vector $\vec{M}$ parallel to the line of supports ($\vec{M} \parallel Oy$) (Tables 2.4 and 2.12). This is the most frequent case among the categories of designed slab structures (see Sub-sec. 5.1.2.1. — Alternative 1). The case of the moment vector $\vec{M}$ normal to the line of supports ($\vec{M} \parallel Ox$) (Table 2.5 and

3   *Behaviour of plates subjected to a surface couple*

2.13) appears in structural analysis when the condition of zero relative rotation in the slab-column nodes must be set in both directions of the plane (see Sub-sec. 5.1.2.1 — Alternative 2).

3.3.1   *Magnitude and distribution of bending moments along the plate width*

The most important conclusion derived from the study of the distribution of the sectional bending moments M_x and M_y induced in the infinitely long strip ($b = \infty$) by a surface couple refers to the high values of coefficients k_M^x and k_M^y on the boundary of the area subjected to the couple M, and to the very quick decrease of the values — especially those of k_M^x — transverse to the plate (Figs. 3.12, 3.19, 3.20, 3.22, 3.26 and Table 3.10).

The characteristics of a local perturbation specific to the effects induced by a surface couple M is stronger as the loaded area is smaller: $c = d = 0.02a$ (Figs. 3.12, 3.16, 3.25 and 3.26). The quick decrease of the values of stresses from the extreme values on the boundary $x = \xi \pm c$ to negligible values is also illustrated by the three-dimensional representation of the variation of k_M^x, k_M^y and k_M^{xy} which are plotted on the boundary of the loaded area (Fig. 3.13).

The influence of the position parameter ξ on the extreme values of the bending moments is clearly shown throughout this study (Figs. 3.12, 3.13, 3.14, 3.15, 3.19, 3.20, 3.22, 3,25, 3.26 and Tables 3.7, 3.8 and 3.9). The highest values k_M^x and k_M^y *increase very slowly* in the range $0.500a \geqslant \xi \geqslant 0.075a$, whereas the lowest values k_M^x and k_M^y decrease in absolute value. Depending on the position ξ of the surface couple, greater variations of the extreme values are observed in the range $0.075a \leqslant \xi \leqslant c$.

The same approach shows that the sizes $2c \times 2d$ of the area subjected to the couple M bears heavily on the values max (min) k_M^x and max (min) k_M^y (Figs. 3.14 and 3.15). The influence of ξ is even greater over the range $0.01a \leqslant c, d \leqslant 0.03a$, where the extreme values of the stresses decrease to one third for both the bending moments M_x (Tables 3.7 and 3.8) and M_y (Table 3.9).

The conclusions of the study on the influence of the parameters c, d, ξ and μ (Sub-sec. 3.3.4) on the extreme values of the bending moments are systematically presented in diagrams so as to allow a convenient estimation by interpolation of the design stresses corresponding to all ordinary values of these parameters (Figs. 3.14 and 3.15).

Like the state of deformation, the state of stress was examined by referring the behaviour of the infinitely long strip ($b = \infty$)

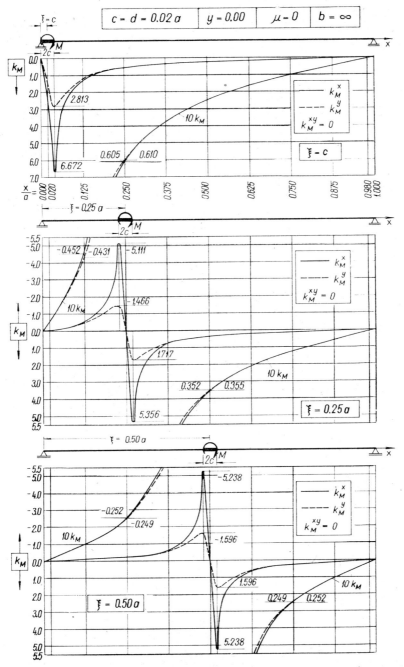

Figure 3.12 Infinitely long plate subjected to a surface couple $\vec{M} \parallel Oy$. Diagrams for sectional moments M_x and M_y (dimensionless coefficients k_M^x and k_M^y) induced in transverse plane $y = 0$ for three pertinent positions $(\xi; 0)$ of surface couple M.

3 *Behaviour of plates subjected to a surface couple*

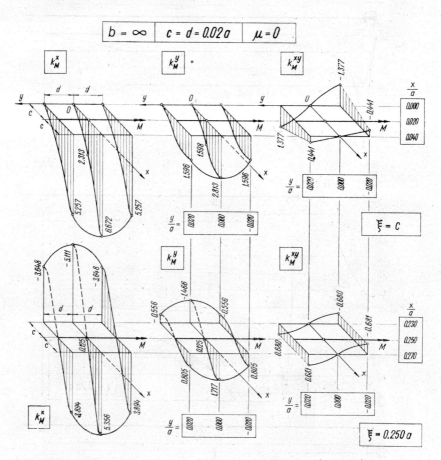

Figure 3.13 Infinitely long plate subjected to a surface couple $\vec{M} \parallel Oy$. Three-dimensional distribution of sectional moments M_x, M_y and M_{xy} (dimensionless coefficients k_M^x, k_M^y and k_M^{xy}) along the boundary of surface $2c \times 2d$ on which couple M is applied.

to the behaviour of finite boundary plates ($a \leqslant b \leqslant 3a$). The results obtained in the earlier numerical study regarding the magnitude and distribution of the bending moments M_x and M_y lead to similar conclusions.

The difference in values between the stresses M_x and M_y (dimensionless coefficients k_M^x and k_M^y) of the infinitely long plate and those of the rectangular plate derive from the different boundary conditions and, obviously, from the different values of the parameter b/a which defines the ratio of spans of the rectangular

3.3 *Magnitude and distribution of stresses*

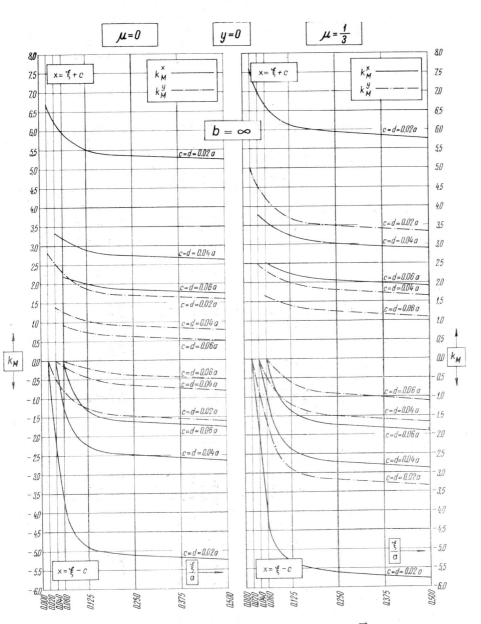

Figure 3.14 Infinitely long plate subjected to a surface couple $\vec{M} \parallel Oy$. Variation of extreme values of bending moments max (min) M_x (coefficients k_M^x) and max (min) M_y (coefficients k_M^y) with position parameter ξ of surface couple M.

Table 3.7

The infinitely long plate subjected to a surface couple $\vec{M} \| Oy$ at $(\xi, 0)$

Highest values of bending moment M_x

The dimensionless coefficients $\max k_M^x$ at $(x = \xi + c; y = 0)$

$c = d$	$\xi = c$	m	$\xi = 2c$	m	$\xi = 0.060$	m	$\xi = 0.075a$	m	$\xi = 0.125a$	m	$\xi = 0.250a$	m	$\xi = 0.500a$	m
$\mu = 0$														
0.01a	13.263	1500	11.719	530	10.956	1110	10.437	410	10.364	510	10.448	1000	10.217	700
0.02a	6.672	1580	6.059	1540	5.817	1610	5.653	650	5.507	1050	5.356	1250	5.238	1350
0.03a	4.458	2370	4.047	2110	4.047	2110	3.930	840	3.769	1170	3.613	1070	3.497	1270
0.04a	3.343	2830	3.030	1950	3.141	1630	3.038	680	2.892	1630	2.742	1430	2.625	1630
0.06a	2.221	2950	2.009	2170	2.221	2950	2.141	1370	1.999	2220	1.861	2170	1.745	2270
$\mu = 1/6$														
0.01a	14.204	1500	12.511	530	11.591	1110	11.053	410	10.947	510	11.003	1000	10.751	700
0.02a	7.142	1580	6.454	1540	6.176	1610	5.996	650	5.820	1050	5.643	1250	5.505	1350
0.03a	4.771	2370	4.309	2110	4.309	2110	4.178	840	3.991	1170	3.810	1070	3.674	1270
0.04a	3.576	2830	3.226	1950	3.352	1630	3.237	680	3.067	1630	2.894	1430	2.757	1630
0.06a	2.375	2950	2.137	2170	2.375	2950	2.286	1370	2.126	2220	1.966	2170	1.831	2270
$\mu = 1/3$														
0.01a	15.139	1500	13.299	530	12.223	1110	11.665	410	11.526	510	11.555	1000	11.282	700
0.02a	7.609	1580	6.847	1540	6.533	1610	6.336	650	6.131	1050	5.982	1250	5.770	1350
0.03a	5.081	2370	4.570	2110	4.570	2110	4.425	840	4.211	1170	4.005	1070	3.850	1270
0.04a	3.808	2830	3.420	1950	3.561	1630	3.435	680	3.241	1630	3.044	1430	2.888	1630
0.06a	2.528	2950	2.264	2170	2.528	2950	2.431	1370	2.252	1220	2.071	2170	1.917	2270

Note. The order of the last summed term in the series is m.

Table 3.8

The infinitely long plate subjected to a surface couple $\vec{M}\|Oy$ at $(\xi, 0)$

Lowest values of bending moment M_x

The dimensionless coefficients $\min k_M^x$ at $(x = \xi - c; \ y = 0)$

$c = d$	$\xi = c$	m	$\xi = 2c$	m	$\xi=0.060a$	m	$\xi=0.075a$	m	$\xi=0.125a$	m	$\xi=0.250a$	m	$\xi=0.500a$	m
\multicolumn{15}{c}{$\mu = 0$}
0.01a	0.000	20	−7.108	380	−9.579	910	−9.388	400	−9.759	500	−10.030	600	−10.217	700
0.02a	0.000	20	−3.828	1650	−4.418	1060	−4.597	800	−4.899	1100	−5.111	1400	−5.238	1350
0.03a	0.000	20	−2.554	1110	−2.554	1110	−2.814	1000	−3.089	340	−3.365	1270	−3.497	1270
0.04a	0.000	20	−1.926	1450	−1.475	1060	−1.846	1000	−2.252	980	−2.489	1430	−2.625	1630
0.06a	0.000	20	−1.235	120	0.000	20	−0.643	600	−1.318	1020	−1.603	2070	−1.745	2270
\multicolumn{15}{c}{$\mu = 1/6$}
0.01a	0.000	20	−7.237	380	−9.994	910	−9.830	400	−10.241	500	−10.543	600	−10.751	700
0.02a	0.000	20	−3.893	1650	−4.557	1060	−4.765	800	−5.111	1100	−5.356	1400	−5.505	1350
0.03a	0.000	20	−2.598	1110	−2.598	1110	−2.887	1000	−3.210	340	−3.520	1270	−3.674	1270
0.04a	0.000	20	−1.959	1450	−1.475	1060	−1.871	1000	−2.236	980	−2.599	1430	−2.757	1630
0.06a	0.000	20	−1.258	120	0.000	20	−0.636	600	−1.344	1020	−1.666	2070	−1.831	2270
\multicolumn{15}{c}{$\mu = 1/3$}
0.01a	0.000	20	−7.364	380	−10.403	910	−10.268	400	−10.720	500	−11.053	600	−11.282	700
0.02a	0.000	20	−3.957	1650	−4.695	1060	−4.932	800	−5.322	1100	−5.599	1400	−5.770	1350
0.03a	0.000	20	−2.641	1110	−2.641	1110	−2.960	1000	−3.330	340	−3.674	1270	−3.850	1270
0.04a	0.000	20	−1.992	1060	−1.475	1060	−1.897	1000	−2.399	980	−2.708	1430	−2.888	1630
0.06a	0.000	20	−1.280	120	0.000	20	−0.628	600	−1.369	1020	−1.730	2070	−1.917	2270

Note. The order of the last summed term in the series is m.

3 Behaviour of plates subjected to a surface couple

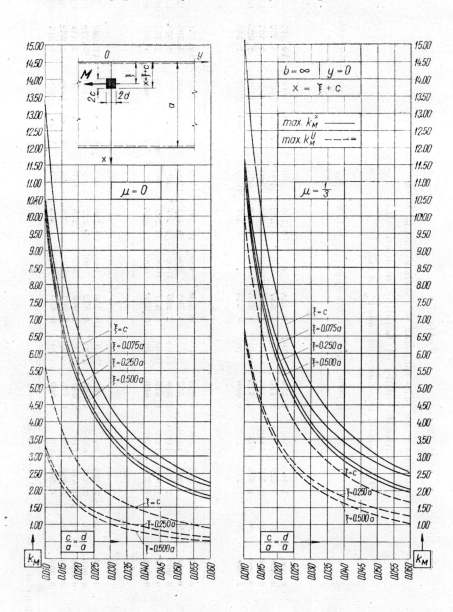

Figure 3.15 Infinitely long plate subjected to a surface couple $\vec{M} \parallel Oy$. Variation of max M_x (max k_M^x) and max M_y (max k_M^y) with parameters c and d defining sizes of surface on which couple M is applied.

3.3 Magnitude and distribution of stresses

Table 3.9

The infinitely long plate subjected to a surface couple $\vec{M} \parallel Oy$ at $(\xi; 0)$

Extreme values of bending moment M_y

The dimensionless coefficients **max k_M^y** at $(x = \xi + c\,;\ y = 0)$

$c = d$	$\xi = c$	m	$\xi = 0.25a$	m	$\xi = 0.50a$	m
			$\mu = 0$			
0.01a	5.633	200	3.321	190	3.198	190
0.02a	2.813	130	1.717	110	1.596	140
0.03a	1.871	90	1.179	100	1.059	100
0.04a	1.399	80	0.908	80	0.790	80
0.06a	0.924	60	0.632	60	0.518	60
			$\mu = 1/6$			
0.01a	7.807	600	5.004	410	4.826	300
0.02a	3.918	680	2.601	550	2.461	600
0.03a	2.613	970	1.776	470	1.639	570
0.04a	1.956	1180	1.363	630	1.226	680
0.06a	1.294	1200	0.942	870	0.809	920
			$\mu = 1/3$			
0.01a	10.007	850	6.746	600	6.554	500
0.02a	5.025	930	3.489	750	3.329	800
0.03a	3.353	1370	2.377	670	2.219	770
0.04a	2.510	1630	1.818	830	1.662	980
0.06a	1.663	1700	1.251	1270	1.098	1320

The dimensionless coefficients **min k_M^y** at $(x = \xi - c\,;\ y = 0)$

$c = d$	$\xi = c$	m	$\xi = 0.25a$	m	$\xi = 0.50a$	m
			$\mu = 0$			
0.01a	0.000	20	−3.071	200	−3.198	190
0.02a	0.000	20	−1.466	110	−1.596	140
0.03a	0.000	20	−0.928	100	−1.059	100
0.04a	0.000	20	−0.657	80	−0.790	80
0.06a	0.000	20	−0.381	60	−0.518	60
			$\mu = 1/6$			
0.01a	0.000	20	−4.677	300	−4.826	300
0.02a	0.000	20	−2.310	600	−2.461	600
0.03a	0.000	20	−1.486	570	−1.639	570
0.04a	0.000	20	−1.070	630	−1.226	680
0.06a	0.000	20	−0.648	870	−0.809	920
			$\mu = 1/3$			
0.01a	0.000	20	−6.342	400	−6.554	500
0.02a	0.000	20	−3.157	810	−3.329	800
0.03a	0.000	20	−2.044	770	−2.219	770
0.04a	0.000	20	−1.483	830	−1.662	980
0.06a	0.000	20	−0.914	1170	−1.098	1320

Note. 1. The order of the last summed term in the simple series is m.
2. The value of Poisson's ratio used in computations was $\mu = 1/6 = 0.167$; $\mu = 1/3 = 0.333$.

3 Behaviour of plates subjected to a surface couple

plates. Such differences agree with predictions, the extreme absolute values of the coefficient k_M^x increasing with the longitudinal span b of the plate (Tables 3.10—3.13).

Table 3.10

The infinitely long plate and the very long rectangular plate ($b = 3a$) subjected to a surface couple $\vec{M} \| Oy$ at $(\xi; \eta)$

Comparative values of transverse displacements w and of bending moments M_x and M_y

$$c = d = 0.04a\,;\ \xi = c\,;\ \begin{cases} b = \infty\,;\ y = \eta = 0 \\ b = 3a\,;\ y = \eta = 0.50b \end{cases}$$

x	$10^5 k_w$		k_M^x		k_M^y		μ
	$b = \infty$	$b = 3a$	$b = \infty$	$b = 3a$	$b = \infty$	$b = 3a$	
c	1463	1439	1.153	1.145	0.795	0.797	0.000
	—	—	1.285	1.278	0.987	1.004	0.167
	—	—	1.418	1.411	1.179	1.191	0.333
$2c$	—	—	3.343	3.236	1.399	1.406	0.000
	—	—	3.576	3.470	1.955	1.945	0.167
	—	—	3.808	3.704	2.510	2.484	0.333
$0.125a$	3676	3654	1.560	1.554	1.221	1.229	0.000
	—	—	1.762	1.758	1.480	1.495	0.167
	—	—	1.966	1.963	1.740	1.748	0.333
$0.250a$	4893	4889	0.633	0.632	0.607	0.609	0.000
	—	—	0.734	0.734	0.712	0.713	0.167
	—	—	0.835	0.835	0.818	0.818	0.333
$0.375a$	5056	5048	0.382	0.382	0.376	0.376	0.000
	—	—	0.445	0.444	0.439	0.440	0.167
	—	—	0.507	0.507	0.503	0.503	0.333
$0.500a$	4607	4599	0.253	0.252	0.251	0.251	0.000
	—	—	0.295	0.294	0.293	0.293	0.167
	—	—	0.337	0.336	0.335	0.335	0.333
$0.625a$	3754	3749	0.169	0.168	0.167	0.168	0.000
	—	—	0.196	0.196	0.195	0.196	0.167
	—	—	0.224	0.224	0.224	0.224	0.333
$0.750a$	2640	2634	0.104	0.104	0.104	0.104	0.000
	—	—	0.122	0.121	0.121	0.121	0.167
	—	—	0.139	0.138	0.138	0.139	0.333
$0.875a$	1360	1358	0.050	0.049	0.050	0.051	0.000
	—	—	0.058	0.058	0.058	0.058	0.167
	—	—	0.067	0.066	0.066	0.066	0.333
$a - c$	439	439	0.017	0.016	0.016	0.017	0.000
	—	—	0.020	0.018	0.019	0.018	0.167
	—	—	0.023	0.021	0.022	0.021	0.333
Plane	$y = 0$	$y = 1.5a$	$y = 0$	$y = 1.5a$	$y = 0$	$y = 1.5a$	—

Note. The torsional moments M_{xy} are zero ($k_M^{xy} = 0$) in the planes of symmetry $y = \eta = 0$ (for $b = \infty$) and $y = \eta = 0.50b$ (for $b = 3a$), respectively.

3.3 Magnitude and distribution of stresses

Very small differences exist between the values of the elastic displacements w and bending moments M_x and M_y calculated for the very long plate ($b \geqslant 3a$) and the infinitely long plate subjected to a surface couple M (Table 3.10). The differences are so small that we can state that the behaviour of the two categories o plates is identical with respect to stresses and displacements.

We retain the extreme values for Poisson's ratio ($\mu = 0$ and $\mu = 1/3$), and the parameters defining the area where the surface couple M ($c = d = 0.040a$) acts and also the symmetrical position of the couple along the plate length ($\eta = 0$ for $b = \infty$ and $\eta = 0.50b$ for $b = 3a$, respectively), and extend the study to the entire range $a \leqslant b \leqslant \infty$. In the range $2a \leqslant b \leqslant \infty$, the maximum values (Table 3.11) are very close and so are the minimum ones (Table 3.12). Larger difference in values appears only in the case of plates whose sizes are close to the dimension of the square plate ($a \leqslant b \leqslant 1.5a$). This observation prompted us to approach the two extreme categories: the infinitely long and the square plate (see the auxiliary values tabulated in the Appendix).

Similar conclusions could be derived in the case of an unsymmetrical position of the surface couple ($\xi = 0.25a$; $\eta = 0.25b$) (Table 3.13). Hence, the use of the model for the infinitely long plate in current design or pre-design is justified for any slab where $b \geqslant 2a$ [75]. If local corrections are required at the edge or corner panels, we use the data obtained in the case of the square plate.

The comparative study of the distribution of sectional bending moments transverse to the plate was extended in the case of the square plate (Fig. 3.16) by consideration of basic values for the design parameters (Tables 3.14—3.16):
— the sizes $2c \times 2d$ of the area subjected to the couple M ($c = 0.02a$; $0.04a$; $0.06a$), ($d = 0.02a$; $0.04a$; $0.06a$);
— the position (ξ, η) of the surface couple ($\xi = c$; $0.25a$; $0.50a$), ($\eta = d$; $0.25b$; $0.50b$);
— the value of Poisson's ratio ($\mu = 0$; $1/6$; $1/3$).

The relative values chosen for the sizes of the slab-column area ($0.02a \leqslant c, d \leqslant 0.06a$) cover almost the entire range of flat slabs. The three values for the parameters (ξ) and (η) define nine characteristic positions of the surface couple M in the first quarter of the square plate, thus allowing, for instance, the shaping of slab structures with 1—4 spans in each of the two directions of the plane (Chapts. 4, 5 and 6). Finally, we should mention that the accuracy of computations imposed by the programmes derived

3 Behaviour of plates subjected to a surface couple

Table 3.11

Rectangular plates subjected to a surface couple $\vec{M} \parallel Oy$ at $(\xi; \eta)$

Comparative values of bending moments

max M_x and **max M_y** at $(x = \xi + c; y = \eta)$

$c = d = 0.040a$

b/a	x/a	y/a	Coefficients max k_M^x			Coefficients max k_M^y		
			$\mu = 0$	$\mu = 1/3$	$m = n$	$\mu = 0$	$\mu = 1/3$	$m = n$

$\xi = 0.50a;\quad \eta = 0.50b$

b/a	x/a	y/a	$\mu = 0$	$\mu = 1/3$	$m = n$	$\mu = 0$	$\mu = 1/3$	$m = n$
1.0		0.500	2.357	2.622	63	0.794	1.579	63
1.5		0.750	2.431	2.695	83	0.790	1.601	83
2.0		1.000	2.476	2.740	103	0.791	1.616	103
3.0	0.540	1.500	2.522	2.788	138	0.797	1.638	138
						0.790	—	80
∞^*		0.000	2.625	2.888	1630	—	1.662	980

$\xi = 0.25a;\quad \eta = 0.50b$

b/a	x/a	y/a	$\mu = 0$	$\mu = 1/3$	$m = n$	$\mu = 0$	$\mu = 1/3$	$m = n$
1.0		0.500	2.384	2.702	63	0.956	1.750	63
1.5		0.750	2.521	2.829	83	0.926	1.767	83
2.0		1.000	2.586	2.891	103	0.914	1.775	103
3.0	0.290	1.500	2.640	2.945	138	0.915	1.795	138
						0.908	—	80
∞^*		0.000	2.742	3.044	1430	—	1.818	830

$\xi = c;\quad \eta = 0.50b$

b/a	x/a	y/a	$\mu = 0$	$\mu = 1/3$	$m = n$	$\mu = 0$	$\mu = 1/3$	$m = n$
1.0		0.500	3.030	3.505	63	1.427	2.437	63
1.5		0.750	3.131	3.600	83	1.408	2.452	83
2.0		1.000	3.184	3.652	103	1.402	2.464	103
3.0	0.080	1.500	3.236	3.704	138	1.406	2.484	138
						1.399	—	80
∞^*		0.000	3.343	3.808	2830	—	2.510	1630

* In the case of simple series, we have considered that $\mu = 1/3 = 0.333$.

3.3 Magnitude and distribution of stresses

Table 3.12

Rectangular plates subjected to a surface couple $\vec{M} \| Oy$ at $(\xi; \eta)$

Comparative values of bending moments

min M_x and **min M_y** at $(x = \xi - c; y = \eta)$

$c = d = 0.040a$

b/a	x/a	y/a	Coefficients min k_M^x			Coefficients min k_M^y		
			$\mu = 0$	$\mu = 1/3$	$m = n$	$\mu = 0$	$\mu = 1/3$	$m = n$
			$\xi = 0.50a$;	$\eta = 0.50b$				
1.0		0.500	−2.357	−2.622	63	−0.794	−1.579	63
1.5		0.750	−2.431	−2.695	83	−0.790	−1.601	83
2.0		1.000	−2.476	−2.740	103	−0.791	−1.616	103
3.0	0.460	1.500	−2.522	−2.788	138	−0.797	−1.638	138
						−0.790	—	80
∞*		0.000	−2.625	−2.888	1630	—	−1.662	980
			$\xi = 0.25a$;	$\eta = 0.50b$				
1.0		0.500	−2.294	−2.502	63	−0.623	−1.388	63
1.5		0.750	−2.317	−2.532	83	−0.644	−1.416	83
2.0		1.000	−2.346	−2.564	103	−0.654	−1.436	103
3.0	0.210	1.500	−2.388	−2.609	138	−0.664	−1.460	138
						−0.657	—	80
∞*		0.000	−2.489	−2.708	1430	—	−1.483	830
			$\xi = c$;	$\eta = 0.50b$				
1.0		0.500	0.000	0.000	63	0.000	0.000	63
3.0	0.000	1.500	0.000	0.000	138	0.000	0.000	138
∞*		0.000	0.000	0.000	20	0.000	0.000	20

* In the case of simple series, we have considered that $\mu = 1/3 = 0.333$.

Table 3.13

Rectangular plates $(a \leqslant b \leqslant 3a)$ subjected to a surface couple $\vec{M} \| Oy$ at (ξ, η)

Comparative values of bending moments M_x

$\xi = 0.25a$; $\eta = 0.25b$

$c = d = 0.040 a$

b/a	x/a	y/a	Coefficients max k_M^x		x/a	y/a	Coefficients min k_M^x		$m = n$
			$\mu = 0$	$\mu = 1/3$			$\mu = 0$	$\mu = 1/3$	
1.0		0.250	2.324	2.639		0.250	−2.326	−2.540	63
1.5	0.290	0.375	2.460	2.775	0.210	0.375	−2.363	−2.574	83
2.0		0.500	2.544	2.856		0.500	−2.380	−2.592	103
3.0		0.750	2.627	2.935		0.750	−2.399	−2.618	138

3 Behaviour of plates subjected to a surface couple

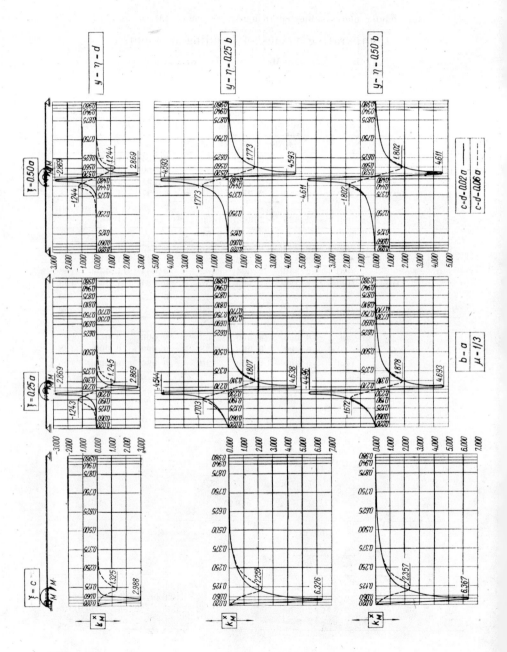

3.3 Magnitude and distribution of stresses

Table 3.14

The square plate subject to a surface couple $\vec{M} \| Oy$ at (ξ, η)

Extreme values of bending moment M_x

$\eta = d$

	The dimensionless coefficient max k_M^x at $(x = \xi + c;\ y = \eta)$			The dimensionless coefficient min k_M^x at $(x = \xi - c;\ y = \eta)$		
$c = d$ \ $\xi =$	c	$0.25\,a$	$0.50\,a$	c	$0.25\,a$	$0.50\,a$
			$\mu = 0$			
$0.02\,a$	2.083	2.234	2.235	0.000	−2.236	−2.235
$0.04\,a$	1.435	1.451	1.454	0.000	−1.459	−1.454
$0.06\,a$	1.025	1.033	1.039	0.000	−1.048	−1.039
			$\mu = 1/6$			
$0.02\,a$	2.535	2.552	2.552	0.000	−2.553	−2.552
$0.04\,a$	1.661	1.607	1.609	0.000	−1.611	−1.609
$0.06\,a$	1.175	1.139	1.142	0.000	−1.146	−1.142
			$\mu = 1/3$			
$0.02\,a$	2.988	2.869	2.869	0.000	−2.896	−2.869
$0.04\,a$	1.887	1.764	1.764	0.000	−1.763	−1.764
$0.06\,a$	1.325	1.245	1.244	0.000	−1.243	−1.244

Extreme values of bending moment M_y

$\eta = d$

	The dimensionless coefficient max k_M^y at $(x = \xi + c;\ y = \eta)$			The dimensionless coefficient min k_M^y at $(x = \xi - c;\ y = \eta)$		
$c = d$ \ $\xi =$	c	$0.25\,a$	$0.50\,a$	c	$0.25\,a$	$0.50\,a$
			$\mu = 0$			
$0.02\,a$	2.716	1.906	1.903	0.000	−1.900	−1.903
$0.04\,a$	1.355	0.938	0.929	0.000	−0.914	−0.929
$0.06\,a$	0.901	0.635	0.616	0.000	−0.583	−0.616
			$\mu = 1/6$			
$0.02\,a$	3.063	2.278	2.276	0.000	−2.273	−2.276
$0.04\,a$	1.595	1.180	1.171	0.000	−1.158	−1.171
$0.06\,a$	1.071	0.807	0.789	0.000	−0.758	−0.789
			$\mu = 1/3$			
$0.02\,a$	3.410	2.651	2.648	0.000	−2.645	−2.648
$0.04\,a$	1.834	1.422	1.413	0.000	−1.401	−1.413
$0.06\,a$	1.242	0.979	0.962	0.000	−0.932	−0.962

Note. The order of the last summed term in the double series is $m = n = 63$

Figure 3.16 Square plate subjected to a surface couple $\vec{M} \| Oy$. Diagrams for sectional bending moments $M_x (k_M^x)$ in transverse planes $y = \eta$ for nine pertinent positions (ξ, η) of couple M. Influence of the relative size of contact area $(2c \times 2d)$.

3 Behaviour of plates subjected to a surface couple

Table 3.15

The square plate subjected to a surface couple $\vec{M}\|Oy$ at $(\xi;\eta)$

Extreme values of bending moment M_x

$\eta = 0.25\ b$

	The dimensionless coefficient max k_M^x at $(x = \xi + c; y = \eta)$			The dimensionless coefficient min k_M^x at $(x = \xi - c; y = \eta)$		
$c = d$ $\xi =$	c	$0.25\ a$	$0.50\ a$	c	$0.25\ a$	$0.50\ a$
$\mu = 0$						
$0.02\ a$	5.374	4.046	4.050	0.000	−4.052	−4.050
$0.04\ a$	2.944	2.324	2.332	0.000	−2.326	−2.332
$0.06\ a$	1.924	1.583	1.594	0.000	−1.580	−1.594
$\mu = 1/6$						
$0.02\ a$	5.850	4.342	4.322	0.000	−4.298	−4.322
$0.04\ a$	3.186	2.481	2.466	0.000	−2.433	−2.466
$0.06\ a$	2.089	1.695	1.684	0.000	−1.642	−1.684
$\mu = 1/3$						
$0.02\ a$	6.326	4.638	4.593	0.000	−4.544	−4.593
$0.04\ a$	3.428	2.639	2.601	0.000	−2.540	−2.601
$0.06\ a$	2.255	1.807	1.773	0.000	−1.703	−1.773

Extreme values of bending moment M_y

$\eta = 0.25\ b$

	The dimensionless coefficient max k_M^y at $(x = \xi + c; y = \eta)$			The dimensionless coefficient min k_M^y at $(x = \xi - c; y = \eta)$		
$c = d$ $\xi =$	c	$0.25\ a$	$0.50\ a$	c	$0.25\ a$	$0.50\ a$
$\mu = 0$						
$0.02\ a$	2.856	1.777	1.630	0.000	−1.477	−1.630
$0.04\ a$	1.455	0.945	0.806	0.000	−0.643	−0.806
$0.06\ a$	0.993	0.673	0.538	0.000	−0.369	−0.538
$\mu = 1/6$						
$0.02\ a$	3.751	2.452	2.305	0.000	−2.152	−2.305
$0.04\ a$	1.945	1.333	1.195	0.000	−1.030	−1.195
$0.06\ a$	1.314	0.936	0.804	0.000	−0.633	−0.804
$\mu = 1/3$						
$0.02\ a$	4.647	3.126	2.980	0.000	−1.827	−2.980
$0.04\ a$	2.436	1.720	1.583	0.000	−1.418	−1.583
$0.06\ a$	1.634	1.200	1.070	0.000	−0.896	−1.070

Note. The order of the last summed term in the double series is $m = n = 63$.

3.3 Magnitude and distribution of stresses

Table 3.16

The square plate subjected to a surface couple $\vec{M} \| Oy$ at $(\xi; \eta)$

Extreme values of bending moment M_x

$\eta = 0.50\ b$

	The dimensionless coefficient max k_M^x at $(x = \xi + c; y = \eta)$			The dimensionless coefficient min k_M^x at $(x = \xi - c; y = \eta)$		
$c = d$ $\quad \xi =$	c	$0.25\ a$	$0.50\ a$	c	$0.25\ a$	$0.50\ a$
			$\mu = 0$			
$0.02\ a$	5.421	4.097	4.068	0.000	−4.010	−4.068
$0.04\ a$	3.030	2.384	2.357	0.000	−2.294	−2.357
$0.06\ a$	2.037	1.651	1.628	0.000	−1.555	−1.628
			$\mu = 1/6$			
$0.02\ a$	5.894	4.395	4.339	0.000	−4.253	−4.339
$0.04\ a$	3.267	2.543	2.489	0.000	−2.398	−2.489
$0.06\ a$	2.197	1.764	1.715	0.000	−1.614	−1.715
			$\mu = 1/3$			
$0.02\ a$	6.367	4.693	4.611	0.000	−4.496	−4.611
$0.04\ a$	3.505	2.702	2.622	0.000	−2.502	−2.622
$0.06\ a$	2.357	1.878	1.802	0.000	−1.672	−1.802

Extreme values of bending moment M_y

$\eta = 0.50\ b$

	The dimensionless coefficient max k_M^y at $(x = \xi + c; y = \eta)$			The dimensionless coefficient min k_M^y at $(x = \xi - c; y = \eta)$		
$c = d$ $\quad \xi =$	c	$0.25\ a$	$0.50\ a$	c	$0.25\ a$	$0.50\ a$
			$\mu = 0$			
$0.02\ a$	2.840	1.788	1.629	0.000	−1.456	−1.629
$0.04\ a$	1.427	0.956	0.794	0.000	−0.623	−0.794
$0.06\ a$	0.960	0.680	0.522	0.000	−0.349	−0.522
			$\mu = 1/6$			
$0.02\ a$	3.743	2.471	2.307	0.000	−2.124	−2.307
$0.04\ a$	1.932	1.353	1.187	0.000	−1.005	−1.187
$0.06\ a$	1.300	0.956	0.793	0.000	−0.608	−0.793
			$\mu = 1/3$			
$0.02\ a$	4.646	3.154	2.985	0.000	−2.792	−2.985
$0.04\ a$	2.437	1.750	1.579	0.000	−1.388	−1.579
$0.06\ a$	1.639	1.231	1.064	0.000	−0.867	−1.064

Note. The order of the last summed term in the double series is $m = n = 63$

3   *Behaviour of plates subjected to a surface couple*

from the exigencies of our study. This accuracy is higher than that required in the calculation involved in the design of structures (see Sec. 3.4).

3.3.2   Magnitude and distribution of bending moments along the plate length

That the surface couple acts like a local perturbation is also obvious along the plate length. Thus, the bending moments M_x and M_y decrease quickly in absolute value from the highest (lowest) values, calculated on the perimeter of the area subjected to the couple M, to negligible values along the plate length. For instance, in the case of the infinitely long plate, the values of k_M^x and k_M^y are so small at a distance $y \geqslant 0.75a$ from the transverse plane of symmetry, that they can be overlooked in practice (Figs. 3.17–3.20) *.

All the diagrams k_M^y show points of zero moment. It is clear that the points where M_y changes sign correspond to those where the curves k_w show a change of sign of the curvature of the elastic surface (Figs. 3.17 and 3.18).

An analysis of the manner in which the bending moments M_x and M_y vary along the plate length shows that the diagrams k_M^y and k_M^x tend to be symmetrical to the axis of symmetry $x = 0.50a$ of the plate (Fig. 3.19). The tendency towards a symmetrical distribution of values in planes sufficiently far away from the loaded area, which was also noticed in the case of elastic displacements, is in compliance with the physical phenomenon described by the established mathematical equations. This tendency constitutes the *response of an elastic symmetrical structure* to an unsymmetrical load.

In the case of an anti-symmetrical load, the diagrams of M_x and M_y remain antisymmetric in any plane transverse to the plate. The characteristics of a local perturbation of the effects induced by a surface couple becomes substantially weaker as the distance y from the loaded area increases (Fig. 3.20).

We pass now to the case of finite boundary plates ($a \leqslant b \leqslant 3a$) and observe that under a load M, their behaviour is similar to that of the infinitely long plate. Of course, several quantitative differences appear between the two categories of plates, but these

---

* However, in the structural analysis carried out in Chaps. 4—6, only those values of the dimensionless coefficients which correspond to the distances $y > 2a$ were neglected (see Sub-secs. 4.1.1.1 and 4.1.2.1).

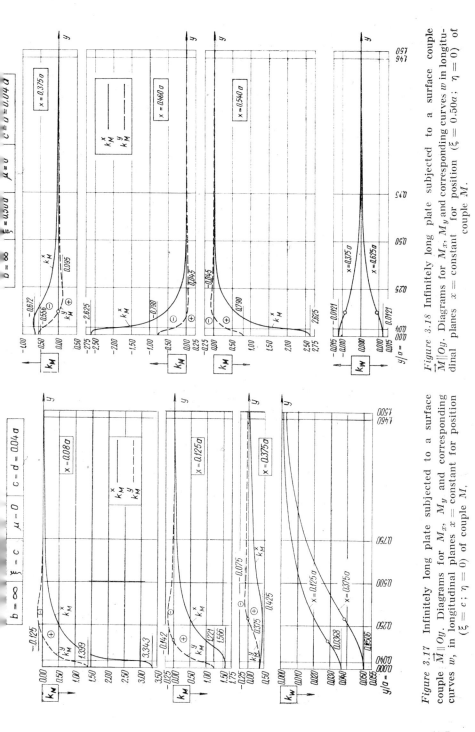

Figure 3.17 Infinitely long plate subjected to a surface couple $\overline{M} \parallel Oy$. Diagrams for M_x, M_y and corresponding curves w, in longitudinal planes $x = $ constant for position ($\xi = c$; $\eta = 0$) of couple M.

Figure 3.18 Infinitely long plate subjected to a surface couple $\overline{M} \parallel Oy$. Diagrams for M_x, M_y and corresponding curves w in longitudinal planes $x = $ constant for position ($\xi = 0.50a$; $\eta = 0$) of couple M.

175

3 Behaviour of plates subjected to a surface couple

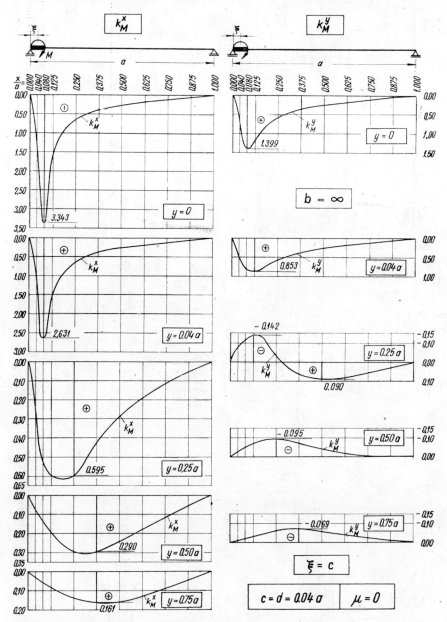

Figure 3.19 Infinitely long plate subjected to a surface couple $\vec{M} \parallel Oy$. Diagrams for M_x and M_y in transverse planes $y =$ constant, located at a gradually increasing distance away from loaded area, for position ($\xi = c$; $\eta = 0$) of couple M.

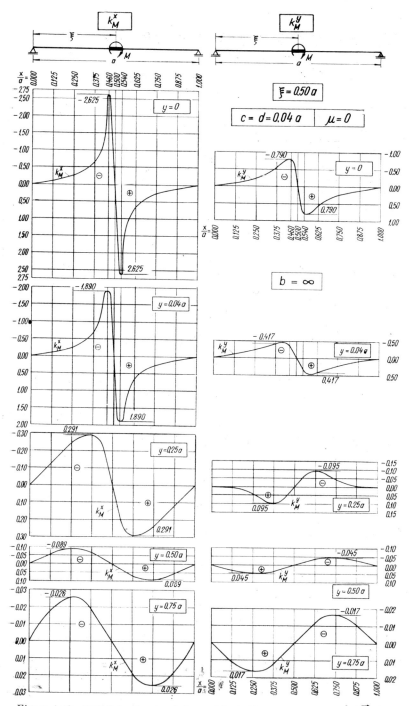

Figure 3.20 Infinitely long plate subjected to a surface couple $\vec{M} \parallel Oy$. Diagrams for M_x and M_y in transverse planes $y =$ constant located at a gradually increasing distance away from loaded area, for position ($\xi = 0.50a$; $\eta = 0$) of couple M.

3 Behaviour of plates subjected to a surface couple

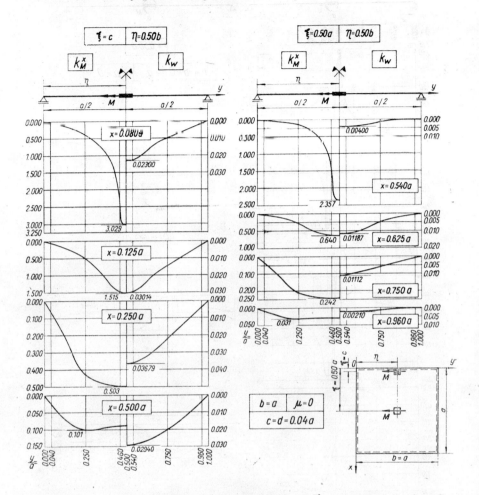

Figure 3.21 Square plate subjected to a surface couple $\vec{M} \parallel Oy$. Diagrams for $M_x(k_M^x)$ and corresponding curves w (k_w) in longitudinal planes $x =$ constant, for two pertinent positions (ξ, η) of couple M.

differences are in compliance with expectations (Figs. 3.21 and 3.22) and are determined by the different boundary conditions used, including also the ratio b/a. Hence, the similar behaviour of the two categories of plates is also confirmed by the magnitude and distribution of the bending moments along the plate length.

3.3 Magnitude and distribution of stresses

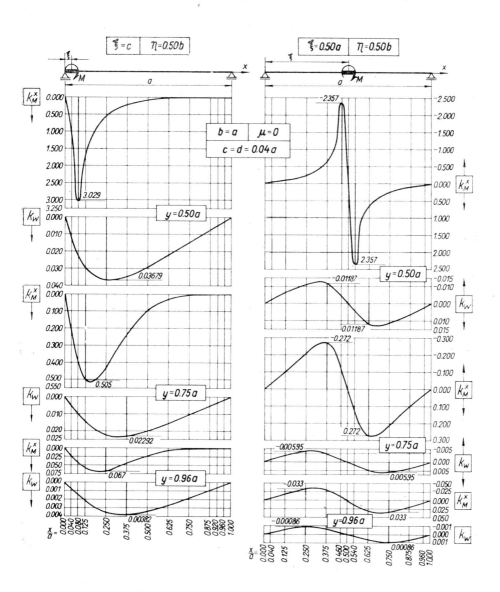

Figure 3.22 Square plate subjected to a surface couple $\vec{M} \parallel Oy$. Diagrams for $M_x(k_M^x)$ and corresponding $w(k_w)$ curves in transverse planes $y =$ constant, located at gradually increasing distances away from loaded area, for two pertinent positions (ξ, η) of couple M.

3 *Behaviour of plates subjected to a surface couple*

3.3.3 Magnitude and distribution of torsional moments

An examination of the torsional moments M_{xy} can rarely be found in the literature dealing with slabs. Moreover, the particular case of plates subjected to a surface couple M is first approached in [70], [75]. The results obtained in this book allow an examination of the distribution of the values of M_{xy} throughout the plate surface under this load (Figs. 3.23, 3.24 — Tables 3.17, 3.18).

The diagrams k_M^{xy} plotted for the infinitely long plate clearly show that the characteristic of a local perturbation is also obvious for to the torsional moments (Fig. 3.23). In the transverse plane of symmetry $y = 0$, the value of the torsional moments is zero ($k_M^{xy} = 0$), irrespective of the position (ξ, 0) of the surface couple M.

However, the tendency towards symmetrization is different from that observed in the case of bending moments. Thus, the diagrams k_M^{xy}, which for $\xi = c$; $\xi = 0.25a$ were non-symmetrical, tend to become *antisymmetric*, as the distance y from the loaded area increases. The symmetry, and the antisymmetry respectively, of the moment diagrams M_x, M_y and M_{xy} in transverse planes that are sufficiently far from the loaded area is clearly noticed in the planes located from $y = 0.75a$ farther on (Figs. 3.19, 3.20, 3.23 and 3.24). In this plane we plotted the diagrams k_M^x, k_M^y and k_M^{xy} corresponding to the three characteristic positions of the couple M under consideration: the edge couple ($\xi = c$), the couple at a quarter of the span ($\xi = 0.25\ a$) and the couple at the mid-span ($\xi = 0.50a$).

Table 3.17

The infinitely long plate subjected to a surface couple $\vec{M} \| Oy$

Extreme comparative values of bending and torsional moments

$c = d = 0.04\ a;\ \mu = 0$

ξ/a	x/a	y/a	k_M^x	k_M^y	k_M^{xy}
0.04	0.04	0.04	1.123	0.470	**0.669**
	0.08	0	3.343	1.399	0
0.25	0.25	0.04	0.127	0.121	**0.783**
	0.29	0	2.742	0.908	0
0.50	0.50	0.04	0	0	**0.790**
	0.54	0	2.625	0.790	0

Table 3.18

The infinitely long plate and the square plate subjected to a surface couple $\vec{M} \parallel Oy$

Comparative values of the sectional torsional moments M_{xy}. The dimensionless coefficients $k_M^{xy} = M_{xy} / \dfrac{M}{a}$

$c = d = 0.04\,a$

	$b = \infty$					$b = a$				$b = \infty$			$b = a$		
y	x	μ	$\xi=c,\ \eta=0$	$\xi=c,\ \eta=0.5a$		x	y		x	$\xi=0.50a,\ \eta=0$	$\xi=0.50a,\ \eta=0.50a$	μ	x	y	
$-d$	0	0	−0.684	−0.702		0			0.375a	0.129	0.124	0	0.375a		
		1/3	−0.456	−0.468						0.086	0.083	1/3			
	c	0	−0.669	−0.683		c	$0.50a-d$		0.500a	−0.790	−0.795	0	0.500a	$0.50a-d$	
		1/3	−0.446	−0.455						−0.527	−0.530	1/3			
	0.125a	0	0.223	0.213		0.125 a			0.625a	0.129	0.124	0	0.625a		
		1/3	0.149	0.142						0.086	0.083	1/3			
0	—	—	0.000	0.000		—	0		—	0.000	0.000	—	—	0.50a	
$+d$	0	0	0.684	0.702		0			0.375a	−0.129	−0.124	0	0.375a		
		1/3	0.456	0.468						−0.086	−0.083	1/3			
	c	0	0.669	0.683		c	$0.50a+d$		0.500a	0.790	0.795	0	0.500a	$0.50a+d$	
		1/3	0.446	0.455						0.527	0.530	1/3			
	0.125a	0	−0.223	−0.213		0.125a			0.625a	−0.129	−0.124	0	0.625a		
		1/3	−0.149	−0.142						−0.086	−0.083	1/3			
0.25a	0	0	0.549	0.635		0			c	−0.110	−0.129	0	c		
		1/3	0.366	0.423						−0.074	−0.086	1/3			
	c	0	0.513	0.598		c	0.75a		0.500a	0.255	0.275	0	0.500a	0.75a	
		1/3	0.342	0.399						0.170	0.183	1/3			
	0.625a	0	−0.101	−0.136		0.625a			$a-c$	−0.110*	−0.129	0	$a-c$		
		1/3	−0.067	−0.091						−0.074	−0.086	1/3			

Note: For the square plate, the order of the last term in the double series is $m = n = 61$.

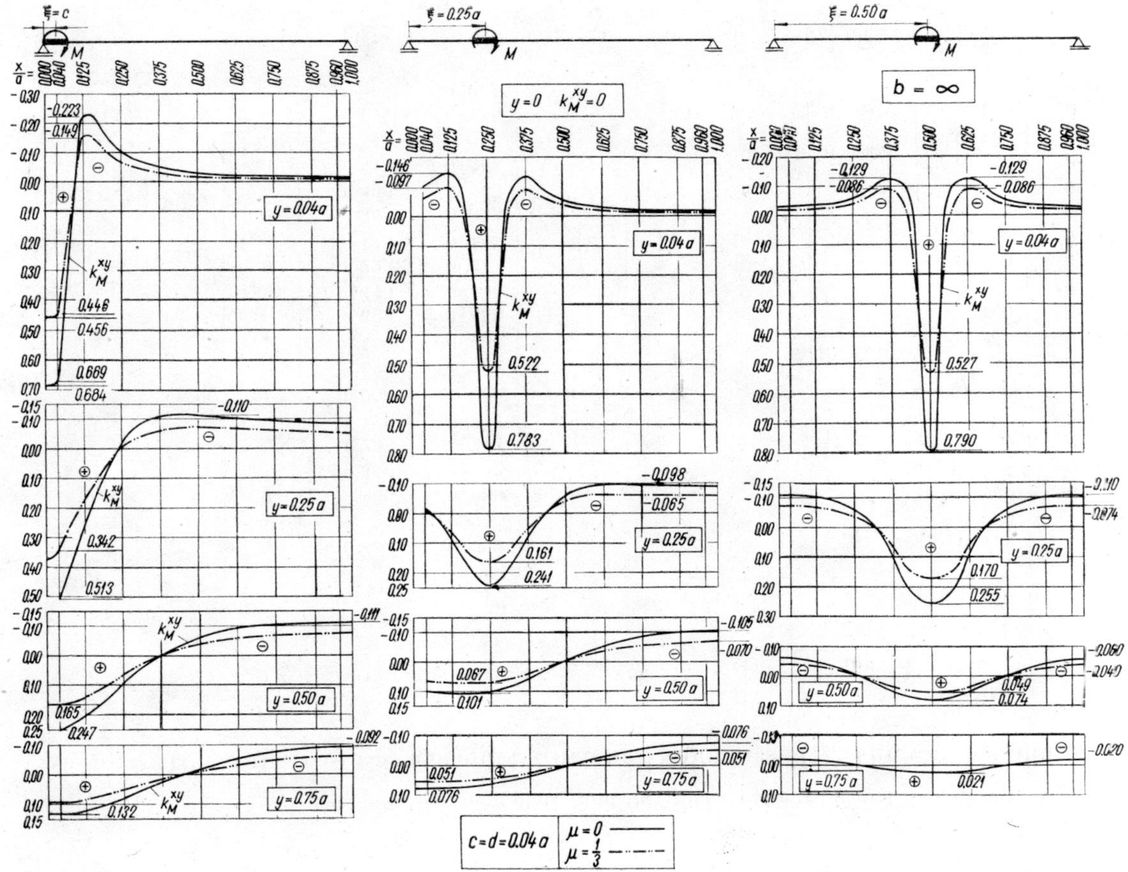

Figure 3.23 Infinitely long plate subjected to a surface couple $\vec{M} \parallel Oy$. Diagrams of sectional torsional moments $M_{xy}(k_M^{xy})$ in transverse planes $y =$ constant located at gradually increasing distances away from loaded area, for three pertinent positions $(\xi, 0)$ of couple M.

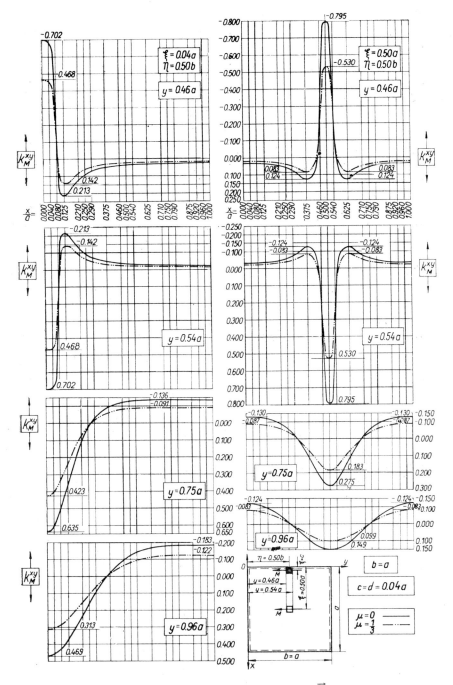

Figure 3.24 Square plate subjected to a surface couple $\vec{M} \| Oy$. Diagrams of sectional torsional moments M_{xy} (k_M^{xy}) in transverse planes $y =$ constant, located at gradually increasing distances away from loaded area, for two pertinent positions (ξ, η) of couple M.

3 Behaviour of plates subjected to a surface couple

Depending on the type of the elastic displacement or stress considered, and for distances lying between $y = 1.25\,a$ and $y = 1.75\,a$ and more, all the curves representing the dimensionless coefficients (k_w, k_φ^x, k_φ^y, k_M^x, k_M^y and k_M^{xy}) become strictly symmetric and strictly antisymmetric, respectively.

A comparative examination of the diagrams k_M^{xy} (Figs. 3.23 and 3.24) and the corresponding diagrams k_M^x and k_M^y (Figs. 3.19—3.22) shows several differences and similarities. The extreme values of the three moments appear in the area loaded with the couple M, whereas the highest and lowest values of the bending moments M_x and M_y are not recorded at the points (x, y), where the highest values of the torsional moments M_{xy} appear (Table 3.17). We also notice that for any position ξ of the surface couple M, the order of magnitude of the highest value k_M^{xy} is clearly lower than the highest value k_M^x corresponding to the same position of the couple.

The magnitude and distribution of the torsional moments M_{xy} was also studied in the case of finite boundary plates ($a \leqslant b \leqslant 3a$) subjected to the action of a surface couple M (Fig. 3.24). The diagrams k_M^{xy}, plotted for the square plate ($b = a$), show that the torsional moments have also the characteristics of a local perturbation. The value of the distribution of the torsional moments confirm the previous statement regarding the similar behaviour of the rectangular plate and the infinitely long plate (Table 3.18 and Figs. 3.23, 3.24).

The diagrams plotted in the comparative study (Sub-secs. 3.3.1—3.3.3) for k_M^x, k_M^y and k_M^{xy} give a clear image of the behaviour of plates under a surface couple. The region of maximum stresses in the elastic range are particularly obvious.

Identification of such stress concentration is of importance for further developments in threes distinct fields which were not approached in the present work:

a) it facilitates tracing of the influence lines for the moments M_x, M_y and M_{xy};

b) it allows a more convenient use of the finite-element method (in specifying the portions for which the density of the computation grid must be increased or reduced);

c) it may lead to the study of the post-elastic behaviour of plates subjected to surface couples (i.e. identification of the regions where the plastic stage is reached).

3.3 Magnitude and distribution of stresses

3.3.4 Influence of Poisson's ratio on the values of bending and torsional moments

The discussion of the parameters which influence the magnitude and distribution of the stresses in plates subjected to a surface couple, revolved around the influence of the geometrical parameters a, b, c, d, ξ, η (Sub-secs. 3.3.1—3.3.3). Comparative diagrams describing the influence of Poisson's ratio, which is the only physical parameter in the expressions of the dimensionless coefficients, were only sporadically given (Figs. 3.14, 3.15, 3.23, and 3.24). However, most of the tables listing values for k_M^x, k_M^y and k_M^{xy} contain the comparative data computed for two or three characteristic values of this parameter. Thus, $\mu = 0$ $\mu = 1/6$; $\mu = 1/3$ (Tables 3.7—3.16, 3.18, 3.20). This sub-section gives several supplementary data as well as the observations and conclusions concerning the influence of Poisson's ratio on the magnitude and distribution of the stresses M_x, M_y and M_{xy} [70], [75].

The bending moments M_x and M_y increase *linearly* in absolute value with the value of Poisson's ratio (Figs. 3.25—3.29 and Tables 3.7—3.16). The absolute values of the twisting moments M_{xy} decrease linearly as μ increases (Figs. 3.23, 3.24 — Table 3.18).

As was predicted, the influence of Poisson's ratio is stronger for the values max (min) k_M^y than for the values max (min) k_M^x (Tables 3.7—3.9, Fig. 3.28). For the values k_M^y this influence is stronger in the central position $\xi = 0.50\,a$ (Fig. 3.26), whereas the values k_M^x are influenced by the increase of μ in the edge position $\xi = c$ (Figs. 3.25 and 3.27) of the couple M. That the values max k_M^{xy} depend on μ is barely influenced by the position ξ of the surface couple M (Figs. 3.23 and 3.24).

In the study of the influence of Poisson's ratio on the stresses we have additionally considered the comparative diagrams of max (min) k_M^y and k_M^x, drawn as functions of μ (Fig. 3.27). These diagrams which were plotted for four characteristic positions of the couple $M(\xi = c\,;\ 0.125a\,;\ 0.25a$ and $0.50a)$, reinforce the small influence of the parameter ξ on the extreme values of coefficients k_M^x and k_M^y, over the entire range $0.125\,a \leqslant \xi \leqslant 0.50\,a$ (Sub-secs. 3.3.1, Fig. 3.14).

We also notice that the influence of Poisson's ratio on the extreme values of the moments M_x and M_y increases when the sizes $2c \times 2d$ of the loaded area are smaller. In the absolute value, the coefficients k_M^y increase more rapidly with the values of μ than do the coefficients k_M^x, especially when the values of the parameters c and d are small. The diagrams max(min) k_M^x and k_M^y were drawn so

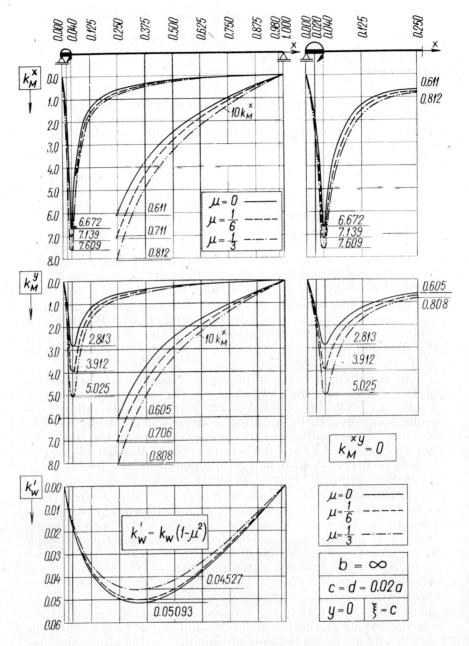

Figure 3.25 Infinitely long plate subjected to a surface couple $\vec{M} \parallel Oy$. Influence of Poisson's ratio on value of bending moments M_x, M_y and on ordinates $w(k'_w)$ of deflected mid-surface for edge-position ($\xi = c$) of couple M.

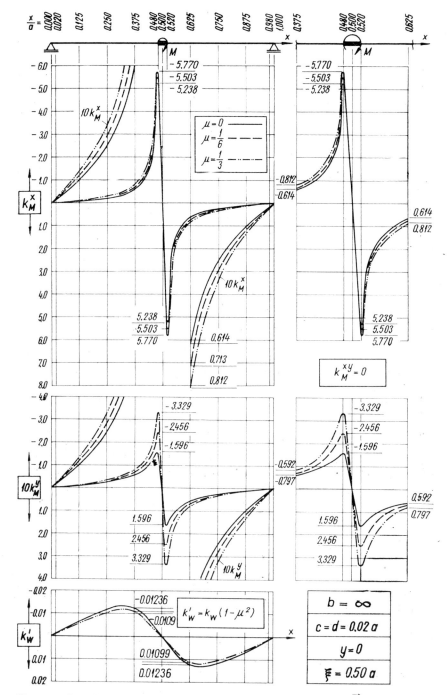

Figure 3.26 Infinitely long plate subjected to a surface couple $\vec{M} \parallel Oy$. Influence of Poisson's ratio on value of bending moments M_x, M_y and on ordinates w (k'_w) of deflected mid-surface for central position ($\xi' = 0.50a$) of couple M.

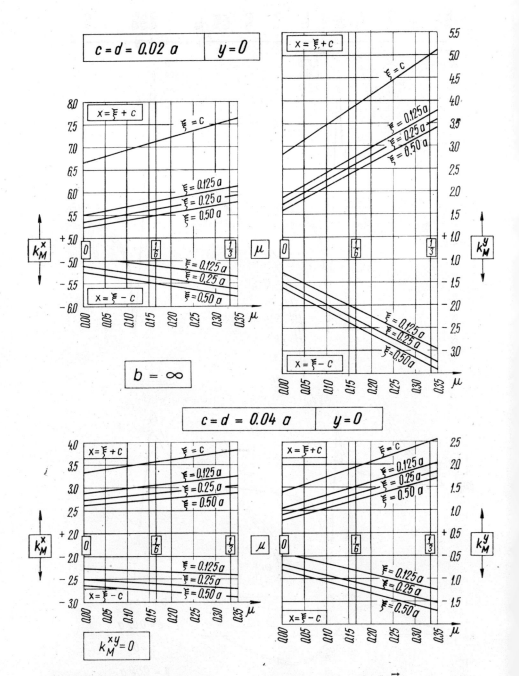

Figure 3.27 Infinitely long plate subjected to a surface couple $\vec{M} \parallel Oy$. Variation of extreme values of sectional bending moments M_x and M_y with Poisson's ratio.

3.3 Magnitude and distribution of stresses

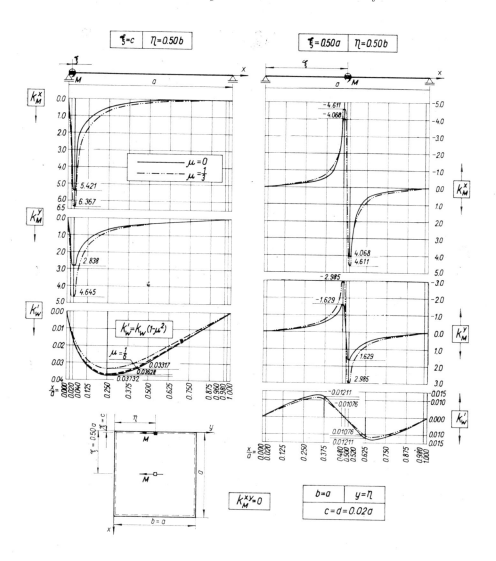

Figure 3.28 Square plate subjected to a surface couple $\vec{M} \parallel Oy$. Influence of Poisson's ratio on value of bending moments M_x, M_y and on ordinates $w\,(k'_w)$ of deflected mid-surface, for two pertinent positions (ξ, η) of couple M.

3 Behaviour of plates subjected to a surface couple

as to allow direct or interpolated reading for any value of Poisson's ratio and of the geometrical parameters c, d and ξ which may appear in practice.

The influence of the parameter μ on the moments M_x, M_y and M_{xy} was studied in the case of the infinitely long plate (Figs. 3.14, 3.15, 3.23, 3.25, 3.26 and 3.27 and 3.29) and finite boundary plate ($a \leqslant b \leqslant 3a$) subjected to a surface couple (Figs. 3.24, 3.28, 3.29). The study of the influence of the parameter μ also includes a comparison of the mid-surfaces of plates deflected by the load M. Thus, in the same transverse planes of symmetry $y = 0$ and $y = 0.5a$, respectively, we plotted the curves $k'_w = k_w(1-\mu^2)$, where k_w is the dimensionless coefficient defining the elastic displacements of plates, independenty of their cylindrical rigidity D (see Table 2.2 and formulae 2.51). The ordinates k'_w, i.e. the transverse displacements w of the plate, decrease as D grows, or as the square of Poisson's ratio increases (Figs. 3.25, 3.26, 3.28).

The large number of comparative data obtained from the numerical study (Tables 3.2—3.18) validate the idea of the similar behaviour of the two categories of plates. Whence the accurate modelling of the physical phenomenon by the two types of solutions to the fundamental equation of plates.

3.4 Accuracy of numerical computations

3.4.1 Influence of the number of summed terms in the series on the accuracy of numerical computations

Calculation of the values of the elastic displacements w, φ_x, φ_y, stresses M_x, M_y, M_{xy}, T_x, T_y and boundary reactions R_x and R_y is reduced to the computation of the corresponding dimensionless coefficients k_r^{st}, which are expressed by series of functions with different structures (Secs. 2.2 and 2.3). In order to establish the accuracy condition of the computation programs, we have considered the specific aspects of these series of functions, which appear as simple (Sub-secs. 2.3.2.1 and 2.3.2.2) or as double series (Sub-sec. 2.3.2.3).

The requirements of our comparative study implied a computation precision superior to what is usually admitted in design work. This is why the numerical results given by the author were computed so as to become reference values in research work, including tests carried out on models in the elastic range [75]. However, a strict condition of precision would lead to the

3.4 Accuracy of numerical computations

summation of an exaggerated number of terms and, hence, to a long time required for computations.

★

The accuracy condition was expressed in the computer programmes by the magnitude ER of the admissible error — an arbitrarily small value. For the same accuracy condition ER, the total number N of summed terms *is substantially larger in the case of double series* than it is for simple series. As the convergence of the series decreases with every step of differentiation, the order m of the last summed term depends also on the nature of the static quantity to be determined. Thus, for the same boundary conditions and the same type of load, N increases significantly from k_w to k_φ and to k_M, the convergence of k_R being the slowest.

Amongst the three types of loads M, P and q considered in the analysis of plates (Sub-sec. 2.1.4), the algorithms with the slowest convergence are obtained in the case of the surface couple M, assuming that the category of plate and the static quantity are not changed. Hence, for the extreme case of a plate simply supported on the entire boundary, subjected to a surface couple, the exaggerated increase of N prohibited the numerical computation of the boundary reactions k_R^x and k_R^y.

Finally, for the same value ER of the allowed error, the number N of summed terms varies substantially with the value of the geometrical parameters b/a, c/a, d/a, ξ/a, η/a and the coordinate parameters $(x/a, y/a)$, in the plane of the slab.

It is clear from what was stated above that the use of a unique number N of terms in computation programs is unacceptable. Indeed, if the value of N were constant, irrespective of the boundary conditions, type of load, nature of the static quantity and value of the geometrical parameters, then the distribution of computation error would become non-uniform. This inconvenience could be removed by adopting different values for N depending upon the category of plate and the nature of the static quantities used in computation. Hence, the accuracy of the numerical computations would only vary with the type of load and the position of the node (x, y) of the computation grid.

Bearing this in mind, a unique accuracy condition was used in computer programs for all the static quantities and alternatives compared, namely,

$$\left| \sum_{10} \right| \leqslant ER \tag{3.1}$$

where $\left| \sum_{10} \right|$ denotes the absolute value of the sum of the last ten summed terms in the simple or double series. Hence, for the same value of the error ER admitted in computation, the total number N of summed terms is different for almost each computed value.

However, the programmes used allow us to set an arbitrary order of magnitude for the error ER, i.e. 10^{-4}, 10^{-5}, 10^{-6}, which is constant for all values or only for a certain category of static quantities (elastic displacements, moments and boundary reactions). The greatest number of static quantities were computed assuming that $ER = 10^{-6}$. This accuracy condition is severely restrictive if referred to the conditions adopted by other authors (Pfaffinger [48], Stiglat [60], Scholz [66]).

For each computed value, we have written the order m — $m = n$ in the case of double series — of the last summed term. For simple series, $N = m$. For double series, where the summation was done in triangle,

$$a_{11} + (a_{12} + a_{21}) + (a_{31} + a_{22} + a_{13}) +$$
$$(a_{14} + a_{23} + a_{32} + a_{41}) + \ldots,$$

the total number N of summed terms, corresponding to the order $m = n$ of the last term, is given by

$$N = \frac{m(m+1)}{2}. \tag{3.2}$$

When the load is a surface couple M, a comparison between the values of the elastic displacements w and φ_x, computed first with $ER = 10^{-5}$ and then with $ER = 10^{-6}$, leads to the conclusion that the accuracy condition for the coefficients k_w and k_φ^x can be less severe (Table 3.19). For instance, if in the case of elastic displacements we admit that the accuracy of computation is 10^{-5} instead of 10^{-6}, then the time required by work on computer would be substantially shortened. However, by modifying the accuracy condition ER (i.e. the order $m = n$ of the last summed term), far greater changes of the computed values are noticed for the bending moments M_x and M_y (Table 3.20).

The diagrams showing the variation of the extreme values of the coefficient k_M^x were plotted for the infinitely long and rectangular ($a \leqslant b \leqslant 3a$) plates, subjected to a surface couple M and to a locally distributed load P. An examination of the diagrams shows that the order m ($m = n$) of the last term in the series grows with the parameter b/a, assuming that the accuracy condition $ER = 10^{-6}$ is not changed. For the square plate under the load M, the modification of the accuracy condition from $ER = 10^{-6}$ to $ER = 10^{-5}$ (dotted line) leads to a lower order of the

3.4 Accuracy of numerical computations

Table 3.19

The infinitely long plate subjected to a surface couple M

Influence of accuracy condition ER on the value of elastic displacements

$c = d = 0.02\,a$

ξ/a	x/a	y/a	ER	k_w	$N = m$	k_φ^x	$N = m$
0.02	0.02	0.00	10^{-5}	0 00952	—	0 46136	74
			10^{-6}	0 009524	50	0 461090	130
		0.75	10^{-5}	0 00110	—	0 05481	5
			10^{-6}	0 001098	20	0 054809	20
		1.48	10^{-5}	0.00017	—	0.00864	4
			10^{-6}	0.000173	20	0.008640	20
0.25	0.25	0.00	10^{-5}	0.02318	—	0.26244	74
			10^{-6}	0.023185	50	0 261697	200
		0.75	10^{-5}	0 00804	—	0 02552	5
			10^{-6}	0.008035	20	0.025521	20
		1.48	10^{-5}	0 00137	—	0 00430	4
			10^{-6}	0.001369	20	0.004300	20

last term, i.e. $m=n=42$ instead of $m=n=63$ (Fig. 3.29). In this case, differences from 8 per cent to 10 per cent appear between the values of max (min) k_M^x. It follows that for this load a similar lowering of the accuracy condition is not recommended for the computation of the values M_x and M_y. However, for the load P, the same modification of the accuracy condition does not induce large bariations of the values max (min) k_M^x and hence these values may ve computed with $ER = 10^{-5}$.

★

The question of shortening the time required for computation does not arise in the determination of the static quantities defining the behaviour of the infinitely long plate, since the simple series of the hyperbolic and exponential functions are sufficiently rapidly convergent. In the case of plates simply supported along the entire boundary, the behaviour of which is mathematically modelled by expressions containing double series of very slowly convergent trigonometric functions, we have to consider any

3 *Behaviour of plates subjected to a surface couple*

Table

The square plate subjected

The influence of the order $m = n$ of the last summed term

$c = d = 0.04a$; $\xi = c$;

$10^5 k_w$	$m = n$	N	$10^5 k_\varphi^x$	$m = n$	N	k_M^x
1247	16	136	29475	18	171	1.176 1.421
1251	18	171	29774	38	741	—
1248	29	435	29702	87	3828	1.159 1.427
1245	54	1485	29693	194	18915	1.126 1.399

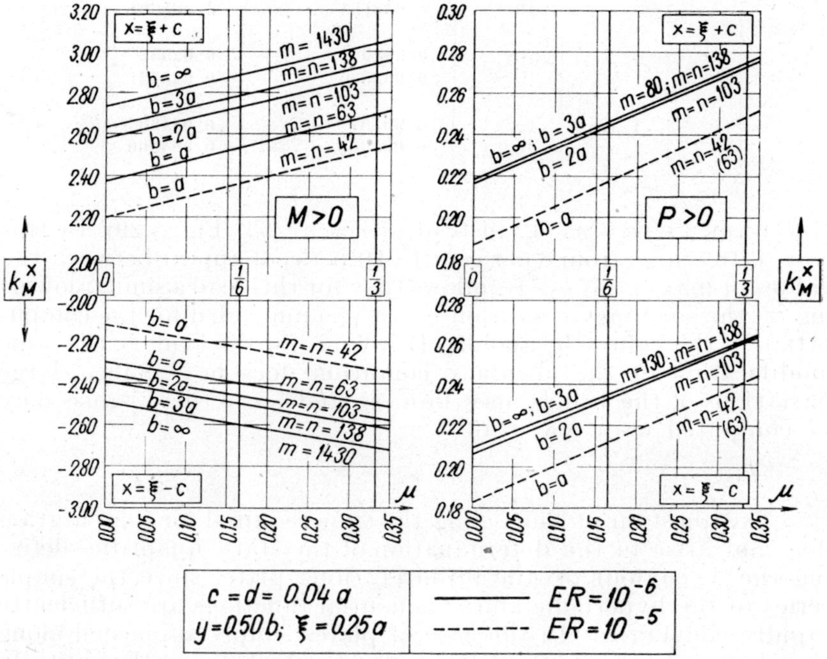

Figure 3.29 Plates subjected to a surface couple M and a locally distributed force P. Variation of the magnitude max (min) k_M^x and order $m = n$ of last term in series as a function of parameter b/a.

194

3.4 Accuracy of numerical computations

3.20

to a surface couple M

in the double series on the value of the static quantities

$\eta = 0.50\ b$; $x = \xi$; $y = \eta$

$m = n$	N	k^y_M	$m = n$	N	μ
32	528	0.747	17	153	0
34	595	1.290	25	325	1/3
—		—	—		0
71	2556	0.825	35	630	0
74	2775	1.157	53	1431	1/3
252	31878	0.809	113	6441	0
261	34191	1.186	181	16471	1/3

means possible to increase the computation convergence rate. This is important for the computation of the stresses and especially the values max (min) M_x which require the largest number N of summed terms.

A higher computation rate can be eventually obtained assuming that the order $m_0 = n_0$ of the last summed term corresponding to a given computation error ER is known. Substituting in the programmes condition (3.1) the accuracy condition

$$m \leqslant m_0, \qquad (3.3)$$

the summation of terms in the series is limited to the term of order $m = n$ corresponding to ER. Through this substitution, all the operations required by the successive use of condition (3.1) are cancelled. However, if $m_0 = n_0$ is the order of the last term in the series defining k^x_M, and condition (3.3) is imposed on the computation of all static quantities, then the values k^x_w, k^x_φ, k^y_φ, k^y_M and k^{xy}_M have a degree of accuracy higher than ER. Moreover, the order $m_0 = n_0$ of the last summed term in the series depends, as was mentioned before, on several other factors, one of which is the ratio b/a, defining the relative dimensions of the rectangular plate (Table 3.21).

A more efficient manner of increasing the rate of numerical computation of the static quantities modelled by double series of trigonometric functions is to transform the expressions using double series into expressions with simple series of exponential

3   *Behaviour of plates subjected to a surface couple*

Table 3.21

Rectangular plates subjected to a surface couple M

Computation of the dimensionless coefficients max(min) k_M^x

The order $m_0 = n_0$ of the last term in the series, corresponding to the accuracy condition $ER = 10^{-6}$

$c = d = 0.04\,a\,;\,\xi = 0.25\,a\,;\,\eta = 0.50\,b\,(\eta = 0)$

b/a	$m_0 = n_0$	N	Observations
1.00	63	1953	double series
2.00	103	5253	double series
3.00	138	9453	double series
∞	$m = 1430$	1430	simple series

functions. This procedure required that one of the indefinite series is expressed by its finite sum. For particular cases, such transformations were carried out by Nádai [1], Girkmann — Tungl [101] and Sonier [139].

3.4.2   Checking the accuracy of computations by use of the reciprocity theorem of unit displacements in the case of plates

Maxwell's theorem on the reciprocity of unit elastic displacements is known in the theory of reticular structures [8], [22] in the form

$$\delta_{ij} = \delta_{ji}. \tag{3.4}$$

An extension to the case of plates allowed us to check the accuracy of the mathematical models established by the author (Secs. 2.2 and 2.3) and the accuracy of the numerical calculations (Chap. 3). The extension of Maxwell's theorem requires not only the transition from a reticular static system to a mixed system consisting of bars and plates, but also that from concentrated forces P_i, P_j (and couples M_i, M_j, respectively) to *locally distributed loads* P_i, P_j (and surface couples M_i, M_j, respectively) [75].

The use of the reciprocity theorem of unit elastic displacements in the case of mixed systems shows that the value of the displacement δ_{ij} produced at point i, in the direction of the generalized force P_i by the generalized force $P_j = 1$, is equal to the value

3.4 Accuracy of numerical computations

of the displacement δ_{ji} produced at point j, in the direction of the generalized force P_j by the generalized force $P_i = 1$.

Checking the accuracy of computations indicates that equality (3.4) holds (Table 3.22) when the plate is under the locally distributed loads $P_i = P_j = P = 1$, in the form of an equality of linear displacements

$$\delta_{ij}^P = \delta_{ji}^P \tag{3.4a}$$

and when the plate is subjected to the surface couples $M_i = M_j = M = 1$, in the form of an equality of angular displacements

$$\theta_{ij}^M = \theta_{ji}^M, \tag{3.4b}$$

respectively.

Using the notation given in Table 2.2, equalities (3.4a) and (3.4b) can be written in the forms

$$k_{wi}^{P_j} = k_{wj}^{P_i} \tag{3.5a}$$

and

$$k_{\varphi_i}^{xM_j} = k_{\varphi_j}^{xM_i}, \tag{3.5b}$$

respectively.

The values of the dimensionless coefficients k_w and k_φ^x, computed for the infinitely long plate and for rectangular plates supported on the entire boundary, satisfy the equalities (3.5a) and (3.5b) with a precision corresponding to the condition $ER = 10^{-6}$ (Table 3.22).

The most interesting verifications were achieved by applying Maxwell's theorem in the form

$$\delta_{ij}^M = \theta_{ji}^P. \tag{3.6}$$

Extending this equation to mixed structures, we have

$$k_{wi}^{M_j} = k_{\varphi_j}^{xP_i}, \tag{3.7}$$

which indicates that the transverse unit displacement w produced at i by a surface couple $M_j = 1$ at j is equal to the rotation φ_x produced at j by a locally distributed load $P_i = 1$ at i. This reciprocity of unit displacements of different kinds and due to different causes (Table 3.23) validates the correctness of the computation algorithms employed and the precision of the numerical computations.

3 Behaviour of plates subjected to a surface couple

Table 3.22

Plates subjected to a surface couple M_j, and a locally distributed force P_j, respectively

Check of accuracy of numerical computations by reciprocal theorem of unit displacements *

Infinity long plate

Position j	Position i	Load $M_j = 1$ $10^6 \, k_\varphi^{xM}$		Load $P_j = 1$ $10^6 \, k_w^P$	
		$c = d = 0.01a$	$c = d = 0.06a$	$c = d = 0.02a$	$c = d = 0.04a$
$\xi = 0.25a$, $\eta = 0$	$x = 0.50a$, $y = 0$	-27469	-25683	10919	10891
$\xi = 0.50a$, $\eta = 0$	$x = 0.25a$, $y = 0$	-27469	-25683	10919	10891
$\xi = 0.25a$, $\eta = 0$	$x = 0.50a$, $y = 0.25a$	-6853	-6558	9134	9115
$\xi = 0.50a$, $\eta = 0.25a$	$x = 0.25a$, $y = 0$	-6853	-6558	9134	9115
$\xi = 0.25a$, $\eta = 0$	$x = 0.75a$, $y = 0.25a$	-67724	-67267	5581	5572
$\xi = 0.75a$, $\eta = 0.25a$	$x = 0.25a$, $y = 0$	-67724	-67267	5581	5572
$\xi = c$, $\eta = 0$	$x = 0.50a$, $y = 0.125a$	-49399	-47055	889	1773
$\xi = 0.50a$, $\eta = 0.125a$	$x = c$, $y = 0$	-49399	-47055	889	1773

Square plate

Position j	Position i	Load $M_j = 1$ $10^8 \, k_\varphi^{xM}$		Load $P_j = 1$ $10^8 \, k_w^P$	
		$c = d = 0.02a$	$c = d = 0.06a$	$c = d = 0.02a$	$c = d = 0.06a$
$\xi = c$, $\eta = d$	$x = 0.25a$, $y = 0.25b$	-147978	-105020	5563	48022
$\xi = 0.25a$, $\eta = 0.25b$	$x = c$, $y = d$	-147978	-105020	5563	48022
$\xi = 0.25a$, $\eta = d$	$x = 0.50a$, $y = 0.25b$	-184954	-714410	44782	131739
$\xi = 0.50a$, $\eta = 0.25b$	$x = 0.25a$, $y = d$	-184954	-714410	44782	131739
$\xi = 0.25a$, $\eta = 0.50b$	$x = 0.50a$, $y = 0.75b$	-675306	-652580	476278	472569
$\xi = 0.50a$, $\eta = 0.75b$	$x = 0.25a$, $y = 0.50b$	-675306	-652580	476278	472569
$\xi = 0.50a$, $\eta = 0.50b$	$x = c$, $y = 0.50b$	-5275523	-5076505	59074	175633
$\xi = c$, $\eta = 0.50b$	$x = 0.50a$, $y = 0.50b$	-5275523	-5076505	59074	175633

* See Eqns. (3.4.) and (3.5).

Table 3.23

Plates subjected to a surface couple M_j, and a locally distributed force P_i, respectively

Check of the accuracy of numerical computations by reciprocal theorem of unit displacements*

Infinitely long plate   Square plate

	Load $M_j = 1$			Load $P_i = 1$				Load $M_j = 1$			Load $P_i = 1$	
Position j	Position i	$10^6 k_{wi}^{M_j}$	$10^6 k_{\varphi j}^{P_i}$	Position i	Position j	$\mathbf{d} =$	Position j	Position i	$10^8 k_{wi}^{M_j}$	$10^8 k_{\varphi j}^{P_i}$	Position i	Position j
$\begin{cases}\xi=0.25\,a\\\eta=0\end{cases}$	$\begin{aligned}x&=0.50\,a\\y&=0\end{aligned}$	38091 37715	38091 37715	$\begin{cases}\xi=0.50\,a\\\eta=0\end{cases}$	$\begin{aligned}x&=0.25\,a\\y&=0\end{aligned}$	$\begin{aligned}0.02\,a\\0.06\,a\end{aligned}$	$\begin{cases}\xi=c\\\eta=d\end{cases}$	$\begin{aligned}x&=0.25\,a\\y&=0.25\,b\end{aligned}$	277330 782630	277330 782630	$\begin{cases}\xi=0.25\,a\\\eta=0.25\,b\end{cases}$	$\begin{aligned}x&=c\\y&=d\end{aligned}$
$\begin{cases}\xi=0.25\,a\\\eta=0\end{cases}$	$\begin{aligned}x&=0.50\,a\\y&=0.25\,a\end{aligned}$	30247 30048	30247 30048	$\begin{cases}\xi=0.50\,a\\\eta=0.25\,a\end{cases}$	$\begin{aligned}x&=0.25\,a\\y&=0\end{aligned}$	$\begin{aligned}0.02\,a\\0.06\,a\end{aligned}$	$\begin{cases}\xi=0.25\,a\\\eta=0.25\,b\end{cases}$	$\begin{aligned}x&=c\\y&=d\end{aligned}$	−6331 −19576	−6331 −19576	$\begin{cases}\xi=c\\\eta=d\end{cases}$	$\begin{aligned}x&=0.25\,a\\y&=0.25\,b\end{aligned}$
$\begin{cases}\xi=0.25\,a\\\eta=0\end{cases}$	$\begin{aligned}x&=0.75\,a\\y&=0.25\,a\end{aligned}$	19798 19698	19798 19698	$\begin{cases}\xi=0.75\,a\\\eta=0.25\,a\end{cases}$	$\begin{aligned}x&=0.25\,a\\y&=0\end{aligned}$	$\begin{aligned}0.02\,a\\0.06\,a\end{aligned}$	$\begin{cases}\xi=0.25\,a\\\eta=0\end{cases}$	$\begin{aligned}x&=0.50\,a\\y&=0.25\,b\end{aligned}$	179328 521882	179328 521882	$\begin{cases}\xi=0.50\,a\\\eta=0.25\,b\end{cases}$	$\begin{aligned}x&=0.25\,a\\y&=d\end{aligned}$
$\begin{cases}\xi=c\\\eta=0\end{cases}$	$\begin{aligned}x&=0.50\,a\\y&=0.125\,a\end{aligned}$	44438 43760	44438 43760	$\begin{cases}\xi=0.50\,a\\\eta=0.125\,a\end{cases}$	$\begin{aligned}x&=c\\y&=0\end{aligned}$	$\begin{aligned}0.02\,a\\0.06\,a\end{aligned}$	$\begin{cases}\xi=0.25\,a\\\eta=d\end{cases}$	$\begin{aligned}x&=0.50\,a\\y&=0.25\,b\end{aligned}$	−99871 −286875	−99871 −286875	$\begin{cases}\xi=0.25\,a\\\eta=d\end{cases}$	$\begin{aligned}x&=0.50\,a\\y&=0.25\,b\end{aligned}$
$\begin{cases}\xi=0.50\,a\\\eta=0\end{cases}$	$\begin{aligned}x&=0.25\,a\\y&=0\end{aligned}$	−11565 −11299	−11565 −11299	$\begin{cases}\xi=0.25\,a\\\eta=0\end{cases}$	$\begin{aligned}x&=0.50\,a\\y&=0\end{aligned}$	$\begin{aligned}0.02\,a\\0.06\,a\end{aligned}$	$\begin{cases}\xi=0.50\,a\\\eta=0.75\,b\end{cases}$	$\begin{aligned}x&=0.25\,a\\y&=0.75\,b\end{aligned}$	1646599 1632420	1646599 1632420	$\begin{cases}\xi=0.25\,a\\\eta=0.75\,b\end{cases}$	$\begin{aligned}x&=0.25\,a\\y&=0.50\,b\end{aligned}$
$\begin{cases}\xi=0.50\,a\\\eta=0.25\,a\end{cases}$	$\begin{aligned}x&=0.25\,a\\y&=0\end{aligned}$	−6685 −6579	−6685 −6579	$\begin{cases}\xi=0.25\,a\\\eta=0\end{cases}$	$\begin{aligned}x&=0.50\,a\\y&=0.25\,a\end{aligned}$	$\begin{aligned}0.02\,a\\0.06\,a\end{aligned}$	$\begin{cases}\xi=0.25\,a\\\eta=0.50\,b\end{cases}$	$\begin{aligned}x&=0.25\,a\\y&=0.75\,b\end{aligned}$	−597635 −589598	−597635 −589598	$\begin{cases}\xi=0.25\,a\\\eta=0.50\,b\end{cases}$	$\begin{aligned}x&=0.50\,a\\y&=0.75\,b\end{aligned}$
$\begin{cases}\xi=0.75\,a\\\eta=0.25\,a\end{cases}$	$\begin{aligned}x&=0.25\,a\\y&=0\end{aligned}$	−19798 −19698	−19798 −19698	$\begin{cases}\xi=0.25\,a\\\eta=0\end{cases}$	$\begin{aligned}x&=0.75\,a\\y&=0.25\,a\end{aligned}$	$\begin{aligned}0.02\,a\\0.06\,a\end{aligned}$	$\begin{cases}\xi=0.50\,a\\\eta=0.50\,b\end{cases}$	$\begin{aligned}x&=c\\y&=0.50\,b\end{aligned}$	−105750 −309845	−105750 −309845	$\begin{cases}\xi=0.25\,a\\\eta=0.50\,b\end{cases}$	$\begin{aligned}x&=0.50\,a\\y&=0.50\,b\end{aligned}$
$\begin{cases}\xi=0.50\,a\\\eta=0.125\,a\end{cases}$	$\begin{aligned}x&=c\\y&=0\end{aligned}$	−986 −2887	−986 −2887	$\begin{cases}\xi=c\\\eta=0\end{cases}$	$\begin{aligned}x&=0.50\,a\\y&=0.125\,a\end{aligned}$	$\begin{aligned}0.02\,a\\0.06\,a\end{aligned}$	$\begin{cases}\xi=0.50\,a\\\eta=0.50\,b\end{cases}$	$\begin{aligned}x&=0.50\,a\\y&=0.50\,b\end{aligned}$	2953001 2916469	2953001 2916469	$\begin{cases}\xi=0.50\,a\\\eta=0.50\,b\end{cases}$	$\begin{aligned}x&=c\\y&=0.50\,b\end{aligned}$

* See Eqns. (3.6) and (3.7).

3 Behaviour of plates subjected to a surface couple

3.4.3 Checking the accuracy of computations by reference to the results obtained by other authors

No data exist concerning the elastic displacements and stresses produced in finite boundary plates by a surface couple M. That is why, for this type of load, only the values obtained in the case of the infinitely long plate can be referred to the data given elsewhere (Tables 1.7, 1.8, 3.24 and 3.25). Several minor differences exist between the values established by Stiglat [60] and those given herein (Table 3.24), but this small disagreement derives from the less accuracy conditions used by Stiglat (Sub-sec. 1.2.4.3).

Table 3.24

Infinitely long plate subjected to a surface couple $\vec{M} \parallel Oy$

Comparative values of rotations $\varphi_x / \dfrac{M}{D}$ after Stiglat, Scholz and Negruțiu

$x = \xi$

ξ	y	$c = d$	Stiglat *	Scholz **	Negruțiu	$N = m$
c	0.00	0.02a	0.461	0.4561	0.46109	130
	0.50a	0.02a	—	0.0082	0.00864	20
0.50a	0.00	0.01a	—	0.2536	0.28878	200
		0.02a	0.234	0.2306	0.23409	200
		0.04a	0.179	0.1818	0.17914	130
		0.06a	0.147	0.1498	0.14702	170
		0.08a	—	0.1272	0.12437	170
0.50a	1.50a	0.02a	—	0.000067	0.000075	20

* $N = 160$.
** $N = 60$.

Greater differences exist between the author's data and those given by Scholz [66], who restricts the number of summed terms to $N = 60$ (Sec. 1.2.5.1). Unlike the values found by Scholz for the rotations φ_x, which are satisfactory, those for the bending moments and, especially for max k_M^x, are not. The lack of precision of Scholz' computations leads to differences of 22 per cent — 32 percent (Table 3.25).

As the distribution of elastic displacements and stresses generated in finite boundary plates by a surface couple M has

3.4 Accuracy of numerical computations

not been studied in the literature, the numerical data obtained herein were verified by virtue of the *similar behaviour* of long rectangular plates ($2a \leqslant b < 3a$) and infinitely long plates. A com-

Table 3.25

Infinitely long plate subjected to a surface couple $\vec{M} \parallel Oy$

Comparative values of sectional bending moments M_x and M_y, after Stiglat, Scholz and Negruțiu

$y = 0$; $\mu = 0$

$c = d$	ξ	x	Stiglat *	Scholz **	Negruțiu	$N = m$
			Coefficients k_M^x			
0.02a	c	c $2c$ $0.50a$	— 6.68 —	2.425 4.575 0.243	2.3130 6.6718 0.2513	1080 1580 280
0.02a	0.50a	$\xi + c$	5.25	4.088	5.2384	1350
0.04a	c 0.25a 0.50a	$\xi + c$ $\xi + c$ $\xi + c$	3.33 2.75 2.63	— — —	3.343 2.742 2.625	2830 1430 1630
			Coefficients k_M^y			
0.02a	c	c $2c$ $0.50a$	— 2.82 —	1.750 2.650 0.248	1.5984 2.8129 0.2502	130 130 110
0.02a	0.50a	$\xi + c$	1.60	1.575	1.5957	140
0.04a	c 0.25a 0.50a	$\xi + c$ $\xi + c$ $\xi + c$	1.40 0.91 0.79	— — —	1.399 0.908 0.790	80 80 80

* $N = 160$.
** $N = 60$.

parison between the values found for these two categories of plates highlights their identical behaviour under any load M, P or q (Secs. 3.2 and 3.3; Chaps. 4 and 5).

4

Elastic plates as structural members

The static and dynamic analysis of mixed systems consisting of bars and plates can be approached by any of the general methods used in the analysis of hyperstatic reticular structure. The computation algorithms and the systematized treatment presented in Chap. 2 prompted the use of the general force method (Sub-secs. 4.1, 5.1 and Chap. 6). This chapter also gives several preliminary elements of the structural analysis of slab structures through the general displacement method (Secs. 4.2 and 5.2).

4.1 Slab structures subjected to transverse loads
Effect of M- and P-type loads

Owing to the geometrical and mechanical symmetry of most slab structures, the analysis can be substantially simplified by grouping the statically indeterminate unknowns. The basic system is chosen here so as to allow the identification of three types of generalized interacting forces (Fig. 4.1) at the level of the plate-column nodes (Sub-sec. 5.1.1):

X_i — force-reactions P normal to the mid-surface of the plate;

X_j, X_l — couple-reactions, the moment vector $\vec{M}$ of which is in the plane of the plate (the components of this vector develop along any Ox and Oy directions defining the plane);

X_k — force-reactions in the plane xOy of the plate.

The last type of interacting force, which appears in the compatibility equations, acts on the slab as axial forces in one or both directions of the plane. The corresponding stresses σ_x and σ_y are relatively small if referred to the normal stresses caused by loads

4 Elastic plates as structural members

acting transverse to the slabs. Hence, in the general case and within the approximations derived from the basic hypotheses (Sub-sec. 2.1.1), the slab deformations due to the force reactions of type X_k are negligible. This simplified treatment resembles the approxima-

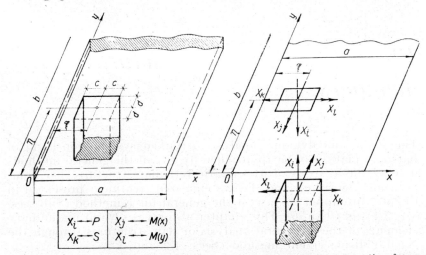

Figure 4.1 Plates as structural members. Types of generalized interacting forces as defined by use of general force method.

tions obtained in the analysis of the reticular system by disregarding the longitudinal deformations of bars. Hence, in order that the states of stress and deformation be thoroughly defined, only the first two types of generalized interacting forces must be introduced in the analysis of plates considered as structural members: surface couples M and locally distributed forces P. Since loads of type M and P occur frequently as given loads, too, the effect of these sets of transverse loads should be duly investigated.

★

The study of the sets of surface couples will rely on the results obtained from the analysis of plates subjected to only one surface couple M (Chap. 3). By virtue of the superposition of effects, we can model two types of loads which are frequently used in design practice: symmetrical and antisymmetrical sets of surface couples. Thus, we can obtain a picture of the distribution of the elastic displacements and moments generated throughout the plate in the case when the load is induced by sets of surface couples M.

The sets of locally distributed forces P are examined comparatively for the same value of the calculation parameters, namely: the sizes $2c \times 2d$ of the area subjected to the force P (the couple

4.1 Slab structures subjected to transverse loads

M, respectively), the location (ξ, η) of the centroid of this area in the plate plane and the value of Poisson's ratio μ.

If the associated surface couples M and locally distributed forces P are assumed to be sets of unknowns, then the slab structures can be treated in the analysis as a multiple statically indeterminate system (Sec. 5.1). The elastic displacements δ_{ij} of the slab, corresponding to these generalized interacting forces (the coefficients of the associated unknowns) will be calculated with the aid of the dimensionless coefficients K_w^M and K_w^P in the case of transverse displacements and using K_φ^M and K_φ^P in the case of rotations. Similarly, the unit flexural and torsional moments generated in the plate will be computed using the coefficients K_M^M and K_M^P. *

4.1.1 Symmetrical and antisymmetrical sets of surface couples M

The study of the effects of sets of surface couples on plates relies on the following conclusions concerning the effect of only one surface couple M (Secs. 3.2 and 3.3):

a. The size $2c \times 2d$ of the area over which the couple is applied bears on the magnitude of the elastic displacements, i.e. the values of the coefficients k_w, and particularly those of the coefficients k_φ^x and k_φ^y, become smaller as this area becomes larger. This observation holds in the case of long plates and those with finite boundaries.

That the extreme — maximum and minimum — values of the bending moments M_x and M_y decrease in proportion to the increase of the parameters c and d is even more evident;

b. The influence of the value of Poisson's ratio is more moderate in the case of elastic displacements, which decrease with the increase of μ. The values of the bending moments M_x and M_y tend to grow with the increase of μ. The absolute value of the coefficients max. (min.) k_M^y indicate a definite increase with the increasing value of μ, if referred to the coefficients max. (min.) k_M^x;

c. The similar behaviour of the infinitely long plate and finite boundary plates ($a \leqslant b \leqslant 3a$) under the load M is one of the main conclusions of the comparative study.

By virtue of the above conclusions, a simplified approach will be described in the sequel. This approach, in which a smaller number of alternatives is examined, leads to a reduced amount

---

* We must discriminate between the notations K_w, K_φ and K_M, introduced for the dimensionless coefficients corresponding to sets of unknowns and the dimensionless coefficients k_w, k_φ and k_M corresponding to the load induced by only one surface couple M or only one locally distributed force P (Sec. 2.3).

4 *Elastic plates as structural members*

of processed data and a substantial time saving for work on computers.

Thus, for instance, if the intermediate value of $c = d = 0.04a$ is retained, then we can relinquish the alternatives $c = d = 0.02a$ and $c = d = 0.06\,a$. For the relative values b/a of the in-plane dimensions of the plate, account was taken of the extreme cases alone, namely the infinitely long plate (Sub-sec. 2.3.2.1) and the square plate (Sub-sec. 2.3.2.3). For Poisson's ratio only the extreme values $\mu = 0$ and $\mu = 1/3$ were considered. However, the computation programmes developed allow the consideration of any ordinary value for any of the design parameters [73], [75].

4.1.1.1 *The infinitely long plate.* The effect of the associated surface couples on infinitely long plates was studied in two stages: 1 — an infinite row of associated surface couples, parallel to the line of supports (Fig. 4.2 and 4.3); 2 — alternatives of double, triple or more rows of associated symmetrical or antisymmetrical surface couples, obtained by superposing the effects of two, three or more simple rows of couples (Figs. 4.4—4.9). Keeping the abscissa ($\xi = 0.25\,a$) of the simple row of couples M constant, the value of equidistance e between the locations of the surface couples (i.e. the equidistance between the columns of the slab structure) bears heavily on the elastic displacements, and only slightly on the value of the bending moments M_x (Fig. 4.2). In this case, we can also observe the *high stiffness of the square plate* $b = a$ (the curves A) as against the stiffness of the infinitely long plate (curves B and C). The extreme values of the bending moments are less affected by the valuue of parameter b/a.

The procedure used to study the superposition of effects enables us to establish precisely the number of couples M included in the one-row set in the case of the infinitely long plate. We have already shown in the foregoing chapter that the effect induced in plates by a surface couple has the characteristic of a local perturbation (Sec. 3.3). Since the values of k_w, and particularly those of k_M, decrease steeply with the increasing distance from the loaded area, for a distance exceeding $1.5a - 2a$, the influence of the couples can be neglected. This is why, admitting that $y_{max} = \pm 2a$ in the estimation of the effects of the three compared sets, we had to consider only 9 couples in the row where $e = 0.50a$ (curves C), and only 5 couples in the row where $e = 0.75a$ (curves B) (see also the note on p. 225).Obviously, in the limiting case of the square plate $b = a$, for $e = 0.50\,a$, the simple row includes only two couples M (curves A).

In order to have a faithful picture of the effect of grouping the surface couples, the curves A_1 for the square plate and the

4.1 Slab structures subjected to transverse loads

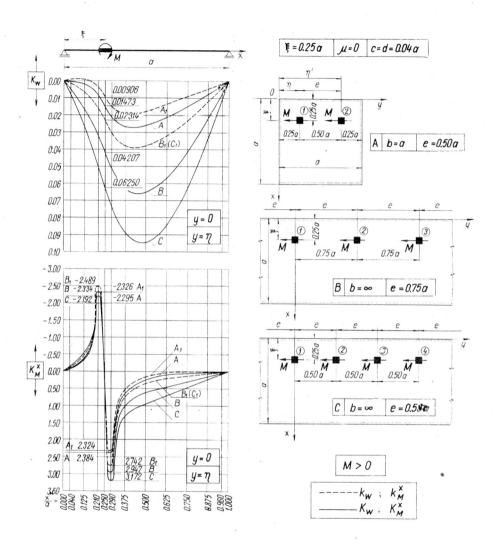

Figure 4.2 Plates as structural members. Sets of surface couples M associated in simple row. Influence of parameter b/a and of equidistance e between location point of couples in row.

4 *Elastic plates as structural members*

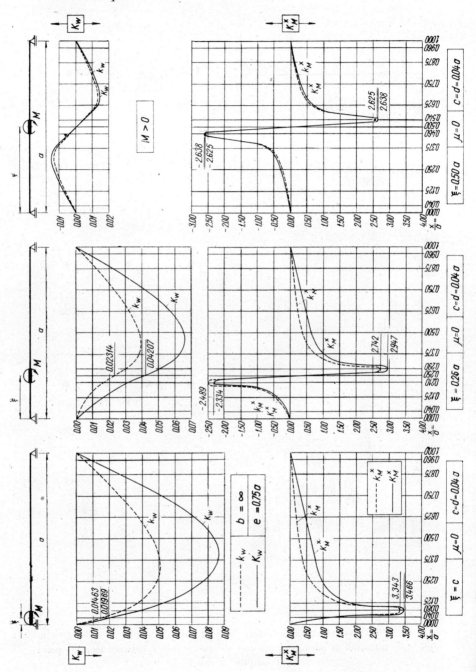

208

4.1 Slab structures subjected to transverse loads

Figure 4.3 Infinitely long plate as structural member. Sets of surface couples M associated in simple row. Effect of grouping pattern and influence of position parameter ξ of row of couples.

curves $B_1 = C_1$ for the infinitely long plate were also plotted in the diagrams of K_w and K_M^x; these curves correspond to load induced in the three types of plates under study by only one surface couple M (Fig. 4.2). If the equidistance e is large enough (for instance $e = 0.75a$), small differences appear between the values of k_w and k_M^x, corresponding to the effect of only one couple M on an infinitely long plate (curves B_1) and the respective values of K_w and K_M^x, corresponding to the set of couples associated in simple row (curves B).

The results of the comparative study allow us to state that for the equidistance $e > 0.75\,a$ (for flexural moments) and the equidistance $e > a$ (for elastic displacements) respectively, the effect of grouping surface couples may be neglected in practice. Thus, in the first approximation step, it is sufficient to consider only the effect of a surface couple M. This means that much of the calculations (elastic displacements, bending moments, etc.) required by the elastic analysis of a large category of slab structures can be saved. *

In order to have a clear idea of the range in which the above conclusions are valid, the effect of the location ξ of the couple row was also examined. Thus, account was taken of three characteristic values of this parameter ($\xi = c$, $\xi = 0.25a$ and $\xi = 0.50a$), the value of the equidistance $e = 0.75a$ being constant (Fig. 4.3).

The change of the location ξ of the simple row of surface couples M shows that the grouping of couples bears heavily on the magnitude of the elastic displacements in the range $c \leqslant \xi \leqslant 0.25a$ (diagrams K_w). Again, the diagrams K_M^x show that the value of the bending moments is little influenced by the effect of grouping couples associated in a simple row. For the central location $\xi = 0.50\,a$ of the row of couples M, the effect of the set is almost negligible for both elastic displacements (diagram K_w) and bending moments (diagram K_M^x). For all these three locations, the effect of a simple row of couples (solid lines) was compared to that of a couple M applied on an infinitely long plate (dashed lines).

---

* For a rigorous comparative study, throughout the book, the set of surface couples associated in a simple row was investigated taking into account the superposition of the effects of surface couples located at a distance up to $y = \pm 2a$ from the origin (see also Sub-sec. 4.1.2.1).

4 Elastic plates as structural members

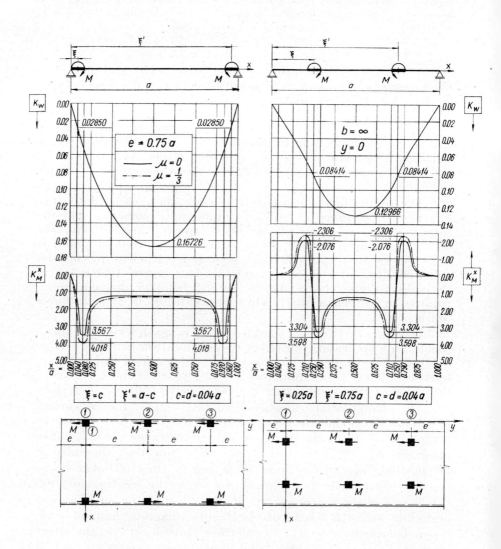

Figure 4.4 Infinitely long plate. Sets of double symmetrical row of surface couples M. Deflected mid-surface w and diagrams M_x. Influence of parameters ξ and μ.

4.1 Slab structures subjected to transverse loads

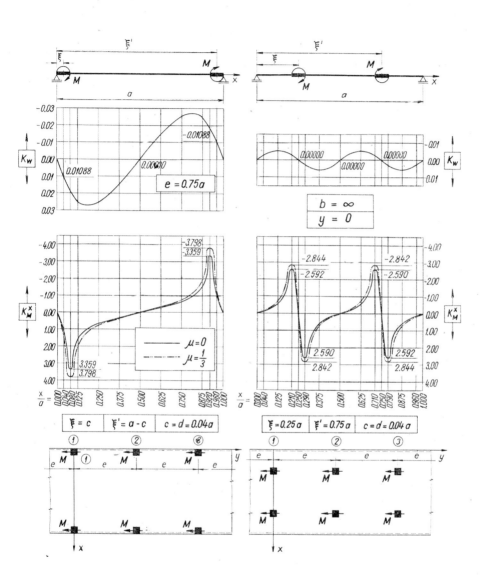

Figure 4.5 Infinitely long plate. Sets of double antisymmetrical rows of surface couples M. Deflected mid-surface w and diagrams M_x. Influence of parameters ξ and μ.

4 Elastic plates as structural members

In the second step of the approach to sets of couples, by superposing the effects of two or more couple rows, we obtained computation schemes similar to those effective in the design practice of slab structures (Figs. 4.4—4.9). That the *physical phenomenon* is correctly modelled is illustrated by four simple schemes: sets of double symmetrical (Fig. 4.4) and antisymmetrical (Fig. 4.5) rows of surface couples, each with two alternatives (edge rows and rows at the quarter of the span).

The similar behaviour of plates and the simply supported beam, subjected to concentrated symmetrical and antisymmetrical couples, is relevant. Thus, the inflection points of the deflected midsurface of the plate correspond to the points of zero bending moment, a flattened portion exists between the location ξ and ξ' of the symmetrical couples, etc.

We can notice the influence of the location ξ of the couple rows on the elastic displacements and on the extreme values of the bending moments, as well as the effect produced by introducing Poisson's ratio in these values (Table 4.1). The characteristics of a local perturbation induced in the plates by the surface couples is also apparent in the case of associated couples, though slightly attenuated.

Table 4.1

Set of surface couples M associated in a double row, acting on the infinitely long plate

$c = d = 0.04\,a\,;\ e = 0.75\,a\,;\ y = \lambda e$

Highest and lowest values of K_M^x in the case of symmetrical and antisymmetrical sets of couples

ξ	Symmetrical couples		Antisymmetrical couples	
	$\mu = 0$	$\mu = \dfrac{1}{3}$	$\mu = 0$	$\mu = \dfrac{1}{3}$
Values max $K_M^x(x = \xi + c)$				
c	3.567	4.018	3.359	3.798
$0.25a$	3.304	3.598	2.590 *	2.842 *
Values min $K_M^x(x = \xi - c)$				
c	0.000	0.000	0.000	0.000
$0.25a$	−2.076	−2.306	−2.592 *	−2.844 *

* In the case of sets of antisymmetrical couples, we can check the precision of the values determined for the dimensionless coefficients K_M^x using the superposition of effects. The computation error for the set under study is below 0.08 per cent. The values K_w were calculated with better precision.

4.1 Slab structures subjected to transverse loads

In the study of the effect of antisymmetrical sets of couples on the infinitely long plate, account was taken of more complex schemes consisting of three rows of surface couples (Fig. 4.6): one row along the axis of the plate ($\xi' = 0.50\ a$) and two edge rows ($\xi = c;\ \xi'' = a - c$) or two rows at the quarter of the span ($\xi = 0.25\ a;\ \xi' = 0.75\ a$). The first of these two sets shows that the central row of couples M barely affects the general shape of the deflected mid-surface of the plate, which is similar to the diagram K_w, corresponding to the double row of antisymmetrical edge couples. In the case of the triple row of couples, the design values of the elastic displacements are of the same order of magnitude, but are however smaller (diagrams $\xi = c$). In the case of the second set of triple rows of antisymmetrical couples, both the general shape of the deflected mid-surface and the design values of the coefficients K_w are substantially different from those obtained in the case of a double row of antisymmetrical couples located at the quarter of the span (Fig. 4.6, diagrams $\xi = 0.25\ a$).

By superposing the effects of sets of double symmetrical and antisymmetrical rows of edge couples ($\xi = c$) and those located at the quarter of the span ($\xi = 0.25\ a$) (Fig. 4.4 and 4.5), we obtain three complex schemes shaping the effect of the reaction-couples in the case of slab structures with four column rows (Fig. 4.7).

The comparative study of the effect of symmetrical and antisymmetrical sets of couples M leads to original conclusions that are effective in the design practice. Thus, for both sets of three rows of antisymmetrical couples, the diagrams K_w show that the transverse displacements w are not zero at the points ξ and ξ'' where the couples M pertaining to the edge rows are applied (Fig. 4.6). An examination of diagrams K_w for the double rows of antisymmetrical edge couples (Fig. 4.5, $\xi = c$) and the complex set of four couple rows (Fig. 4.7B) leads to the same conclusion. Consequently, these examples allow us to state that the assumption of zero transverse displacement w along the axis of the columns is not valid for any position ξ of the row of antisymmetrical couples.

The above conclusion is of immediate practical interest in the static analysis of slab structures (Sec. 5.1). In the case of horizontal loads, we must write two types of compatibility equations for each slab-column node: to the conditions of zero relative rotation we associate the conditions of zero relative displacement along the column axes.

★

For a complete image of the effect of sets of surface couples on the infinitely long plate, we must show the distribution of the elastic displacements and flexural moments along the

4 Elastic plates as structural members

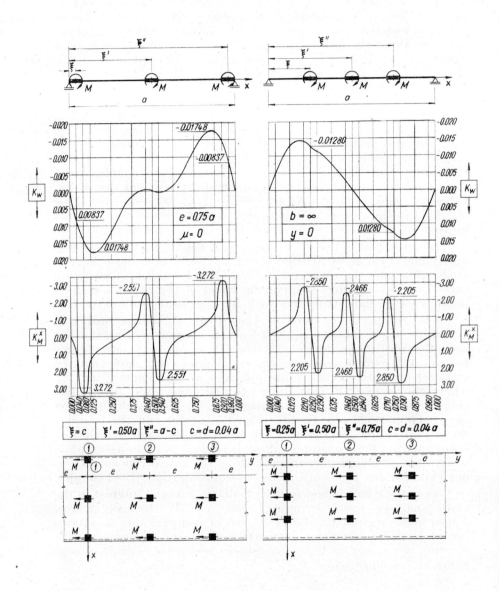

Figure 4.6 Infinitely long plate. Sets of three antisymmetrical rows of surface couples M. Influence of position parameters ξ and ξ'' of lateral rows.

4.1 Slab structures subjected to tranverse loads

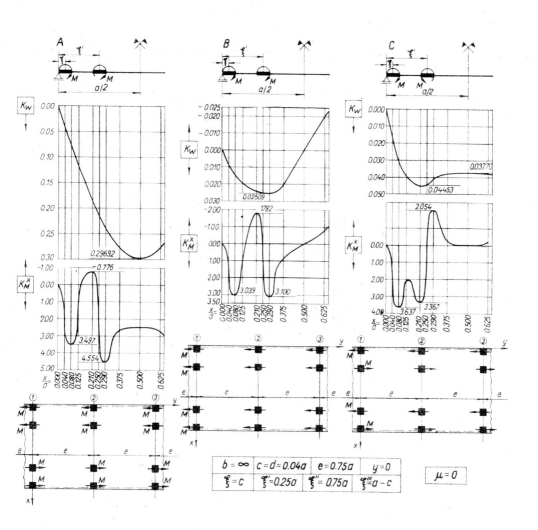

Figure 4.7 Infinitely long plate. Complex sets of four symmetrical and antisymmetrical rows of surface couples M. Deflected mid-surface w and diagrams for moments M_x.

4 *Elastic plates as structural members*

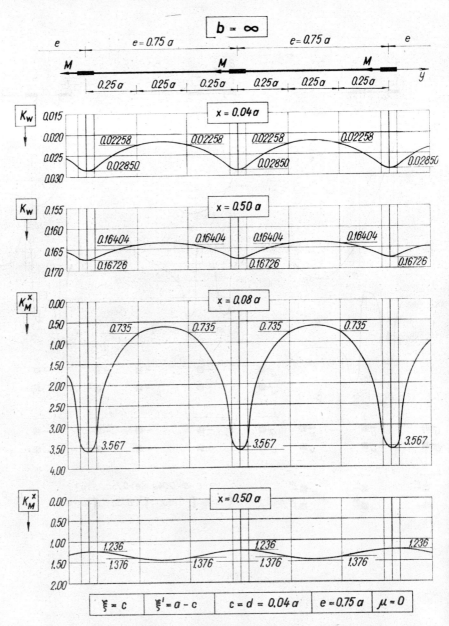

Figure 4.8 Infinitely long plate. Set of symmetrical edge-rows of surface couples M ($\xi = c$; $\xi' = a - c$) associated in two rows. Deflected mid-surface w and diagrams for M_x along the plate length.

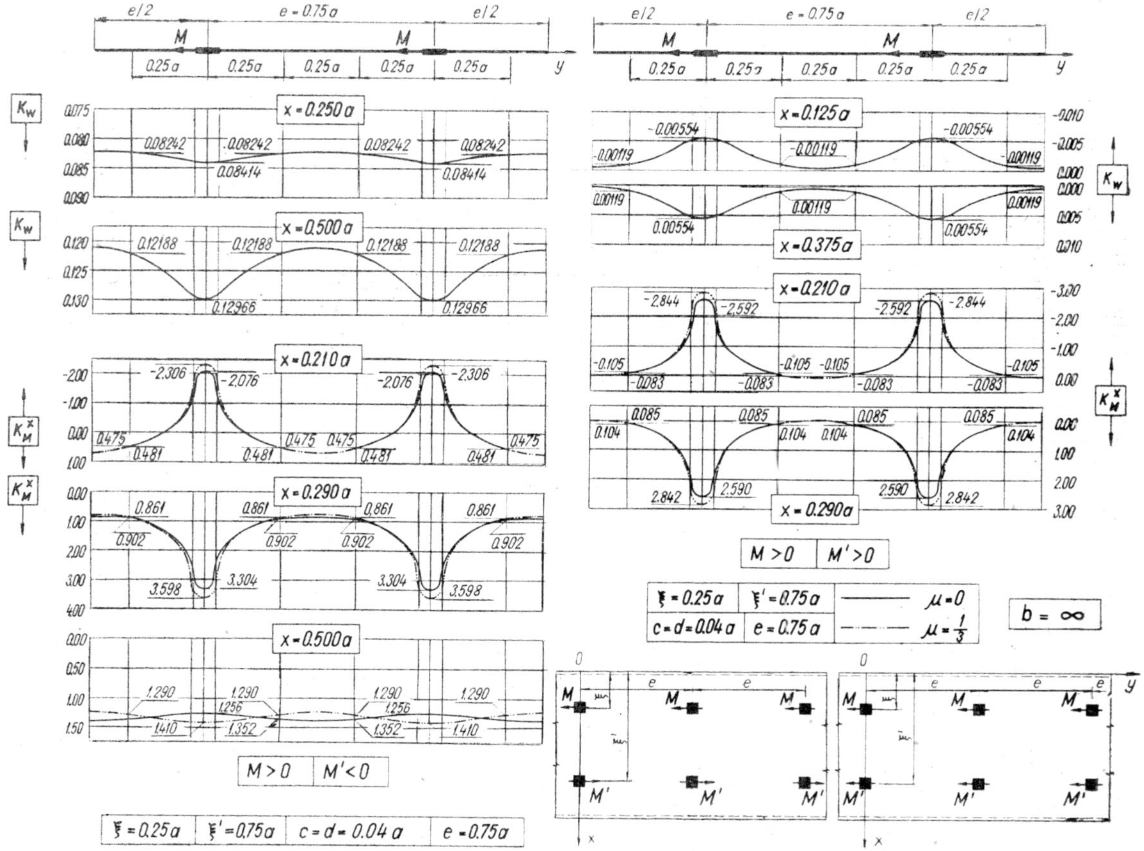

Figure 4.9 Infinitely long plate. Sets of symmetrical and antisymmetrical surface couples M associated in two rows at quarter of the span ($\xi = 0.25a$; $\xi' = 0.75a$). Deflected mid-surface w and diagrams for M_x along the plate length.

4 *Elastic plates as structural members*

length of the plate. Diagrams were drawn for three different sets of couples M : a double row of symmetrical edge couples ($\xi = c$; $\xi' = a - c$) (Fig. 4.8); a double row of symmetrical and antisymmetrical couples at the quarter of the span ($\xi = 0.25a$; $\xi' = 0.75\ a$) The value $e = 0.75a$ of the equidistance was preserved (Fig. 4.9).

After having noted the influence of the parameters ξ and μ, we observe that all the loading alternatives under examination leads to the same conclusions, namely :

— the deflected mid-surface remains practically level along any plane $x = $ constant (the diagrams K_w were plotted to an intentionally exaggerated scale);

— along the axis of the plate ($x = 0.50\ a$), the values of the bending moments are almost constant and are practically independent of the location ξ of the rows of couples associated in symmetrical sets (Table 4.2);

Table 4.2

Sets of symmetrical surface couples M associated in a double row, acting on the infinitely long plate

$c = d = 0.04\ a$; $e = 0.75\ a$; $x = 0.50\ a$; $\mu = 0$

Comparative values of dimensionless coefficients K_M^x in plate axis

Type of set of couples	ξ	ξ'	$y = \lambda e$	$y = (3\lambda+1)e/3$
Edge rows	c	$a-c$	1.236	1.376
Rows at the quarter of span	$0.25a$	$0.75a$	1.256	1.352

— in the particular planes $x = \xi \pm c$, large variations appear between the values of M_x calculated in the area subjected to the surface couples, and those calculated at the free span of each panel. (Thus, the characteristics of a local perturbation of the load are also apparent along the plate length both in the case of symmetrical and antisymmetrical sets of associated couples);

— for any sets of antisymmetrical surface couples, the transverse displacements w and the sectional flexural moments M_x are equal to zero along the axis $x = 0.50\ a$.

We have seen in previous cases that Poisson's ratio bears little on the values of the flexural moments, and so this influence was noted only for the extreme values of K_M^x (Fig. 4.9). By virtue of these conclusions, the analysis of slab structures can be simplified and a substantially smaller amount of computations would be required.

4.1 Slab structures subjected to transverse loads

4.1.1.2 *The square plate*. In the foregoing sub-section, we have approached the effect of the load induced in the infinitely long plate by sets of surface couples M. The results obtained regarding the magnitude and distribution of elastic displacements w and flexural moments M_x, will be compared now with those derived in the case of the square plate ($b = a$). This comparative study deserves special attention as the mathematical models of the mid-surface deflected by a surface couple are structurally different: simple series of hyperbolic functions (Sec. 2.2.1) and double series of trigonometric functions (Sec. 2.2.3).

Under the action of some sets of four symmetrical (Figs. 4.10, 4.12) and antisymmetrical (Fig. 4.11, 4.13) surface couples, applied at the edge and at the quarter of the span a and $b = a$, the behaviour of the square plate is identical with that of the infinitely long plate under similar loads: the shape of the deflected mid-surface w, as well as the shape of the diagram of the bending moments M_x are almost identical (Figs. 4.4 and 4.5). As distinct from the case of the infinitely long plate, where the diagrams K_w and K_M^x were drawn only in the axes of the columns (transverse planes $y = \lambda e$), in the case of the square plate these curves were drawn for two characteristic planes: in the axes of the columns ($y = 0.25a$) and at mid-span ($y = 0.50\ a$).

A comparative examination of these two limit cases indicates that the stiffness of the square plate is significantly higher (2.5 times higher, in the examples under study) than that of the infinitely long plate. The higher deformability of the infinitely long plate subjected to sets of symmetrical couples was also observed in the previous study of the set of simple rows (Fig. 4.2).

The similar behaviour of the square plate and the infinitely long plate subjected to the sets of associated couples under consideration is more clear if we compare the extreme values of the flexural moments (Table 4.3). That the values for max. (min.) K_M^x, calculated for the two limit values of the parameter b/a are of the same order of magnitude, is not disentangled from pratice. A comparison of these extreme values shows that the quantitative differences are more moderate in the case of antisymmetrical sets. A comparison of the mid-surface deflected by antisymmetrical sets of surface couples shows that throughout the range $a \leqslant b \leqslant \infty$ the same values are obtained for the displacements w (coefficients K_w).

The diagrams K_M^y, drawn also for the characteristic transverse planes $y = 0.25\ a$ ($y = 0.75\ a$) and $y = 0.50\ a$, show that the variation of the bending moments M_y is more moderate than that of the moments M_x. However, in compliance with expectations, the influence of Poisson's ratio is more pronounced in the diagrams K_M^y (Figs. 4.12–4.13).

4 *Elastic plates as structural members*

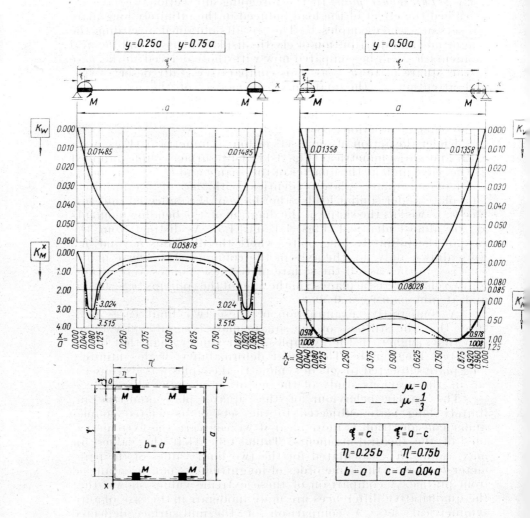

Figure 4.10 Square plate as structural member. Symmetrical set of four edge surface couples M. Deflected mid-surface w and diagrams M_x.

4.1 Slab structures subjected to transverse loads

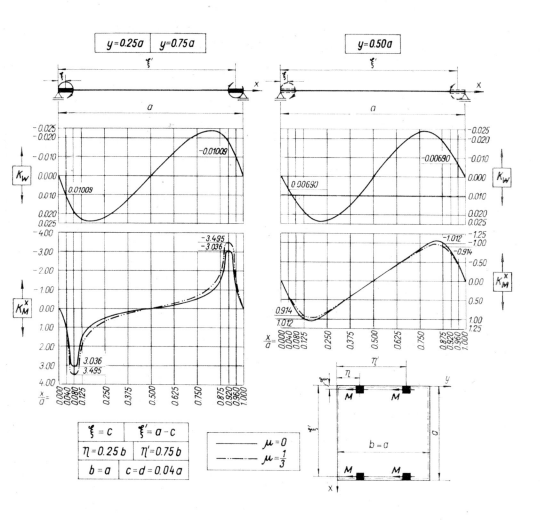

Figure 4.11 Square plate as structural member. Antisymmetrical set of four edge surface couples M. Deflected mid-surface w and diagrams M_x.

4 *Elastic plates as structural members*

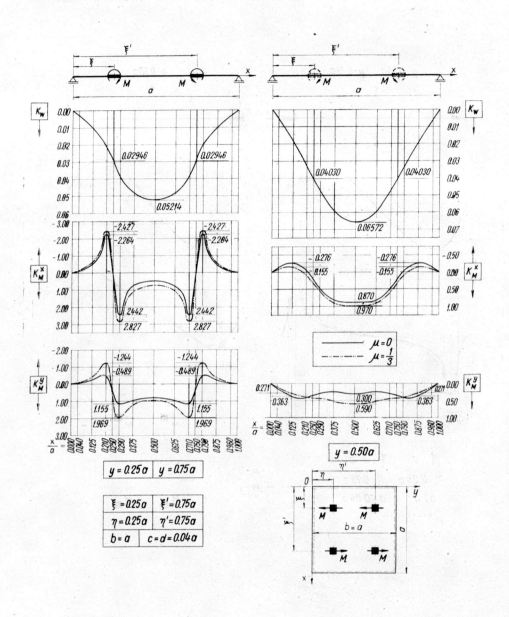

Figure 4.12 Square plate. Symmetrical set of four surface couples M applied at quarter of the span. Deflected mid-surface w and diagrams M_x and M_y.

4.1 Slab structures subjected to transverse loads

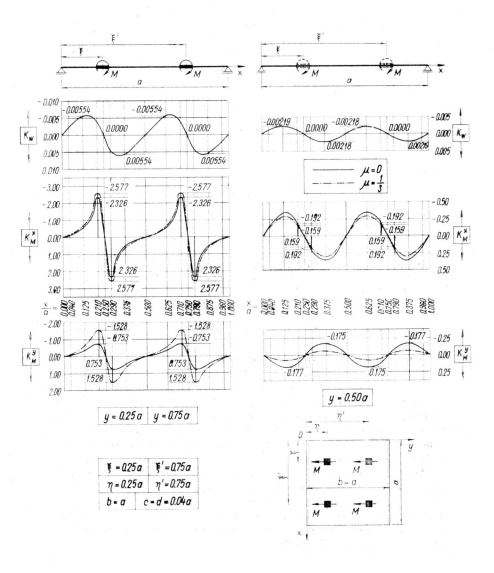

Figure 4.13 Square plate. Antisymmetrical set of four surface couples M applied at quarter of the span. Deflected mid-surface w and diagrams M_x and M_y.

Table 4.3

Sets of surface couples associated in a double row, acting on infinitely long and square plates

$c = d = 0.04\,a$; $\xi = 0.25\,a$; $\xi' = 0.75\,a$; $e = 0.75\,a$; $y = \eta\,(y = \lambda e)$

Comparative values of coefficients max (min) K_M^x

x/a	$b = \infty$ *		$b = a$ **		Observations
	$\mu = 0$	$\mu = 1/3$	$\mu = 0$	$\mu = 1/3$	
	Symmetrical couples				*
0.290	3.304	3.598	2.442	2.827	An infinite double row of
0.210	−2.076	−2.306	−2.264	−2.427	couples M (Figs. 4.4 and 4.5)
	Antisymmetrical couples				**
0.290	2.590	2.842	2.326	2.577	Four couples M (Figs. 4.12
0.210	−2.592	−2.844	−2.326	−2.577	and 4.13)

Note. An eventual verification of the precision with which computations were performed for the antisymmetrical sets of couples was suggested in Sub-sec. 4.1.1.1 — Table 4.1. In the case of the infinitely long plate, the computation error is below 0.08%, whereas in the case of the square plate, the computation error is zero.

Finally, we observe that under antisymmetrical sets of couples applied at the quarter of the span ($\xi = 0.25\,a$ and $\xi' = 0.75a$) of the infinitely long and square plates, there exist three longitudinal planes ($x = 0.25\,a$; $x = 0.50\,a$; $x = 0.75\,a$) along which the transverse displacements w and the bending moments M_x and M_y are zero throughout the longer side b of the plate. This observation is very important, especially for certain slab structures which can be assimilated with similar schemes (for instance those subjected to seismic loads), where resort can be made to the general displacement methods (Sec. 5.2).

All the observations regarding the behaviour of slabs subjected to sets of symmetrical and antisymmetrical surface couples M lead to the conclusion that the elastic displacements and the stresses are distributed throughout the plate in compliance with expectations. Hence, the algorithms suggested represent a correct mathematical model.

4.1.2 Symmetrical and antisymmetrical sets of locally distributed forces P

The solutions of the fundamental equation of plates subjected to a locally distributed load P are known in the classical literature. These solutions cover both the rectangular plate simply supported

4.1 Slab structures subjected to transverse loads

along the boundary (Sub-sec. 2.2.3.1) and the infinitely long plate (Sub-sec. 2.2.1.1). However, the mathematical models of the mid-surface of these two categories of plates deflected under the load due to a surface couple M are not known. (In this respect see Sub-secs. 2.2.1.2, 2.2.1.3 and 2.2.3.2, 2.2.3.3, respectively).

This is why the numerical experiment which is the object of Chapter 3 deals only with the analysis of the behaviour of plates subjected to only one surface couple M. However, for the elastic analysis of slab structures — in which the plates are treated as structural members — through the general force method, it is necessary to study the behaviour of plates subjected to sets of locally distributed forces P. As in the case of sets of surface couples, the sets of forces P can be given loads or inter-acting forces [73], [75].

4.1.2.1 *The infinitely long plate.* The effect of sets of locally distributed forces P was studied in two stages : 1 — a set in a simple row, parallel to the line of supports (Figs. 4.14 and 4.15), and 2 — more complex sets, in double, triple or more rows of symmetrical and antisymmetrical locally distributed forces P (Figs. 4.16 and 4.17, respectively). The parameters differentiating the alternatives compared are the ratio b/a between the two sizes of the plate and the equal distance e between the points subjected to the forces P of a row (i.e. the equal span between the columns of the slab structure).

For a given location of the simple row of locally distributed forces P — for instance, $\xi = 0.25a$ (Fig. 4.14) — the effect of the grouping is paramount on the values of the transverse displacements w and on those of the sectional flexural moments M_x *.

The influence of the parameters e and ξ was comparatively studied for a simple row of locally distributed forces P (Figs. 4.14 and 4.15) taking as reference values the diagrams k_w and k_M^x, respectively (dashed curves A_1, $B_1 = C_1$), which show the effect of only one force P on the plate under study (see also Table 4.5).

Changing the location ξ of the simple row of locally distributed forces P on the infinitely long plate, allows us to establish several analogies with a similar set of surface couples M (Fig. 4.15). However, the effect of the set of forces P is different (Table 4.4).

---

* The actual number of locally distributed forces P associated in a simple row, acting on the infinitely long plate, is limited to five ($e = 0.75a$) and nine forces P ($e = 0.50a$), respectively. This is due to the rapid decrease of the two values of k_w and especially of k_M^x with the distance y from the point where the force is applied. The effect of superposition was limited in all our calculations to a maximum distance $y = \pm 2a$ as in the case of the simple row of surface couples M (see also Sub-sec. 4.1.1.1).

4 Elastic plates as structural members

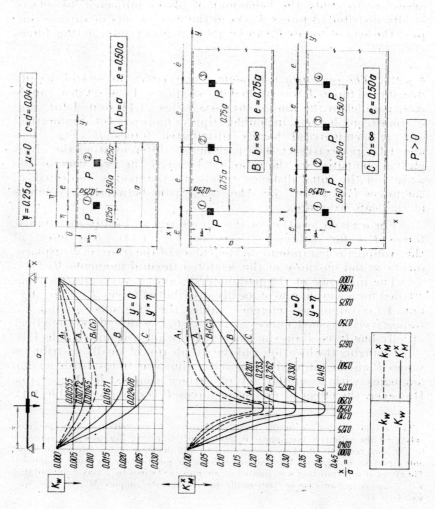

Figure 4.14 Plates as structural members. Sets of locally distributed forces P associated in simple row. Influence of parameter b/a and equidistance e between the location points of forces in row.

4.1 Slab structures subjected to transverse loads

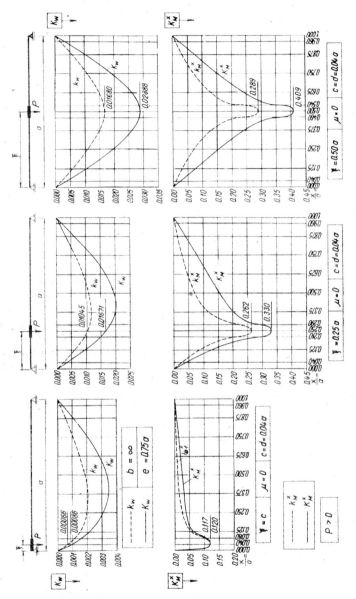

Figure 4.15 Infinitely long plate as structural member. Sets of locally distributed forces associated in simple row. Effect of grouping pattern and influence of position parameter x_p of row of forces.

227

4 Elastic plates as structural members

Table 4.4

Sets of locally distributed forces P and surface couples M associated in a simple row, acting on square and infinitely long plates

$c = d = 0.04 \ a$; $\xi = 0.25 \ a$; $\mu = 0$; $x = \xi$; $(x = \xi + c)$; $y = \eta \ (y = \lambda e)$

Design values of dimensionless coefficients K_w and K_M^x

Category of plate	b/a	e/a	Set P		Set M	
Elastic displacements			K_w	K_w/K_{wA}	K_w	K_w/K_{wA}
A	1	0.50	0.00772	1.00	0.01473	1.00
B	∞	0.75	0.01671	2.17	0.04207	2.85
C	∞	0.50	0.02406	3.12	0.06250	4.24
Bending moments			max K_M^x	K^x/K_{MA}^x	max K_M^x	K^x/K_{MA}^x
A	1	0.50	0.233	1.00	2.384	1.00
B	∞	0.75	0.330	1.42	2.947	1.24
C	∞	0.50	0.419	1.80	3.172	1.33

We give below the main conclusions derived from the comparative study of both types of loads P and M, associated in a simple row:

— The significant influence of the parameter b/a on the value of the elastic displacements (curves K_w), i.e. the great difference in stiffness between the two limiting cases compared: the square plate (curves A) and the infinitely long plate (curves B and C). This influence is decidedly stronger in the case of sets of surface couples M (Fig. 4.2) than it is in the case of loads of type P (Fig. 4.14);

— The influence of the parameter e, which defines the equal distance between the columns of the slab structure is also more moderate on the value of the elastic displacements in the case of the associated forces P;

— The influence of the parameters b/a and e on the value of the bending moments M_x (curves K_M^x), which is generally moderate, becomes more pronounced in the case of sets of P-type loads (Figs. 4.2, 4.14 and Table 4.4);

— Changing the location ξ of the row of forces P, the effect of the set of associated forces P is stronger in the central region of the plate (within the range $0.25a \leqslant \xi \leqslant 0.50a$) than at the edges $\xi = c$ (Fig. 4.15 and Table 4.5); in the case of the simple row of couples M, the effect of this set is stronger at the edges of the plate (the range $c \leqslant \xi \leqslant 0.25a$) and is almost equal to zero when

4.1 Slab structures subjected to transverse loads

the couples are applied at the center $\xi = 0.50a$ of the plate (Fig. 4.3).

— The influence of Poisson's ratio is stronger for the edge position of the set M (Fig. 4.4) and for the central position of a set P (Fig. 4.16).

Table 4.5

Sets of locally distributed forces P and surface couples M associated in a simple row, acting on infinitely long plate

$c = d = 0.004\ a\ ;\quad e = 0.75\ a\ ;\quad y = \lambda e\ ;\quad \mu = 0$

Comparative values of dimensionless coefficients max K_M^x

Set	Locally distributed forces P			Surface couples M		
ξ	max K_M^x	max k_M^x	K_M^x / k_M^x	max K_M^x	max k_M^x	K_M^x / k_M^x
	$x = \xi$			$x = \xi + c$		
c	0.120	0.117	1.03	3.466	3.343	1.04
$0.25a$	0.330	0.262	1.26	2.947	2.742	1.07
$0.50a$	0.409	0.289	1.42	2.638	2.625	1.01

We notice several similarities and differences between the sets of double rows of locally distributed symmetrical and antisymmetrical forces P acting on the infinitely long plate (Figs. 4.16 and 4.17) and the corresponding sets of couples M (Figs. 4.4 and 4.5). Thus,

— the influence of the parameter ξ on the value of the elastic displacements w and bending moments M_x, which increase more than 3.3 times in the range $c \leqslant \xi \leqslant 0.25a$; in the same range, the values max. M_x decrease 1.1—1.3 times in the case of symmetrical and antisymmetrical surface couples M associated in double rows;

— the existence in the symmetrical diagrams K_M^x of a flattened portion between the locations ξ and ξ' of the two rows of forces P; this flattened portion appeared also in the case of symmetrical rows of couples M (Figs. 4.4 and 4.12);

— the characteristics of a local perturbation of the effect of the locally distributed forces P, which is, however, more moderate than in the case of the surface couples M;

— the influence of the parameter μ is similar for both types of sets under study (Figs. 4.16, 4.17 and 4.18).

The investigation of the infinitely long plate subjected to a set of forces P associated in symmetrical double rows shows that the deflected mid-surface is practically horizontal in any

4 Elastic plates as structural members

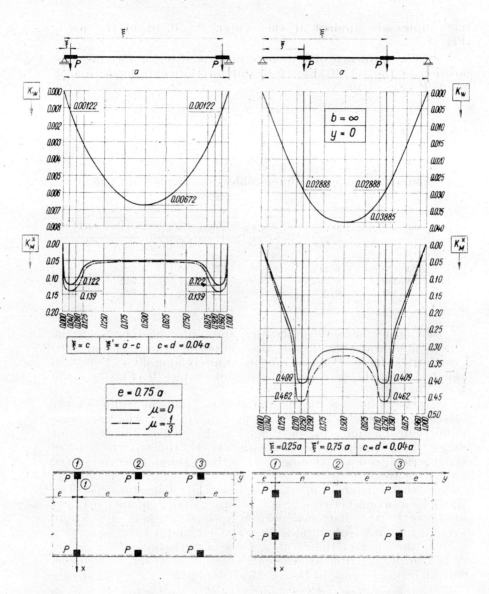

Figure 4.16 Infinitely long plate. Sets of symmetrical locally distributed forces P associated in double row. Deflected mid-surface w and diagrams M_x. Influence of parameters ξ and μ.

4.1 Slab structures subjected to transverse loads

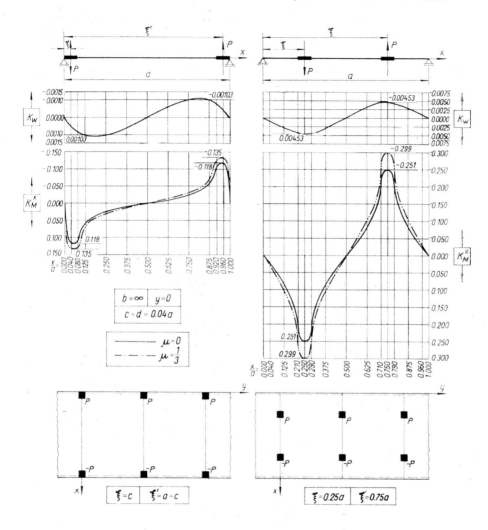

Figure 4.17 Infinitely long plate. Sets of antisymmetrical locally distributed forces P associated in double row. Deflected mid-surface w and diagrams M_x. Influence of parameters ξ and μ.

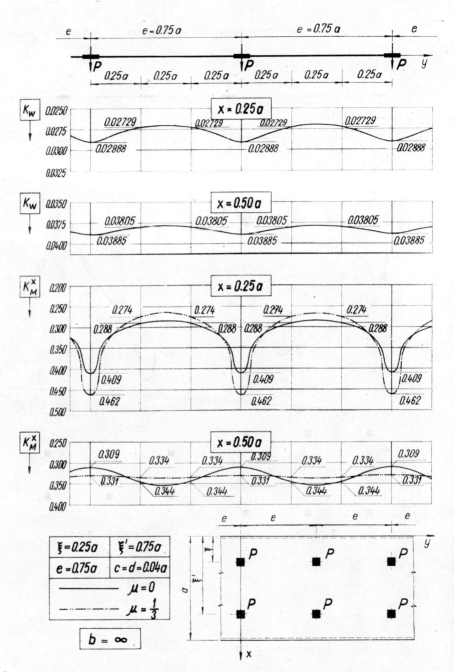

Figure 4.18 Infinitely long plate. Set of symmetrical locally distributed forces P associated in double row, at quarter of the span ($\xi = 0.25a$; $\xi' = 0.75a$). Deflected mid-surface w and diagrams for M_x along the plate length.

4.1 Slab structures subjected to transverse loads

longitudinal plane $x =$ constant. In this case, this characteristic is more pronounced than in the case of the corresponding of surface couples M (Fig. 4.18).

4.1.2.2 *The square plate.* The comparative approach of the effect of the symmetrical and antisymmetrical sets of locally distributed forces P on finite-boundary plates indicates that the square and the infinitely long plate behave similarly under this type of load. In the case of the very long rectangular plates ($b \geqslant 3a$), we observe an identical behaviour with respect to elastic displacements and stresses (Table 4.6).

Table 4.6

Set of two locally distributed symmetric forces P acting on infinitely long and finite-boundary plates

$c = d = 0.04\,a\,;\ \xi=0.25\,a\,;\ \xi' = 0.75\,a\,;\ \eta = 0\,(\eta = 0.50\,b)\,;\ y = \eta\,;\ \mu = 0$

Comparative values of coefficients K_M^x

x/a	$b = \infty$ K_M^x	$b = 3a$ K_{M3}^x	$b = a$ K_{M1}^x	K_{M3}^x / K_{M1}^x
0.210	0.232	0.231	0.185	1.25
0.250	0.290	0.289	0.236	1.22
0.290	0.252	0.251	0.193	1.30
0.710	0.252	0.251	0.193	1.30
0.750	0.290	0.289	0.236	1.22
0.790	0.232	0.231	0.185	1.25

Whereas for $3a \leqslant b \leqslant \infty$ the maximum values of M_x are unaffected by the variation of the parameter b/a, we observe that in the range $a \leqslant b \leqslant 3a$ the variation of max. K_M^x cannot be discarded (22 per cent — 30 per cent). Notwithstanding these quantitative differences, the qualitative similarity is closer in the case of the set of forces P than it is in the case of the set of surface couples M, in the entire range $a \leqslant b \leqslant \infty$.

In the study of finite-boundary plates, we have also considered the effect of sets of four symmetrical and antisymmetrical locally distributed forces P acting at the quarters of the spans a and $b = a$ of the square plate (Figs. 4.19 and 4.20). Here too, we observe that the square plate (Figs. 4.14, 4.19 and 4.20) and the infinitely long plate (Figs. 4.14, 4.16 and 4.17) behave in a similar manner under symmetrical and

4 Elastic plates as structura members

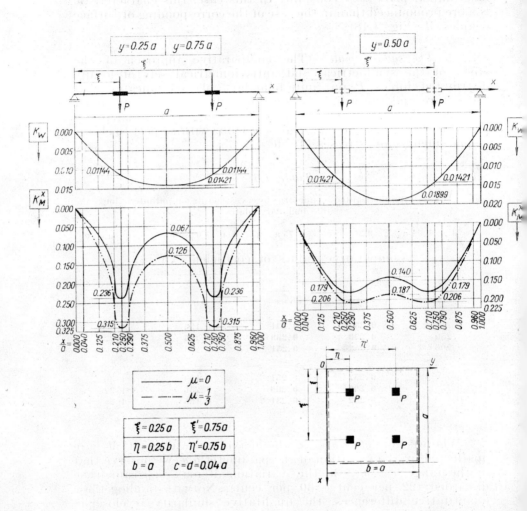

Figure 4.19 Square plate. Set of four symmetrical locally distributed forces P at quarter of the span. Deflected mid-surface w and diagrams M_x.

4.1 Slab structures subjected to transverse loads

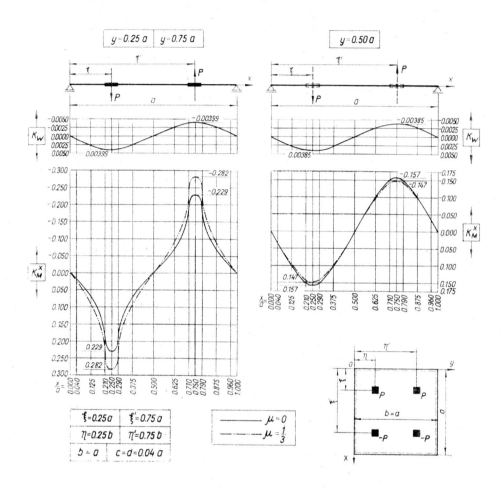

Figure 4.20 Square plate. Antisymmetrical set of four locally distributed forces P at quarter of the span. Deflected mid-surface w and diagrams M_x.

4 Elastic plates as structural members

antisymmetrical sets of locally distributed forces P. Recalling that a similar behaviour of plates was observed in the case of sets of surface couples, the correctness of the mathematical models presented herein is implicit.

The results of the study of the effect of sets of four surface couples M (Sub-sec. 4.1.1.2 — Figs. 4.12 and 4.13) and four locally distributed forces P (Figs. 4.19 and 4.20) on the square plate will be used to illustrate the method suggested for the static analysis of slab structures (Chap. 5). The design values of the coefficients K_w and K_φ^x will serve to express the unit displacements δ_{ij} and the free terms Δ_i required by the compatibility equations. The values of the coefficients K_M^x and K_M^y will allow us to draw the diagrams of the final bending moments M_x and M_y in the slabs of the structure, in the vertical plane of the columns ($y = 0.25a$), in the axis of the mid-span ($y = 0.50a$) or in any other vertical plane $y =$ constant or $x =$ constant.

4.1.3 Associated loads of type M and P applied on partially symmetrical or unsymmetrical slab structures

Using the sets of surface couples M and those of locally distributed forces P, the slab structure could be adequately characterized, and so the method of elastic analysis suggested by the author can more effectively deal with slab structures [64], [68], [70], [73], [75], [82]. Thus, the sets of couples M and of forces P represent generalized interacting forces which are used in the analysis of symmetrical static schemes subjected to symmetrical and antisymmetrical forces (i.e. flat slabs having a two-fold symmetry subjected to a gravity load of the type q, uniformly distributed throughout the floor in strips or in chess-board fashion, or to horizontal forces of the type S). This method can be extended to deal with two categories of slab structures:

1. slab structures (flat slabs, shear-walls, caissions) having no geometrical or mechanical axis of symmetry but displaying a regular shape which allows the use of the superpositioning of effects, and

2. slab structures consisting of bars and plates (flat slabs, installation elements, machine parts, etc.) whose static or dynamic analysis includes couple rows of moment $M_i (i = 1, 2, \ldots, n)$ or rows of forces $P_i (i = 1, 2, \ldots, n)$ which can assume different signs and values from one row to another. Of course, the sets of couples M and of forces P can appear either as generalized interacting forces or as given loads.

4.1 Slab structures subjected to transverse loads

4.1.3.1 *Unsymmetrical sets of couples M and of locally distributed forces P.* An eventual analysis of unsymmetrical static schemes will be illustrated first in the case of slab structures with rhombic or parallelogram-like panels, where the columns are spaced zig-zag (Figs. 4.21, 4.22 and 4.23). Such schemes can be used in the design of oblique slab bridges, exhibition halls, commercial buildings, etc., as well as several types of metal structures.

Using the results obtained from the analysis of sets associated in a simple row, we displace a row off the adjacent rows along the plate length and perform the superposition of the reciprocal effects of the offset rows. Thus, we obtain a simulation of the effect of a set of two or more rows of couples M or forces P applied zig-zag. In this manner, we can approach the elastic analysis of slab structures whose panels have a parallelogram-like or rhombic shape.

Choosing only some of the previously examined alternatives, it is possible to analyse the effect of zig-zag sets where a row of couple M (forces P) is displaced at a distance $e/3$ (for $e = 0.75a$ — Fig. 4.21A, B and C) and $e/2$ (for $e = 0.50a$ — Fig. 4.21D, E and F) respectively, from the alignment of the neighbouring rows. Hence, the displacement of the offset rows in the longitudinal direction of the plate is $0.25a$ in all the cases under study. Obviously, the computation programmes written by the author allow the analysis of any other desired alternative.

For illustration, we shall simulate the effect of an unsymmetrical set of surface couples M acting on the infinitely long plate. This set consists of two rows of associated couples applied at the quarter of the span (Fig. 4.21 E), which are either of opposite (Fig. 4.22) or of the same sign (Fig. 4.23). These models are first approached here. A comparison of these unsymmetrical sets with symmetrical or antisymmetrical couples associated in double rows (Figs. 4.4 and 4.5), from which they derive by displacing some rows, allows the following observations. In the alternative where $M \cdot M' < 0$ (Fig. 4.22), the shape of the deflected midsurface w of the plate resembles closely the symmetrical shape obtained in the case of two rows of symmetrical surface couples (Fig. 4.4). However, the diagrams of the bending moments M_x are no longer symmetrical. In the alternative $M \cdot M' > 0$ (Fig. 4.23), the elastic displacements w and the bending moments M_x are no longer equal to zero at mid- and quarter-span. In the present case, the diagrams are no longer antisymmetrical, as distinct from the diagrams obtained for the set of undisplaced rows of antisymmetrical couples (Fig. 4.5).

The conclusions derived from the comparative analysis of the two sets of zig-zagging couples may become very important

4   Elastic plates as structural members

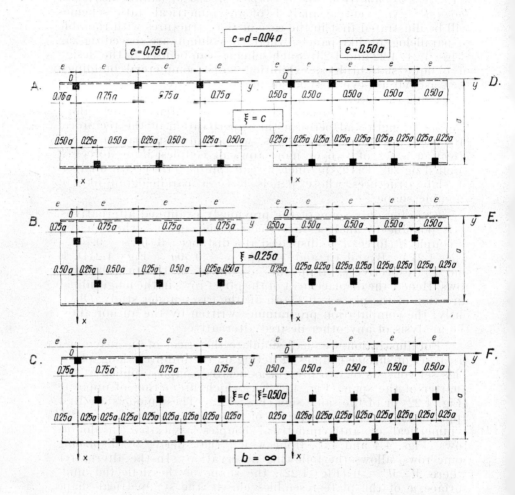

Figure 4.21 Slab structures with parallelogram-like or rhombic panels (zig-zag distributed columns).

4.1 Slab structures subjected to transverse loads

in the design approach to such structures. Thus, the design values of the elastic displacements and bending moments are found to be unaffected by the offset rows of couples M. This means that these values are almost equal to those obtained in the case of symmetrical (Table 4.7) or antisymmetrical sets of surface couples. This conclusion allows us to state that for certain types of flat slabs, the structures with symmetrical rows of columns (rectangular panels) can be replaced by structures with offset rows (parallelogram-like panels) without implying economic disadvantages.

Table 4.7

Sets of two rows of surface couples M ($M \cdot M' < 0$) acting on infinitely long plate with rectangular and rhombic panels, respectively (the offset distance is $e/3$)

$c = d = 0.04\ a$; $e = 0.50\ a$; $\xi = 0.25\ a$; $\xi' = 0.75\ a$; $\mu = 0$

Comparative values of dimensionless coefficients K_w and max K_M^x

x/a	Symmetrical couples (square panels)		Zig-zag couples (rhombic panels)		Observations
	$y = \lambda e$	$y = (2\lambda+1)e/2$	$y = \lambda e$	$y = (2\lambda+1)e/2$	
Coefficients K_w					
0.250	0.12500	0.12396	0.12488	0.12488	
0.500	0.18882	0.18460	0.18671	0.18671	max K_w
0.750	0.12500	0.12396	0.12488	0.12488	
Coefficients K_M^x					
0.210	−1.780	0.659	−1.762	0.641	min K_M^x
0.290	3.735	1.372	3.762	1.345	max K_M^x
0.500	1.894	2.072	1.983	1.983	
0.710	3.735	1.372	1.345	3.762	max K_M^x
0.790	−1.780	0.659	0.641	−1.762	min K_M^x

The range of validity of the approximate calculation methods of slab structures, recommended by the existing code requirements, is restricted to the case of slabs with rectangular panels. This explains why flat slabs with parallelogram-like or rhombic panels are decidedly outside the scope of these methods.

★

4 Elastic plates as structural members

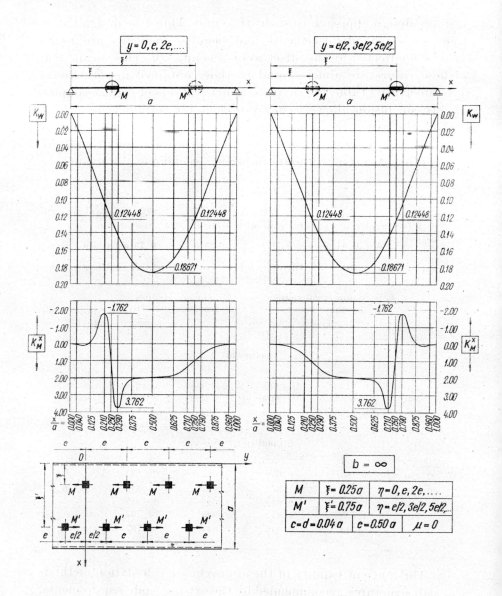

Figure 4.22 Infinitely long plate. Set of zig-zag surface couples M associated in two rows offset by $e/2$ $(M \cdot M' < 0)$.

4.1 Slab structures subjected to transverse loads

Another category of slab structures which cannot be treated with any of the known methods of elastic analysis * are the very unsymmetrical structures in the transverse direction of the slab (Fig. 4.24). We give below as relevant examples several variants

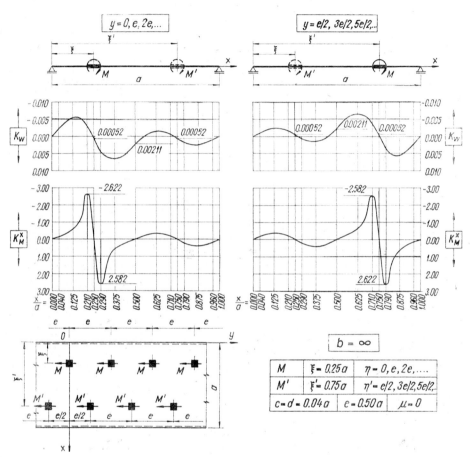

Figure 4.23 Infinitely long plate. Set of zig-zag surface couples M associated in two rows offset by $e/2$ ($M \cdot M' > 0$).

for which the static analysis can be easily carried out using only the numerical data obtained herein [75]: a column row spaced unsymmetrically (Fig. 4.24A); two and three unsymmetrical rows of identical columns (Figs. 4.24 B, C and D); three rows of

---

* With the exception of Pfaffinger's latest study [71].

4   *Elastic plates as structural members*

columns of different cross-sections (Fig. 4.24 E); and four unsymmetrical rows of columns (Fig. 4.24 F).

We shall illustrate the effect of the set of loads applied on unsymmetrical slab structures by considering a set of three rows

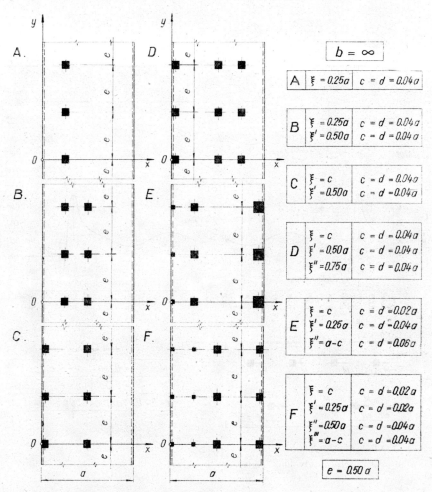

Figure 4.24 Transversally unsymmetric slab structures.

of transversally unsymmetric locally distributed forces P acting on an infinitely long plate (Fig. 4.24 D). The mid-surface w deflected under this load assumes an almost cylindrical shape. The coefficients K_w do not vary between the plane of the columns

242

4.1 Slab structures subjected to transverse loads

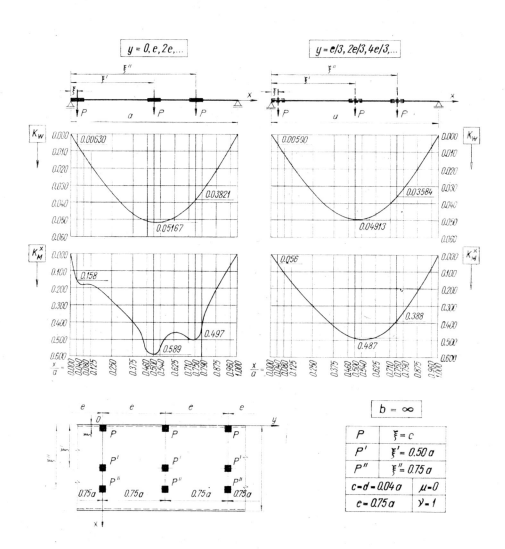

Figure 4.25 Infinitely long plate. Set of locally distributed forces P associated in three transversally unsymmetric rows.

4   Elastic plates as structural members

($y = 0$; e; $2e$; ...) and the intermediate plane ($y = e/3$; $2e/3$; $4e/3$; ...) (Fig. 4.25). However, the diagram of the bending moments M_x drawn in the plane of the columns (the axis of the forces P) has an unusual shape.

4.1.3.2 *Sets of couples νM and forces νP of different value and sign.* In the study of sets of surface couples M or of locally distributed forces P, account was taken of the eventual occurrence in the structural analysis of the loads $M_i = \nu_i M$, and $P_i = \nu_i P$, respectively, ($i = 1, 2, 3, \ldots n$) which, being identical along one row i, assume different values from one row to another. Applying the superposition of effects, we obtain — in the first calculation step — the model of the sets M_i and P_i associated in a simple row (Sub-secs. 4.1.1 and 4.1.2). In the second calculation step we simulate several geometrically symmetric static schemes, subjected this time to unsymmetrical loads, and also schemes with no symmetry.

Among the structures which can be treated with this procedure, special attention should be given to the slab structures subjected to locally distributed forces, also structures where one or more rows of columns are withdrawn from a certain level upward against the lower levels. It is evident that since the factor ν_i can take any value, it may change in sign from one row to another.

Consider now the simple case of an unsymmetrical set of three surface couples of different values and signs ($M \times 2$; $-2M$) applied on the square plate at quarter-span (Fig. 4.26). Referred to the effect of the set of four symmetrical couples M (Fig. 4.12), the mid-surface deflected by the set of three couples is quite different in the axis of the plate ($y = 0.50a$). Larger qualitative and quantitative differences can be observed in the diagrams M_x.

Finally, we consider also the case of the square plate treated as a structural member, subjected to a symmetrical set of five locally distributed forces of different value and opposite signs. The model for this set of loads is obtained by superposing the effects of the set of four forces P applied at quarter-span (Sub-sec. 4.1.2.2 — Fig. 4.19), taken with changed sign ($-P \times 4$) over the effect of a force ($4P$) applied at the center of the plate (Fig. 4.27).

It is of interest to note the similar behaviour of the square plate with the simply supported beam and the characteristics

4.1 Slab structures subjected to transverse loads

of a local perturbation induced by central load νP in the diagram of the flexural moments M_x. In order to verify the precision of the calculations, we notice that the transverse displacements $w(K_w = 0.01421)$ are equal at the points $(x = 0.50a\,;\ y = 0.25a)$ and $(x = 0.25a\,;\ y = 0.50a)$.

This simple loading scheme for the square plate will be further adapted to illustrate the implementation of the new method in the *dynamic analysis* of slab structures (Chap. 6).

Since all the numerical examples considered deal with the infinitely long plate and the square plate, it is clear that the use of the present method and its eventual extension are by no means affected by the restrictions corresponding to the category of plates under study, i.e. from the given boundary conditions.

★

The simplified treatment of structures using sets of surface couples M and locally distributed forces P allows the study of the behaviour of a wide category of slab structures in the elastic range. Thus,

a. slab structures with two axes of symmetry subjected to the uniformly distributed gravity load q (symmetrical sets of interacting forces), or to horizontal forces S, uniformly distributed at every level (antisymmetrical interacting forces);

b. slab structures with one axis of symmetry transverse to the rows of columns (which are spaced at an equal distance along the slab length); the panels are of arbitrary span, being unsymmetric in the transverse direction;

c. slab structures with no geometrical or mechanical axis of symmetry, with offset rows of columns and rhombic or parallelogram-like panels (zig-zagging sets of reaction couples or reaction forces);

d. slab structures with or without geometrical symmetry, where because of the unequal spans, unsymmetrical loads, withdrawn rows of columns, the rows of couples $M_1, M_2, M_3, \ldots M_n$ and those of locally distributed forces $P_1, P_2, P_3, \ldots P_n$ assume different values from one row to another, and can change sign.

By generalization, this procedure can be used in the analysis of any similar slab structures (raft foundations, shear-walls clamped with isolated beams, foundation caissons, piers, docks and members of metal structure, installations and machine parts) which can be idealized using sets of surface couples M and of locally distributed forces P. The effect of these sets can be also simulated in the case of symmetrical and antisymmetrical rows of couples and forces and when the rows forming the static scheme display a regular distribution without being symmetric (for instance, the equal distance between the positions of the couples and forces

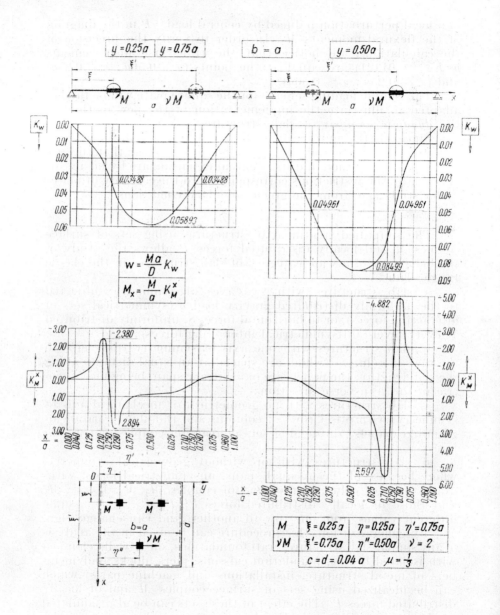

Figure 4.26 Square plate. Unsymmetrical set of three surface couples M of different value and opposite sign ($M \times 2$; $-2M$).

4.1 Slab structures subjected to transverse loads

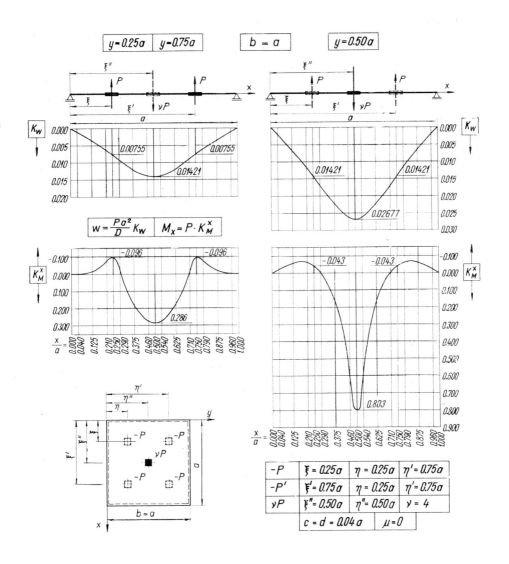

Figure 4.27 Square plate. Symmetrical set of five locally distributed forces P of different value and opposite sign ($-P \times 4$; $4P$).

247

4 *Elastic plates as structural members*

in the longitudinal direction of the plate). Moreover, the modelling procedure using the superposition of effects can be extended to the elastic analysis of any arbitrary slab structure having no geometrical, mechanical or loading symmetry.

However, when the number of spans increases substantially in the transverse or in both directions, calculations may become inconveniently involved. In the case of ordinary slab structures, resort can be made to the approximate methods recommended in standards and code requirements provided that the limits of validity are observed. In the analysis of structures with a relatively small number of panels and large spans, subjected to heavy loads, the approximate methods must be avoided. Also, these approximate methods cannot be used to treat slab structures subjected to seismic or dynamic loads.

4.2 Slab structures subjected to lateral loads
Evaluation of the effective slab width

The behaviour of plates as structural elements was discussed in the previous section in view of using the general force method in the elastic analysis of slab structures. The results obtained from the study of the effect of associated loads of the type M and P will be used in the static (Chap. 5) and dynamic analysis (Chap. 6).

In the condition equations, the symmetrical sets appear as generalized interacting forces, especially in the case of given loads acting transverse to the plate (Sub-sec. 5.1.2.1). The antisymmetrical sets of type M (eventually of type P, too) allow also the approach of the elastic analysis of slab structures subjected to systems of lateral forces (Sub-sec. 5.1.2.2, 6.1.2 and 6.2.2). For the last type of loads *which act in the plane of the slabs*, we give below a static analysis procedure which relies on the *general displacements method*. Considered as a member of the slab structure, the plate is now treated in the analysis as a *frame-beam*.

★

The comparative examination of the behaviour of plates subjected to a surface couple M (Secs. 3.2 and 3.3) or to sets of surface couples M (Sub-secs. 4.1.1 and 4.1.3) showed that the plates behave similarly to a *straight beam* under the same types of loads. On this basis and taking into account the data given elsewhere (see Sub-sec. 1.2.4) the slab structure can be treated as a reticular system for certain categories of structures and types of given loads.

4.2 Slab structures subjected to lateral loads

A type of external load for which we can evaluate the relative slab-column stiffness, assuming that the slab behaves as a frame-beam, is the system of lateral forces S acting in the plane of the slabs. Of course, the slab structure subjected to this system of loads must have a two-fold geometrical and mechanical symmetry and the distribution of the columns should be of a regular pattern in both directions of the plane (Sec. 5.2). Bearing in mind these limitations of applicability, the procedure suggested for the design of symmetrical multistoried slab structures subjected to a system of horizontal forces S, can be considered to be rigorous.

In order to apply the general displacement method in the frame analysis, it is necessary to determine the flexural stiffness of the slab, i.e. the effective plate width subjected to a surface couple, or to a set of couples M.

We present below a rigorous procedure for evaluating the effective plate width relying on the results obtained in the previous chapters from the analysis of the angular elastic displacements of plates subjected to a surface couple M (Sec. 3.2).

4.2.1 The effective width of the plate subjected to only one surface couple M

4.2.1.1 Definition. Calculation procedure. In order to evaluate the relative bar-plate stiffness of a mixed system, particularly the slab-column stiffness of a slab-structure, the existing requirements allow the use of several empirical formulae. However, several exact-calculation procedures for the effective plate width in the elastic range have been suggested elsewhere.

The first use of an *equivalence criterion* for the evaluation of the effective plate width b_y is ascribed to Popov and Mohammed [58] (see Sub-sec. 1.2.4.2), who assimilated the infinitely long plate with the simply supported beam of equal span and height equal to the plate thickness. The same equivalence criterion is reiterated by Stiglat [60] for the case of the infinitely long plate, with the difference that the plate rotation is not induced by a linearly distributed couple but by a *surface couple M* (see Sub-sec. 1.2.4.3).

For a general procedure that would apply to the infinitely long plate and finite-boundary plates, Stiglat's procedure will be extended to the evaluation of the effective width of the rectangular plate [64], [68], [75], etc.

The effective plate width is defined as the working width b_y of an equivalent beam with the same span, the same type of supports and the height equal to the width of the plate (Fig. 4.28). In order

4 Elastic plates as structural members

to estimate the effective plate width, we assume as an equivalence criterion that the rotations are equal: the equivalent beam, subjected to a concentrated couple M, rotates at the loaded point ξ with a value φ_ξ equal to the plate rotation φ_x produced by a

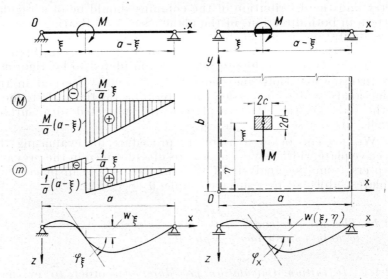

Figure 4.28 Plates as structural members. Evaluation of effective plate width. Schematic diagram.

surface couple M of equal magnitude. The last rotation is calculated at the centroid (ξ, η) of the loaded area.

In the case of a rectangular plate $a \times b$, simply supported on the span a, the equivalent beam is a simply supported beam of span a (Fig. 4.28). According to Maxwell-Mohr, the rotation φ_ξ of this beam, induced by a concentrated couple M applied at ξ is

$$\varphi_\xi = \int Mm \frac{dx}{EI}, \tag{4.1}$$

where
— M represents the diagram of the bending moments induced in the equivalent beam by the couple M acting at ξ;
— m denotes the diagram of the bending moments generated by the unit couple applied at ξ, in the direction of the sought displacement;

4.2 Slab structures subjected to lateral loads

— $I = \dfrac{b_y h^3}{12}$ is the moment of inertia of the equivalent beam of height h;

and

— E is the modulus of elasticity of the equivalent beam. The equivalence equation

$$\varphi_\xi = \varphi_x, \qquad (4.2)$$

where

— $\varphi_x = \dfrac{M}{D} k_\varphi^x$ is the value of the plate rotation at $(x = \xi;\ y = \eta)$;

— k_φ^x is the dimensionless coefficient given by (2.72), (2.82), (2.112), (2.122), (2.152) and (2.162) (Tables 2.4, 2.5, 2.8, 2.9, 2.12 and 2.13);

— D is the cylindrical rigidity of the plate of constant thickness h, given by (1.2) (Sub-sec. 1.2.1),

yields the computation formula for the effective plate width:

$$b_y = \frac{\left(\dfrac{\xi}{a}\right)^3 + \left(1 - \dfrac{\xi}{a}\right)^3}{3(1 - \mu^2) k_\varphi^x}\, a. \qquad (4.3)$$

Once the effective width b_y is determined, we can evaluate the stiffness of the slab, which is treated as a frame beam in the static analysis of slab structures. The values of b_y, computed by means of formula (4.3), may substantially differ from those resulting from the empirical formulae recommended in the code requirements adopted in various countries [154], [155], [157], [159], [160].

The value of b_y has a physical significance only for those points (x, y) of the surface of the plate plane which coincide with the centroid (ξ, η) of the area subjected to the couple M. The values of the dimensionless coefficients k_φ^x for the infinitely long plate and for the rectangular plate (Tables 3.3—3.6 and Figs. 3.5 and 3.6, Sub-sec. 3.2.1), obtained for these points $(x = \xi;\ y = \eta)$ allow a simultaneous computation of the effective plate width b_y using the same programme (Figs. 4.29—4.31) *.

---

* Besides these particular values of the coefficients k_φ^x the computation programmes developed enable us to calculate the values of k_φ^x and k_φ^y at any point (x, y) of the computation grid. Hence, we can also calculate the rotations φ_x and φ_y for any values of the surface couple M, and for any value of the cylindrical rigidity D of the plate, with no restriction.

4 Elastic plates as structural members

The values of the coefficients k_φ^x used for the determination of the effective plate width b_y were calculated assuming that the plate is subjected to only one surface couple M. This assumption is valid in the case of slab structures where the equidistance e between the columns, measured along the floor, is large enough to allow us to ignore the superposition of effects in the longitudinal column row. All the implemented computations lead to the conclusion that the effect of the set of columns associated in a simple row on the rotations φ_x can be neglected if the equidistance e between the columns of a longitudinal row is :

$$e \geqslant 2.0\, a \quad \text{for } c \leqslant \xi \leqslant 0.25a \tag{4.4}$$

$$e \geqslant 1.5\, a \quad \text{for } 0.25a \leqslant \xi \leqslant 0.50a.$$

Under these conditions, the effective plate width b_y can be evaluated assuming the effect of only one surface couple acting at $x = \xi$; $y = \eta$.

For all these types of slab structures, the values b_y given in the diagrams (Figs. 4.29–4.31) can be used for the evaluation of the relative plate-column stiffness.

The study of the effective plate width will be extended to types of slab structures where the value of the equidistance e is smaller than the above mentioned limits (Sub-sec. 4.2.2).

4.2.1.2 *Comparative values of the effective plate width.* The values of b_y given herein for the infinitely long plate and various rectangular plates are presented in the form of diagrams that allow the interpolation of intermediate values.

The diagrams showing the variation of the dimensionless coefficients b_y/a for the infinitely long plate (Figs. 4.29 and 4.30) allow the direct reading of the values of this coefficient, calculated for the three characteristic positions of the surface couple M: $\xi = c$, $0.25a$, $0.50a$. The accuracy of these values is sufficient in structural analysis. In the computations account was taken of most of the sizes of the loaded area that may be related to the design practice of flat slabs with or without capitals : $0.01 \leqslant c/a \leqslant 0.08$; $0.01 \leqslant d/a \leqslant 0.10$.

In the diagrams b_y/a, Poisson's ratio was assumed to be equal to zero only in the case of the infinitely long plate (Fig. 4.29), for which reference can be made to the values given by Stiglat only for $\mu = 0$ [60].

However, when the effective width of the infinitely long plate ($b = \infty$) is compared with the values b_y computed for an ordinary range of rectangular plates ($a \leqslant b \leqslant 3a$), for which no comparable values are available in the literature, the computations are carried

4.2 Slab structures subjected to lateral loads

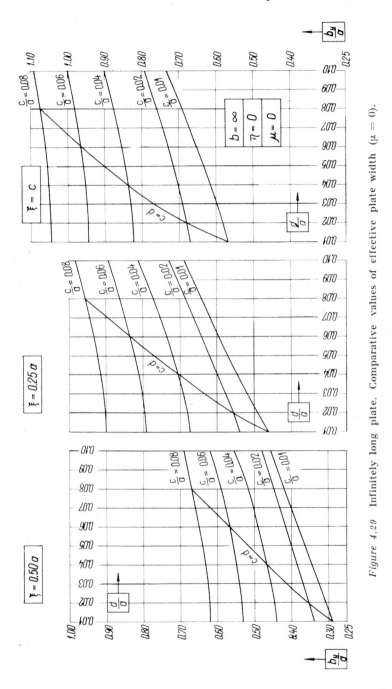

Figure 4.29 Infinitely long plate. Comparative values of effective plate width ($\mu = 0$).

out assuming three characteristic values of Poisson's ratio: $\mu = 0$; $\mu = 1/6$ and $\mu = 1/3$ (Fig. 4.30). For the very long rectangular plate ($b = 3a$), the values of b_y/a and those obtained for the infinitely long plate almost coincide for any value of Poisson's ratio. This result is in compliance with expectations if we keep in mind the similar behaviour of these two kinds of plates subjected to a surface couple M (Sec. 3.2).

In the case of the infinitely long plate (Fig. 4.29), the influence of the parameter ξ on the coefficient b_y/a is obvious; the highest values of the effective width correspond to the edge position of the surface couple ($\xi = c$). Greater differences exist between the center position ($\xi = 0.50a$) and the position at the quarter of the span ($\xi = 0.25a$), than between this last position and the edge location of the couple ($\xi = c$).

The values of b_y/a increase as the parameters c/a and d/a become larger. The influence of the dimension $2c$ of the loaded area, measured normally to the moment vector $\vec{M}$, is more pronounced.

As concerns the precision of computations, the only available comparative data are the values b_y/a computed by Stiglat for the infinitely long plate, assuming that Poisson's ratio is equal to zero. These values inscribe well on the curves plotted in Fig. 4.29, including the extreme values of the parameter d/a ($d = 0.10a$), which exceed the ordinary sizes of the area subjected to the couple M.

The influence of the ratio b/a between the two dimensions of the plate on the value of b_y is clearly illustrated by the comparative diagrams drawn for $b = 3a$ ($b = \infty$), $b = 2a$ and $b = a$ (Fig. 4.30). Besides confirming the identity of behaviour between the infinitely long and the long rectangular plate ($b = 3a$), these diagrams show that in the range $3 > b/a \geqslant 2$ the differences between values are negligible. Larger values of b_y appear however in the range $2 > b/a \geqslant 1$.

The diagrams b_y/a drawn for $0 \leqslant \mu \leqslant 1/3$ are also indicative of the influence of Poisson's ratio on the computations. The values of b_y increase with the growth of this parameter, larger differences appearing in the range $1/6 \leqslant \mu \leqslant 1/3$.

The influence of the position ξ of the surface couple M is in the case of finite boundary plates similar to the case of the infinitely long plate: the increase of b_y is more pronounced in the range $0.50a \geqslant \xi \geqslant 0.25a$ than it is when M is located farther away from the longitudinal axis of symmetry (i.e. for $0.25a > \xi \geqslant c$).

The diagrams showing the variation of the parameter b_y/a (Fig. 4.30) were drawn for the central position of the surface

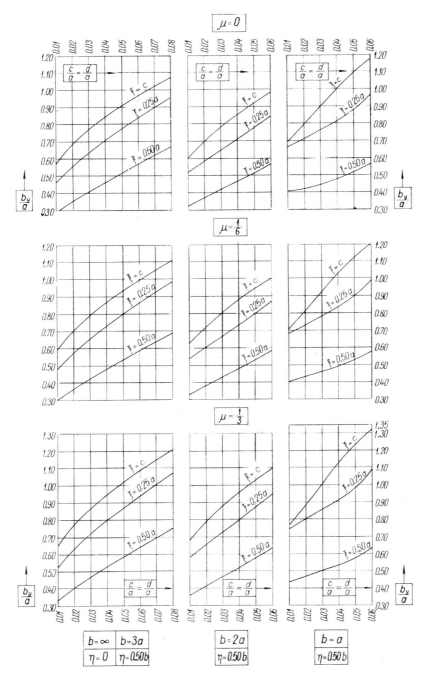

Figure 4.30 Infinitely long and finite-boundary plates. Comparative values of effective plate width ($0 \leqslant \mu \leqslant 1/3$).

4 Elastic plates as structural members

couple $\eta = 0.50b$ which leads to the lowest values of the effective plate width (highest values of the coefficients k_φ^x).

For the case of the square plate ($b = a$), the comparative data include also those resulting from an unsymmetrical position

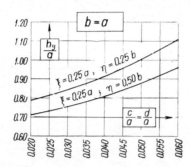

Figure 4.31 Square plate. Comparative values of effective plate width ($\mu=0$).

of the couple M. Thus, the square plate was successively subjected to a couple acting in the transverse plane of symmetry $y = 0.50b$ ($\xi = 0.25a$; $\eta = 0.50b$) and then to a couple acting outside this plane ($\xi = 0.25a$; $\eta = 0.25b$) (Fig. 4.31).

Since the values of the rotations φ_x are smaller for the unsymmetrical position of the couple, the corresponding values of the effective width b_y are higher than those reported for the symmetrical position. This is in compliance with expectations. So, we conclude that in the case of the unsymmetrical position of the couple M, the effective width increases with the distance between this couple and the longitudinal axis of symmetry $y = 0.50b$. Usually, this increase does not exceed 10—15 per cent in the limiting case of the square plate. Here too, we observe that the values b_y/a increase as the dimensions of the area subjected to the couple M are larger. The diagrams showing the variation of the coefficient b_y/a, drawn for the central position of the surface couple ($\eta = 0.50b$), allow the reading and interpolation of the intermediate values of b_y/a with sufficient accuracy in design practice.

4.2.2 The effective width of the plate subjected to sets of surface couples M associated in a simple row

The values of the effective plate width b_y, calculated under the assumption that the plate is subjected to only one surface couple (Figs. 4.29—4.31), can be used in the design of structures where the equidistance e between the columns, measured along the Oy axis, is sufficiently large to allow us to neglect the superposition

4.2 Slab structures subjected to lateral loads

of the effects of the adjacent columns. This simplification is unmistakably effective in the design of slab structures where the equidistance e satisfies conditions (4.4). The differences between the values of the coefficients k_φ^x and K_φ^x, computed for the range $1.5a \leqslant e \leqslant 2a$, are so small that they can scarcely influence the corresponding values of b_y (Table 4.8).

Table 4.8

The infinitely long plate subjected to a set of surface couples M associated in a simple row, acting at ($\xi = 0.50a$; $\eta = 0$)

Design values of rotations φ_x at ($x = \xi$; $y = \eta$)

a. Values of coefficient $10^5 K_\varphi^x$

$c/a = d/a$	K_φ^x			K_φ^x		k_φ^x
	$e = 0.50a$	$e = 0.75a$	$c = a$	$e = 1.50a$	$e = 2a$	
0.01	32070	29715	29096	28892	28879	28878
0.02	26597	24246	23627	23423	23410	23409
0.03	23381	21033	20415	20211	20199	20198
0.04	21090	18748	18131	17928	17915	17914
0.06	17856	15532	14918	14715	14703	17402
0.08	15561	13261	12652	12450	12438	12437

b. Values of coefficient 10^5 mean K_φ^x

$c/a = d/a$	mean K_φ^x			mean K_φ^x		mean k_φ^x
	$e = 0.50a$	$e = 0.75a$	$e = a$	$e = 1.50a$	$e = 2a$	
0.01	31183	28829	28210	28005	27993	27992
0.02	25707	23356	22737	22533	22520	22519
0.03	22483	20136	19517	19314	19301	19300
0.04	20199	17857	17240	17037	17024	17023
0.06	16980	14655	14041	13838	13826	13825
0.08	14702	12402	11793	11591	11579	11578

For the design of slab structures where the equidistance e is smaller than the limit set by (4.4), the values of k_φ^x are replaced in (4.3) by those of K_φ^x, so that the effective plate width is evaluated taking into consideration the superposition of the effects of the adjacent columns [60], [75]:

$$b_y = \frac{\left(\dfrac{\xi}{a}\right)^3 + \left(1 - \dfrac{\xi}{a}\right)^3}{3(1 - \mu^2) K_\varphi^x} a. \qquad (4.5)$$

4   Elastic plates as structural members

For a comprehensive applicability of the analysis procedure of slab structures under seismic loads, by use of the general displacement method, in the computation of the coefficients k_φ^x and K_φ^x at $(x = \xi = 0.50a\,;\ y = \eta = 0)$ a wider range of values

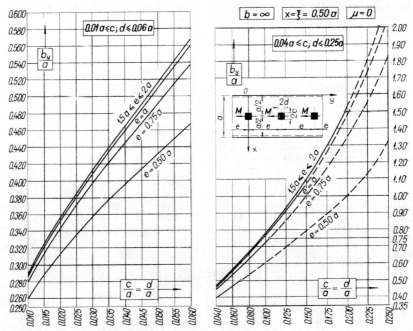

Figure 4.32   Infinitely long plate subjected to set of surface couples M associated in simple row. Design values of effective plate width $(0.01a \leqslant c,\ d \leqslant 0.25a)$.

was considered for the parameters c, d and e than in the foregoing sections (Table 4.8; Fig. 4.32). The sizes $2c \times 2d$ and the values for the equidistance between the columns

$$e = 0.50a\,;\quad 0.75a\,;\quad 1.50a\,;\quad 2a$$

were adopted in order to allow the use of this procedure in the analysis of both categories of structures with beamless floors: slab-structures with or without capitals.

Throughout the book, the values of k_φ^x were computed at the centroid (ξ, η) of the area subjected to the couple M. Large differences may appear between the values of k_φ^x computed at the center $(x = \xi\,;\ y = \eta)$ and along the boundary $(x = \xi\,;\ y = \eta \pm d)$ of the loaded area, especially when the contact area

4.2 Slab structures subjected to lateral loads

is relatively large. That is why we have also calculated the values of mean k_φ^x and mean K_φ^x corresponding to the average between the values computed at the center and on the boundary of the area

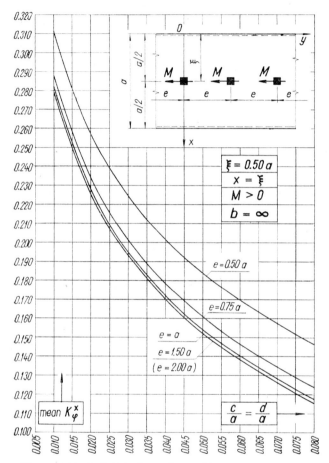

Figure 4.33 Infinitely long plate subjected to set of surface couples M associated in simple row. Values of rotations mean K_φ^x ($0.50a \leqslant e \leqslant 2a$).

subjected to the couple M (Table 4.8b, Fig. 4.33). The diagram mean K_φ^x facilitates eventual interpolations with respect to the parameters c, d and e.

259

5

A new method for the elastic analysis of slab structures

Slab structures and more particularly one- or multi-storied structures consisting of flat slabs are modern structural systems of high versatility. This functional adaptability together with economic, technical and esthetical advantages, explains the use of slabs in the design of reinforced concrete structures over the last thirty years.

Referred to the static scheme, slab structures derive from mixed structures consisting of bars and plates, for which it is possible to relate the states of stress and deformation more accurately to the actual behaviour of the structure by considering in the analysis the effect of space interaction. A rigorous static analysis of these categories of structures leads to new approaches and somewhat complex mathematical models, and requires lengthy calculations that can be performed only on computers. This explains why current design techniques recommend the use of approximate calculation methods in the analysis of ordinary slab structures [154], [155], [158], [160], [162].

The increasing use of slab structures called for new analysis approaches by use of the Theory of Plates and the Theory of Structures. Several such approaches have managed to give new calculation procedures that go beyond the empirical stage of approximate methods.

However, the simplified treatment often adopted here and there takes much from a rigorous analysis and undersirably restricts the range of applications encompassed by the procedures. The main simplifying assumption that is also adopted in recent literature [54], [55] is that of zero transverse displacement in the column axis as the only condition for the compatibility equations (Sec. 1.2).

Disregarding the effect of the rigid connection between slabs and columns of the structure involves several approximations whose order of magnitude is mainly given by the relative slab-column stiffness, the uneven character of spans or the non-uniform distribution of transverse loads on slabs.

In order that the effect of this rigid connection be considered in the analysis, the author suggested, as far back as 1969 [64], [65], the inclusion of a surface couple acting as a generalized interacting force. This allowed a more accurate evaluation of the distribution of elastic displacements and stresses in slabs and throughout the structure [74], [75], [77], [78], [86].

★

The proposed method of analysis is carried out in two steps. Because of the mixed character of slab structures, in the first step it is necessary to establish the calculation elements defining the behaviour of plates under load. In order to do this, the expression of the elastic displacements, stresses and boundary reactions were established for three categories of plates : the infinitely long plate, the rectangular plate with two parallel free edges, and the rectangular plate simply supported on the entire boundary (Sub-sec. 2.1.3). The equations of the deflected mid-surface of the three categories of plates were determined for three types of transverse loads : the locally distributed load P, the surface couple M and the load q uniformly distributed throughout the plate surface. The M-type load, which is first used here in the investigation of plates, is considered under two limiting assumptions concerning the direction of the moment vector $\vec{M}$ in the plate plane : parallel to the line of supports ($\vec{M} \| Oy$) and normal to the line of supports ($\vec{M} \| Ox$) (Sub-secs. 2.1.4 and 2.2).

The computational algorithms developed for all the static quantities defining the behaviour of plates as structural members are expressed as dimensionless coefficients (Sec. 2.3). For a wide range of values of the physical and geometrical parameters that are frequently met in design practice, we have calculated more than 95,000 values for the dimensionless coefficients. These values are given in the form of diagrams and tables in the foregoing comparative approaches (Chapts. 3, 4 and the Appendix).

With the aid of the values obtained in the first calculation step, we can now approach, in the second step, the static analysis of structures consisting of bars and plates, using one of the general methods for solving statically indeterminate reticular systems.

Inserting in the analysis the values of the coefficients k_w and k_φ, which define the elastic point-like displacements of the plates, allows a convenient way of writing the compatibility conditions for the slab-column nodes. The dimensionless coefficients

5.1 General force method. Transverse and in-plane loads

of type K_w and K_φ define the elastic point-like displacements in the cases of associated loads P and M, and associated unknowns, respectively (Sec. 4.1). Based on these facts, we developed a first analysis method for which resort was made to the general force method (Sec. 5.1). This method can be easily adapted to the design requirements for both slabs under gravity loads (transverse loads) and under seismic loads (in-plane loads). Using the same calculation elements, we can approach the static analysis of slab bridges resting of columns subjected to live loads, with the aid of influence surfaces. Finally, in resorting to the dimensionless coefficients k_w and k_φ and K_w and K_φ, respectively, for the description of dynamic point-like displacements, a method of dynamic analysis was obtained (Chap. 6). In this method use is made of the flexibility matrix method for the analysis of slab structures subjected to dynamic loads [83], [86].

In the general displacement method, a different calculation approach proceeds from the evaluation of the effective plate width in order to determine the relative bending stiffness of the plate and the distribution factors of the moments that are out-of-balance at the nodes. In this manner, it is possible, for instance, to design multistoried symmetrical structures subjected to horizontal forces, which cannot be treated with any of the approximate methods (Sec. 5.2). It should be mentioned that, with rare exceptions (Sub-sec. 1.2), recent theoretical investigations deal only with slab structures subjected to transverse (gravity) loads.

★

Considered as multiply statically indeterminate systems, slab structures can be treated with any of the general methods used in the static analysis of reticular structures. However, in the present stage of investigations, the general displacement method leads to over-all design stresses for each strip of the slab. The distribution of stresses transverse to the direction in which calculations are being performed is still empirical, as are all the approximate methods recommended in current prescriptions.

In the author's method of analysis [75], [78], [86] the use of the general force method allows an accurate calculation of the sectional bending moments M_x and M_y and of the sectional torsional moments M_{xy} at any arbitrary point (x, y) of the slabs and throughout the structure. Through data processing the designer can easily solve the systems of conditional equations and the accuracy of the computations is that imposed on the programs written for the computation of the dimensionless coefficients (Sec. 3.4).

5 A new method of elastic analysis

5.1 Use of the general force method. Transverse and in-plane loads

5.1.1 General scheme

5.1.1.1 Basic systems and condition equations. The method given herein for the static analysis of slab structure is rigorous within the limits set by the basic assumptions of the Theory of Plates and Theory of Structures. The theoretical foundation of this method has been developed in Sec. 2.1.

The boundary conditions for the three categories of plates under study were assumed to be free-edge and simply-supported-edge. Based on these basic systems, other boundary conditions, such as the cantilever-free-edge and the rigidly-clamped-edge, can be introduced in the analysis (Fig. 5.1).

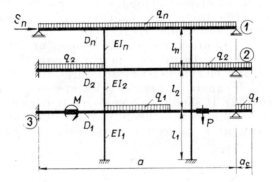

Figure 5.1 Unsymmetrical slab structure subjected to unsymmetrically distributed loads. (1) simply supported edge; (2) cantilever free edge; (3) rigidly clamped edge; q_n — uniformly distributed load throughout floor surface; q_i, q_1 — loads uniformly distributed on limited areas of floors; P — locally distributed force; M — surface couple as load.

The connection between the slabs and columns of the structure is assumed to be rigid (variants V_1 and V_2). In a predesign stage, the calculations can be simplified by disregarding the conditions of zero relative rotation at the slab-column node. Hence, the system of conditional equations reduces to (Sec. 1.2): $w_{c,s} = 0$ (1.7) or $\mathbf{Dp} + \mathbf{l}_k = 0$ (1.42). Such systems of equations, which reduce to the conditions of zero transverse displacement at the slab-column nodes, are used in all the known studies, save for Pfaffinger's latest approach [71]. To this simplified method of

5.1 General force method. Transverse and in-plane loads

analysis — hereinafter referred to as the *approximate variant* (V_0) — there corresponds a real static scheme with which we are seldom confronted in design practice : the slab of the floors is supported by columns of negligible stiffness or the slab-column node is equivalent to a hinge (Sub-sec. 5.1.2.1).

The gravity (transverse) loads are of the type $q\,(\text{N/m}^2)$, $P\,(\text{N})$ or $M\,(\text{Nm})$ and they act directly on the slabs of the structure (Figs. 5.1, 5.2A and 5.2B). In calculations, the load q can be assumed to be uniformly distributed throughout the surface $a \times b$ of the floor, on strips or in chess-board fashion. Account can be also taken of linearly distributed loads of the type $\bar{q}\,(\text{N/m})$ or $\bar{P}\,(\text{N/m})$. As in the case of reticular structures, here too, the horizontal loads of type $S(\text{N})$ are applied at the level of each floor (Figs. 5.1, 5.2A and 5.2C).

The basic system chosen for implementing the present method of analysis is determined by the necessity of suppressing the slab-column interactions (Fig. 5.2). In this way, all the slabs of the floors can be assimilated with one of the previously examined plate diagrams (Chaps. 2, 3 and 4). Irrespective of the type of given load (gravity forces, seismic forces, etc.), its effect on the structure is transmitted through the slabs, which can be assumed to be plates subjected to given loads of type q, P or M.

Besides, the slabs of the basic system are subjected to three types of generalized interacting forces (Fig. 5.2): the P-type force reactions X_i normal to the plane of slabs, the M-type couple reactions X_j and the force reactions X_k acting on the mid-plane of the slabs. The connecting forces of type X_k, which are specific responses of slab structures to the load induced by a set of horizontal forces S (Figs. 5.2A and C), are not directly involved in the calculation of the transverse displacements w and rotations φ_x and φ_y of slabs (Sec. 4.1). Hence, the generalized interacting forces required to express the elastic unit displacements in the compatibility equations (5.1) are either couples M or forces P locally distributed on the slab-column area $2c \times 2d$. The moment vector $\vec{M}$ can develop in either of the main directions Ox or Oy of the slab plane, the surface couple M being either a given load or a generalized interacting force of type X_j (Fig. 5.2).

Hence, by cancelling the slab-column connections, the elements of the basic scheme consisting of the slabs are subjected only to loads q, P and M. The statical effects of these loads are expressed in the form of algorithms for which computational programs are available [75], [78], [82], [86].

★

Since the basic system subjected to the given loads q, P and M or S and the generalized interacting forces of type X_i, X_j

5 A new method of elastic analysis

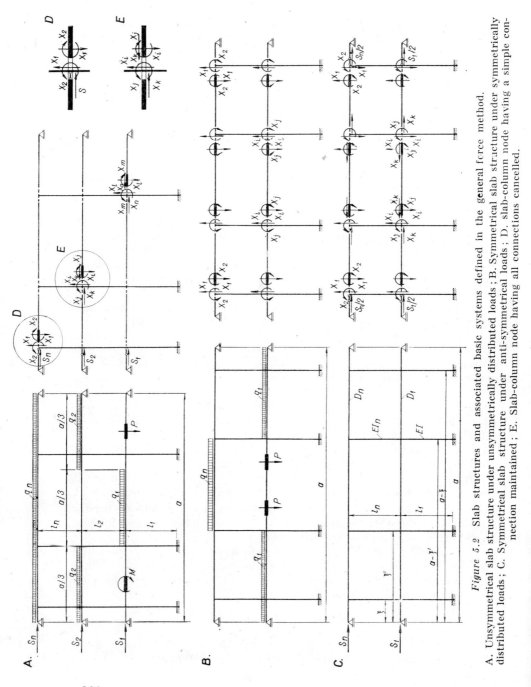

Figure 5.2 Slab structures and associated basic systems defined in the general force method.
A. Unsymmetrical slab structure under unsymmetrically distributed loads; B. Symmetrical slab structure under symmetrically distributed loads; C. Symmetrical slab structure under anti-symmetrical loads; D. slab-column node having a simple connection maintained; E. Slab-column node having all connections cancelled.

5.1 General force method. Transverse and in-plane loads

or X_k corresponding to the cancelled connections, must behave in a manner identical to the initial structure (Fig. 5.2), the compatibility equations express n conditions of zero relative displacement in the direction of each of the n additional unknowns X_i:

$$\mathbf{d X + D + C} = 0, \tag{5.1}$$

where
- **d** represents the square matrix of unit displacements δ_{ij} generated in the direction of the unknown X_i by the load $X_j = 1$ (a locally distributed force or a surface couple) successively applied on the basic system ($i = 1, 2, \ldots, n$; $j = 1, 2, \ldots, n$);
- **X** denotes the vector of the additional unknowns X_j;
- **D** is the vector of the displacements Δ_{is} induced in the direction of the unknown X_i by the given loads q, P, M or S applied on the basic system;

and
- **C** represents the column matrix including the displacements Δ_{ic} induced at the cancelled interaction in the direction of the unknown X_i by the known failure of supports.

The displacements $X_j \cdot \delta_{ij}$, Δ_{is} and Δ_{ic} can be either linear relative displacements or relative rotations. The linear relative displacement at the node i is equal to the transverse elastic displacement w_i of the slab in the axis of the column i. Evidently, it is also possible to introduce in the analysis the elastic deformation of column i in the longitudinal direction. In this case, the linear relative displacements of i is given by the sum of the transverse elastic displacement w_i of the slab and the longitudinal elastic displacement $w_{i\,col}$ of the column i. The relative rotation at the node i is obtained by summating the rotations φ_{xi} and φ_{yi} of the slab, respectively, and the corresponding rotation $\varphi_{i\,col}$ of the column at the cancelled connection i.

The displacements Δ_{ic} induced by known failure of support in the unknown direction X_i are inserted in the analysis also as relative displacements between the slab and the column at the level of the node i. These displacements can be translations or rotations according to whether X_i represents a force reaction or a couple reaction. It follows that any linear or angular total relative displacement Δ_i in the direction of the interacting force X_i at the node i ($i = 1, 2, \ldots, n$) is due to the unknowns X_j ($j = 1, 2, \ldots, n$), the given loads q, P and M and eventually to the failure of the supports.

In design practice, a given slab structure can be analysed by use of several loading diagrams. Meanwhile, several assumptions concerning the failure of supports can be adopted. In this

5 A new method of elastic analysis

case, the system of equations (5.1) becomes

$$\mathbf{dX} + \mathbf{D}_h + \mathbf{C}_k = \mathbf{0}, \qquad (5.1a)$$

where

$\mathbf{D}_h$ denotes one of the columns of the rectangular matrix $\mathbf{D}$ of the displacements Δ_{is}, corresponding to the loading diagram h of the basic system under the given loads ($h = 1, 2, \ldots, l$);

$\mathbf{C}_k$ is one of the columns of the matrix $\mathbf{C}$ of the displacements Δ_{ic}, corresponding to the loading diagram k of the basic system, with support failure of known magnitude ($k = 1, 2, \ldots, m$).

5.1.1.2 Calculation of the elements included in the matrices $\mathbf{d}$ *and* $\mathbf{D}$ *in the general case.* In order to write the square matrix $\mathbf{d}$, in the calculation of the unit displacements δ_{ij}, which are included in this matrix, use is made of the dimensionless coefficients of type k_w, k_φ^x and k_φ^y, corresponding to loads of type P and M acting on the slabs. Thus, for instance, if the effect of rotations in the direction of the Oy axis is neglected in the analysis (i.e. we disregard the effect of the couple reactions X_j, the moment vector of which is parallel to the Ox axis), four cases can be established:

a. If X_i is a force reaction, then the unit displacement δ_{ij} induced in the direction of this force reaction is a linear displacement given by (Table 2.2 — formulae 2.51)

$$(5.2) \quad \begin{cases} \delta_{ij} = \dfrac{a^2}{D} k_w^P, & \text{provided that } X_j = 1 \text{ is an unknown force of type } P; \\ \delta_{ij} = \dfrac{a}{D} k_w^M, & \text{provided that } X_j = 1 \text{ is a couple reaction of type } \vec{M} \parallel Oy; \end{cases}$$

b. If X_i is a couple unknown, then the unit displacement δ_{ij} induced in the direction of this couple is a rotation given by (Table 2.2 — formulae 2.52):

$$(5.3) \quad \begin{cases} \delta_{ij} = \dfrac{a}{D} k_\varphi^{xP}, & \text{provided that } X_j = 1 \text{ is a force reaction of type } P; \\ \delta_{ij} = \dfrac{1}{D} k_\varphi^{xM}, & \text{provided that } X_j = 1 \text{ is a couple unknown of type } \vec{M} \parallel Oy. \end{cases}$$

A similar procedure is used to write the column matrix $\mathbf{D}$ of the displacements Δ_{is} generated by the given loads in the direction of the unknown X_i ($i = 1, 2, \ldots, n$). As a first approximation we assume that the effect of the rotations in the direction Oy is

5.1 General force method. Transverse and in-plane loads

disregarded in the analysis, and consider that the basic system is subjected to only one locally distributed force P, a surface couple $\vec{M} \parallel Oy$ and the load q uniformly distributed throughout the slab surface. Two cases can be established:

c. If X_i is an unknown force, then the displacement Δ_{is} induced in the direction of this force at node i is a liniar displacement given by (Table 2.2—formulae 2.5.1):

$$\Delta_{is} = \frac{Pa^2}{D} k_w^P + \frac{Ma}{D} k_w^M + \frac{qa^4}{D} k_w^q; \qquad (5.4)$$

d. If X_i is a couple reaction, then the displacement Δ_{is} induced in the direction of this reaction at node i is a rotation given by (Table 2.2 — formulae 2.52):

$$\Delta_{is} = \frac{Pa}{D} k_\varphi^{xP} + \frac{M}{D} k_\varphi^{xM} + \frac{qa^3}{D} k_\varphi^{xq}. \qquad (5.5)$$

Expressions (5.2)—(5.5) for the calculation of the displacements δ_{ij} and Δ_{is} will be used to illustrate the implementation of the present method in the study of the approximate variant V_0 and the variant V_1, in which the effect of the rotations in the direction Oy is assumed to be negligible (Sub-sec. 5.1.2.1 — Variant V_1 and Sub-sec. 5.1.2.2). Similar expressions will be used to illustrate the manner in which the method is applied in the dynamic analysis of slab structures (Chap. 6).

In the case of a rigorous static analysis, the effect of the rotations in the direction Oy (i.e. the effect of loads of type $\vec{M} \parallel Oy$ can no longer be neglected (Sub-sec. 5.1.2.1 — Variant V_2). Under this assumption, to the expressions (5.2)—(5.5), we add the following:

e. If $X_j = 1$ is a couple reaction of type $M(y)$, whose moment vector is parallel to the Ox axis, then the displacement δ_{ij} induced in the direction of the unknown X_i is given by (Table 2.2 — formulae 2.51—2.53):

$$(5.6) \begin{cases} \delta_{ij} = \dfrac{a}{D} k_w^{M(y)}, & \text{which is a linear displacement if } X_i \text{ is a force reaction of type } P; \\ \delta_{ij} = \dfrac{1}{D} k_\varphi^{xM(y)}, & \text{which is a rotation in the direction } Ox \text{ if } X_i \text{ is an unknown couple of type } \vec{M} \parallel Oy; \\ \delta_{ij} = \dfrac{1}{D} k_\varphi^{yM(y)}, & \text{which is a rotation in the direction } Oy \text{ if } X_i \text{ is a couple reaction of type } \vec{M} \parallel Ox; \end{cases}$$

f. If the basic system is subjected to a surface couple of type $M(y)$, whose moment vector is parallel to the Ox axis, then the displacement Δ_{is} induced in the direction of the reaction X_i only by this external load, is given by:

$$(5.7) \begin{cases} \Delta_{is} = \dfrac{Ma}{D} k_w^{M(y)}, & \text{which is a linear displacement if } X_i \\ & \text{is an unknown force of type } P; \\[1em] \Delta_{is} = \dfrac{M}{D} k_\varphi^{xM(y)}, & \text{which is a rotation in the direction } Ox \\ & \text{if } X_i \text{ is a couple reaction of type } \vec{M} \parallel Oy; \\[1em] \Delta_{is} = \dfrac{M}{D} k_\varphi^{yM(y)}, & \text{which is a rotation in the direction } Oy \\ & \text{if } X_i \text{ is an unknown couple of type } \\ & \vec{M} \parallel Ox. \end{cases}$$

The set of expressions (5.2)—(5.7) allows the calculation of those elements of the matrices **d** and **D** which define exclusively the relative displacements of the slabs of the basic system. The determination of the displacements δ_{ij} induced by the reactions $X_j = 1$ and the displacements Δ_{is} caused by the external loads applied on the basic system is first discussed here with respect to plates treated as structural members.

Approaching now the relative elastic displacements of the column i at the level of node i we notice that they can be transverse (u_{icol}, v_{icol}) or longitudinal (w_{icol}) linear displacements, or relative rotations (φ_{icol}) in one or both directions of the plane xOy. As the columns are treated as members of the basic system, the values of these elastic displacements are determined by means of the Theory of Structures.

5.1.1.3 *Systematization recommended in the case of associated unknowns.* In design practice we are often confronted with symmetrical slab structures or structures regular in shape where, for instance, the columns in a row are spaced at an equal distance or the loads are uniformly distributed (see Chap. 4). Since the basic system of these structures permits the use of the principle of superposition of effects, it is best to resort to the grouping of unknowns in the elastic—static or dynamic—analysis.

The effect of the symmetrical and antisymmetrical sets of associated couples M and of the locally distributed forces P on plates has been examined with respect of the magnitude and dis-

5.1 General force method. Transverse and in-plane loads

tribution of the elastic displacements w, φ_x, φ_y and to the bending M_x, M_y and twisting M_{xy} moments (Sub-secs. 4.1.1—4.1.3).

In the case of sets of unknowns X_i, the elements of the matrices **d** and **D** are calculated in a manner similar to the unit displacements δ_{ij} and the relative displacements Δ_{is} which occur in the general case (Sub-sec. 5.1.1.2). Indeed, if instead of the dimensionless coefficients k_w, k_φ^x and k_φ^y, corresponding to only one unknown X_i, we introduce the dimensionless coefficients K_w^P, K_φ^{xP}, M_φ^{yP} and M_w^M, M_φ^{xM}, M_φ^{yM} corresponding to associated unknowns of type P or M, then we can establish the expressions of the sets of displacements of type δ_{ij} and Δ_{is}. For the sake of simplicity, we neglect in a first approximation the effect of the sets of couple reactions X_j, the moment vector of which is parallel to the Ox axis, and the effect of the imposed loads of type $M(y)$. By virtue of this approximation, the rotations φ_y induced in the direction of the Ox axis are disregarded in the analysis. This effect is also examined in Sub-sec. 5.1.2.1 (Variant V_2), which illustrates the use of the new method of analysis.

The associated unit displacements δ_{ij} induced in the direction of the unknown X_i by sets of unknowns X_j are calculated by means of expressions (5.8) and (5.9) given in Table 5.1. These formulae, which are similar to expressions (5.2) and (5.3), are derived from (2.51) and (2.52) given in Table 2.2 for particular values $P = 1$ and $M = 1$ assigned to the associated unknown $X_j = 1$.

Table 5.1

Expressions for elements of matrix **d** in the case of associated unknowns X_i, X_j

If X_i represents an	the unit displacement δ_{ij} upon the direction of X_i is an	if $X_j = 1$ represents an
associated force-unknown,	associated linear displacement (5.8) $\begin{cases} \delta_{ij} = \dfrac{a^2}{D} K_w^P, \\ \delta_{ij} = \dfrac{a}{D} K_w^M, \end{cases}$	associated reaction-force ; associated reaction-couple ;
associated couple-unknown,	associated rotation (5.9) $\begin{cases} \delta_{ij} = \dfrac{a}{D} K_\varphi^{xP}, \\ \delta_{ij} = \dfrac{1}{D} K_\varphi^{xM}, \end{cases}$	associated reaction-force ; associated reaction-couple.

The set of relative displacements Δ_{is} induced in the direction of the unknown X_i by subjecting the basic system to associated given loads of type q, P, M or S is calculated by use of expressions

5 A new method of elastic analysis

(5.10)−(5.12) given in Table 5.2. These formulae, which are similar in form to expressions (5.4) and (5.5), stem straightly from (2.51) and (2.52) given in Table 2.2, by replacing the dimensionless coefficients of type k_w and k_φ^x by the coefficients K_w and K_φ^x corresponding to the respective sets of given loads.

Table 5.2

Expressions for the elements of vector **D** in the case of associated given loads and associated unknowns X_i

If the given load represents a	the relative displacement Δ_{is} upon the direction of X_i is an		if X_i represents an
set of locally distributed forces P,	associated linear displacement (5.10)	$\Delta_{is} = \dfrac{Pa^2}{D} K_w^P,$	associated reaction-force;
	associated rotation	$\Delta_{is} = \dfrac{Pa}{D} K_\varphi^{xP},$	associated reaction-couple
set of surface couples M,	associated linear displacement (5.11)	$\Delta_{is} = \dfrac{Ma}{D} K_w^M,$	associated reaction-force;
	associated rotation	$\Delta_{is} = \dfrac{M}{D} K_\varphi^{xM},$	associated reaction-couple;
uniformly distributed load q,	associated linear displacement (5.12)	$\Delta_{is} = \dfrac{qa^4}{D} k_w^q,$	associated reaction-force;
	associated rotation	$\Delta_{is} = \dfrac{qa^3}{D} k_\varphi^{xq},$	associated reaction-couple.

Note. Applied on symmetrical slab structures, the uniformly distributed load q throughout the floor area induces symmetrical elastic displacements in symmetrically located points. Hence, the values of the dimensionless coefficients K_w^q and K_φ^{xq}, corresponding to an associated unknown X_i, are equal to the values of the coefficients k_w^q and k_φ^{xq} in the expressions (5.4) and (5.5), respectively (see Sub-secs. 5.1.1.2 and 5.1.2.1).

Finally, in the case of basic systems which permit the grouping of the unknowns, the displacement Δ_{ic}, induced by known failure of supports, can be an associated linear displacement or an associated rotation depending on whether X_i represents a set of force-reactions or of couple-reactions, respectively.

5.1.1.4 *Expressions for the elastic displacements and stresses in the initial statically indeterminate slab structure.* Once the values of the generalized interacting forces X_i have been determined for

5.1 General force method. Transverse and in-plane loads

a given loading diagram, we can calculate the values of the linear elastic w and angular φ_x, φ_y displacements, moments M_x, M_y, M_{xy} and boundary reactions R_x, R_y of the initial statically indeterminate system at any point (x, y) in the slab plane. This can be done by means of the known dimensionless coefficients and through the superposition of effects.

For instance, if X_i represents an associated force-reaction, X_j denotes a set of couple reactions, P is a set of locally distributed forces P, M denotes a given surface couple, and q is a uniformly distributed load throughout the surface of the floor, using formulae (2.51), (2.52), (2.54) and (2.59), and performing the superposition of effects, we shall have:

$$w(x, y) = \frac{X_i a^2}{D} K_w^P + \frac{X_j a}{D} K_w^M + \frac{Pa^2}{D} K_w^P + \frac{Ma}{D} k_w^M +$$

$$+ \frac{qa^4}{D} k_w^q \text{ (m)} \tag{5.13}$$

$$\varphi_x(x, y) = \frac{X_i a}{D} K_\varphi^{xP} + \frac{X_j}{D} K_\varphi^{xM} + \frac{Pa}{D} K_\varphi^{xP} + \frac{M}{D} k_\varphi^{xM} +$$

$$+ \frac{qa^3}{D} k_\varphi^{xq} \text{ (rad)} \tag{5.14}$$

$$M_x(x, y) = X_i K_M^{xP} + \frac{X_i}{a} K_M^{xM} + P K_M^{xP} + \frac{M}{a} k_M^{xM} +$$

$$+ qa^2 k_M^{xq} \text{ (Nm/m)} \tag{5.15}$$

$$R_x(x, y) = \frac{X_i}{a} K_R^{xP} + \frac{X_j}{a^2} K_R^{xM} + \frac{P}{a} K_R^{xP} + \frac{M}{a^2} k_R^{xM} +$$

$$+ qa k_R^{xq} \text{ (N/m)}. \tag{5.16}$$

Similar expressions can be obtained irrespective of the nature of the generalized interacting forces (associated or not) and of the type of loading diagram which may consist of loads of type q, P, M or S that may be associated or not.

The diagrams showing the variation of the dimensionless coefficients, and their ordinary values in the tables (Chaps. 3 and 4 and the Appendix) facilitate the implementation of the present method of static analysis and the use of expressions such as (5.13)—(5.16) in the calculation of the elastic displacements and stresses generated in the initial statically indeterminate structure.

5 A new method of elastic analysis

5.1.2 Influence of the relative slab-column stiffness on the value of displacements and stresses in slabs and columns

The comparative studies carried out so far [68], [70], [73], [75,] [78], [86] by the author shows that (Chaps. 3 and 4):
- the present analysis method is simplified and efficient;
- the physical phenomenon can be mathematically modelled with good accuracy;
- it is possible to obtain a faithful picture of the distribution of elastic displacements and stresses in slabs and throughout the structure;
- the various parameters in the equations of the deflected mid-surfaces of plates $w(x, y)$ bear specifically on the values of these static quantities;
- by inserting the column stiffness in the analysis, a beneficial effect is obtained which results in a smaller value of displacements and design stresses.

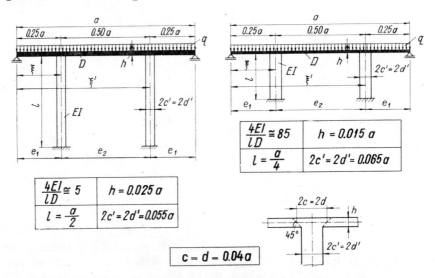

Figure 5.3 Static diagram of comparatively analysed slab structures.

In order to illustrate the implementation of the present computation method let us consider a simple one-storied static diagram having a two-fold topological (geometrical and mechanical) symmetry (Fig. 5.3). The same diagram was used for the static (Chap. 5) and dynamic (Chap. 6) analysis of slab structures.

5.1 General force method. Transverse and in-plane loads

The simplified computation scheme adopted throughout the comparative studies allows a complete and relevant image of most qualitative and quantitative aspects under examination. Additionally, because the type of structure under study is not confined to the applied range of the approximate calculation methods [154], [155], [158], [160], [162], it follows that the method given herein is also valid outside this range. The size b of the slab, measured in the direction of the Oy axis, covers the entire range $a \leqslant b \leqslant \infty$.

In order to have a clear image of the effect of slab-column rigid connection, the analysis was carried out for any relative slab-column stiffness lying within a range

$$10 \leqslant \frac{4\,EI}{l\,D} \leqslant 80 \tag{5.17a}$$

so wide as to include all the cases with which we may be confronted in design practice (Figs. 5.3—5.7, 5.9—5.13 and Chap. 6). In the analysis of slab structures subjected to transverse loads of type q, P or M, a comparative approach was made, including the approximate variant (V_0)

$$\frac{4\,EI}{l\,D} = 0. \tag{5.17b}$$

Additionally, the comparative examination of the values of the unknowns encompassed also the theoretical case of columns that are perfectly rigid with respect to the slab stiffness or the case of the slab of zero stiffness against bending:

$$\frac{4\,EI}{l\,D} = \infty. \tag{5.17c}$$

By virtue of the systematized treatment adopted, the static analysis of a symmetrical structure that is symmetrically (or antisymmetrically) loaded through the general force method leads to a solution of a system having a small number of conditional equations if referred to the complexity of the structure. For instance, for all the one-storey slab structures pertaining to this category which can be schematically assimilated with a plate supported on the boundary and on two rows of interior equidistant columns, system (5.1) reduces to only three compatibility equations for the slab-column node by grouping the unknowns, irrespective of the number of column pairs (Figs. 5.3—5.10):

— an equation for the associated relative linear displacements w in the direction of the Oz axis (5.18a);

— an equation for the associated relative rotations φ_x in the direction of the Ox axis (5.18 b);

— an equation for the associated relative rotations φ_y in the direction of the Oy axis (5.18c);

Under the assumption — generally adopted in the analysis of reticular structures — that the linear elastic deformation of columns in the longitudinal direction is negligible, and assuming that the displacements induced by the failure of supports are zero, the three compatibility equations for one of the two symmetrical nodes of the column pair are *

$$X_1\delta_{11} + X_2\delta_{12} + X_3\delta_{13} + \Delta_{1s} = 0; \quad (5.18a)$$

$$X_1\delta_{21} + X_2\delta_{22} + X_3\delta_{23} + \Delta_{2s} = 0; \quad (5.18b)$$

$$X_1\delta_{31} + X_2\delta_{32} + X_3\delta_{33} + \Delta_{3s} = 0; \quad (5.18c)$$

where X_1 represents the set of force reactions in the axis of the columns that are locally distributed on the loaded area $2c \times 2d$; X_2 denotes the set of couple reactions acting on the same area and having the moment vector $\vec{M} \parallel Oy$; X_3 is the set of couple-interacting forces distributed on the same loaded area, having the moment vector $\vec{M} \parallel Ox$.

In order to make evident the effect of rigid slab-column connections, in the first set of variants (V_1) under comparative study, the effect of the set of couple reactions of type X_3 was assumed to be negligible ($X_3 = 0$). Inserting in the analysis only the first of the two equations for rotations, which corresponds to the moment vector $\vec{M}$ parallel to the longer side of the slab (the Oy axis), system (5.18) reduces to a set of only two conditional equations:

$$X_1\delta_{11} + X_2\delta_{12} + \Delta_{1s} = 0; \quad (5.19a)$$

$$X_1\delta_{21} + X_2\delta_{22} + \Delta_{2s} = 0, \quad (5.19b)$$

where X_1 and X_2 are defined as above (5.18).

Finally, bearing in mind that even in most of the studies published of late (Sec. 1.2) the effect of the slab-column interaction is neglected in both directions ($X_2 = X_3 = 0$), in the present comparative approach we have also considered the approximate variant (V_0), where the condition of zero relative linear displacement w in the column axis represents the only compatibility

---

* No calculation method has so far permitted the writing of the last two compatibility equations.

5.1 General force method. Transverse and in-plane loads

equation:
$$X_1\delta_{11} + \Delta_{1s} = 0, \qquad (5.20)$$

where X_1 is defined as in (5.18) and (5.19). Since in the approximate variant the rigidity of columns must be assumed to be negligible, and hence, the slab-column node is assimilated with a hinge, the solutions obtained with the aid of Eqn. (5.20) are not a faithful image of the real phenomenon.

Summing up what has been stated so far, we can righteously consider that the solutions obtained by use of the system of three equations (5.18) with three unknowns X_1, X_2, X_3 (Variant V_2) are exact for the static diagram used in the comparative approach. Referred to these solutions, those obtained with the aid of the system (5.19) of two compatibility equations (Variant V_1) may be used to advantage as the amount of processed data is smaller. This simplified treatment bears, no doubt, on the accuracy of calculations (Sub-sec. 5.1.2.1).

Assuming that the topological symmetry of the structure is maintained, the expression of the systems (5.18) and (5.19) remains simple irrespective of the boundary conditions (clamped, elastically supported, simply supported or free edges, cantilever, etc.) set for the slab. Additionally, the simplified form of these equations is unaffected by the distribution rule followed by the uniformly distributed load q (throughout the slab surface, on strips or in chessboard fashion) or by the inclusion of given loads of type M, P or $\bar{q}$.

If, in the case of infinitely long symmetrical slabs, the symmetrical (or antisymmetrical) distribution of loads is retained, the number of three (or two, respectively) conditional equations increases only as the number of column rows is greater: 6(4) equations for 3 or 4 rows, 9(6) equations for structures with 5 or 6 column rows. In the case of symmetrical finite-boundary slabs, the number of associated unknowns grows also with the number of columns in a row: 6(4) equations for 3 or 4 columns, 9(6) equations for structures having 5 or 6 columns in a row.

5.1.2.1 *Transverse loads.* Let us consider the class of single-storied symmetrical slab structures supported by two rows of columns at a quarter of the span a ($\xi = 0.25\,a$; $\xi' = 0{,}75\,a$). The size of the longer side of the slab, which is parallel to the Oy axis, ranges between $b = a$ and $b = \infty$ (Figs. 5.3—5.8). The edges of the floor are assumed to be simply supported at $x = 0$ and $x = a$ and the columns in the two rows are assumed to be clamped at the lower end. For the sub-class of structures where the slab of the floor is assimilated with a finite-boundary plate, the transverse edges are also assumed to be simply supported at $y = 0$ and $y = b$.

In the case of structures where the slab of the floor is equivalent to the infinitely long plate ($b = \infty$), the number of columns

5 A new method of elastic analysis

spaced in a row at an equal distance e in the direction of the Oy axis is indeterminate. However, since the values of the elastic displacements and stresses vanish at a distance $y = \pm 2a$ from the loaded area, the effect of the columns located farther away from this area is negligible, irrespective of the degree of accuracy imposed by the programs (Chaps. 3 and 4 and the Appendix). This explains why the number of columns in a row is limited to nine for $e = 0.50a$, and five for an equidistance ranging between $e = 0.75a$ and $e = a$ (Sub-secs. 0.3.2, 4.1.1.1 and 4.1.2.1).

For finite-boundary slab structures, the ratio of the longer-to-shorter span of the rectangular plate with which the slab of the floor has been assimilated was here limited to $1 \leqslant b/a \leqslant 3$ (Secs. 3.1—4.1). In the analysis of this particular sub-class of structures, the distribution of columns in the longitudinal direction was assumed to mimic the distribution in the transverse direction, i.e. at a quarter of the span b ($\eta = 0.25 b$; $\eta' = 0.75b$) (Fig. 5.4). The load q is considered to be uniformly distributed throughout the surface of the floor.

Variant 1

Since the symmetrical structure is subjected to symmetrical loads, a simplified basic system is adopted by grouping the unknowns (Fig. 5.4). The effect of the rigid slab-column connections is des-

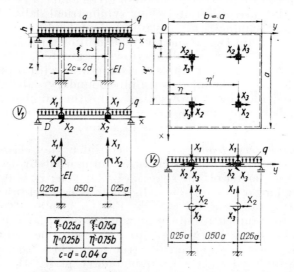

Figure 5.4 Transverse symmetrical loads. Basic system associated to comparatively analysed static diagram.
V_1 — Variant 1. V_2 — Variant 2.

5.1 General force method. Transverse and in-plane loads

cribed by setting the zero relative rotation upon the Ox axis alone. This means that we have to resort to the system of equations (5.19).

In the calculation of the relative displacements of the slab, corresponding to the associated unknowns X_1 and X_2, we insert the expressions (5.8)—(5.12). The expression for the associated relative rotation of the columns of length l, induced at the slab-column nodes by the set of couple-interacting forces X_2 in the direction Ox,

$$\varphi^x_{col} = \frac{X_2 l}{4 EI_x} \text{(rad)}, \tag{5.21}$$

is valid in the case of non-displacing nodes when the far end of the column is clamped. Using these notations, the system of compatibility equations (5.19) becomes

$$X_1 \frac{a^2}{D} K^P_w + X_2 \frac{a}{D} K^M_w + \frac{qa^4}{D} k^q_w = 0 ; \tag{5.22 a}$$

$$X_1 \frac{a}{D} K^{xP}_\varphi + X_2 \left(\frac{1}{D} K^{xM}_\varphi + \frac{l}{4EI_x} \right) + \frac{qa^3}{D} k^{xq}_\varphi = 0. \tag{5.22 b}$$

Denoting the dimensionless factor defining the relative column-slab stiffness in bending by

$$\frac{1}{K^{col}_\varphi} = \frac{4 EI_x}{lD}, \tag{5.23}$$

where

K^{col}_φ represents a relative slab-column flexibility;
EI_x (Nm²) is the bending stiffness of columns in the direction of the Ox axis;

and

D (Nm²/m) denotes the cylindrical stiffness of the slab, given by (1.2),

the system of equations (5.22) assumes a simple form that is more convenient in calculations:

$$X_1 a K^P_w + X_2 K^M_w + qa^3 k^q_w = 0 ; \tag{5.24a}$$

$$X_1 a K^{xP}_\varphi + X_2 (K^{xM}_\varphi + K^{col}_\varphi) + qa^3 k^{xq}_\varphi = 0. \tag{5.24b}$$

In the foregoing comparative static approach, the solutions obtained by solving the system (5.24) (Variant V_1) were referred to those derived in the approximate variant (V_0), where the effect of the rigid slab-column connections is disregarded. This simplifying assumption, which is frequently adopted in classical and modern approaches, leads to only one compatibility equation for the static diagrams under examination. Inserting expressions (5.8) and (5.12) in (5.20) yields the equation

$$X_0 \frac{a^2}{D} K_w^P + \frac{qa^4}{D} k_w^q = 0, \qquad (5.25)$$

which expresses the condition of zero transverse displacement of the slab in the axis of the columns. X_0 above represents the set of force reactions in the variant V_0.

Equations of type (5.25) can be used in the simplified calculation of the associated unknowns X_0 in the axes of the columns of a symmetrical slab structure subjected to a load q uniformly distributed in strips or in chess-board fashion, provided that the loading symmetry is retained.

★

The values of the associated interacting forces X_1 and X_2, obtained in variant V_1, and X_0, obtained in the approximate variant V_o, were calculated for three particular types of slabs (Fig. 5.5):

A — the square plate ($b = a$; $e = 0.50\,a$);

B — the infinitely long plate ($b = \infty$; $e = 0.50\,a$);

C — the infinitely long plate ($b = \infty$; $e = 0.75\,a$).

For the category of finite-boundary plates, the study of the influence of the parameter b/a on the value of the associated unknowns was extended by solving the systems (5.24) and (5.25) for four variants: $b/a = 1$; 1.5; 2; 3 (Table 5.3).

In order to highlight the influence of the relative slab-column stiffness, which has barely been approached so far, the system of equations (5.24) was solved for five values of the stiffness ratio: $4\,EI/lD = 10, 20, 40, 80$ and ∞. The limit values under consideration ($0 \leqslant K_\varphi^{col} \leqslant 0.10$) cover the entire range of ordinary sizes of slabs and columns in the case of the given scheme. A sixth value $4EI/lD = 0$, corresponding to the case of columns of negligible stiffness, was introduced in the analysis by solving Eqn. (5.25) for all the values of the parameters b/a and e/a under consideration (Fig. 5.5 and Table 5.3).

The summed influence of the relative slab-column stiffness and of the parameters b/a and e/a on the values of X_1 and X_2 is illustrated by the sets of curves (Fig. 5.5) drawn for the square plate (curves A) and the infinitely long plate (curves B and C). The results obtained from the comparative examination of the variation of the associated unknowns X_1 and X_2 reinforce the idea

5.1 General force method. Transverse and in-plane loads

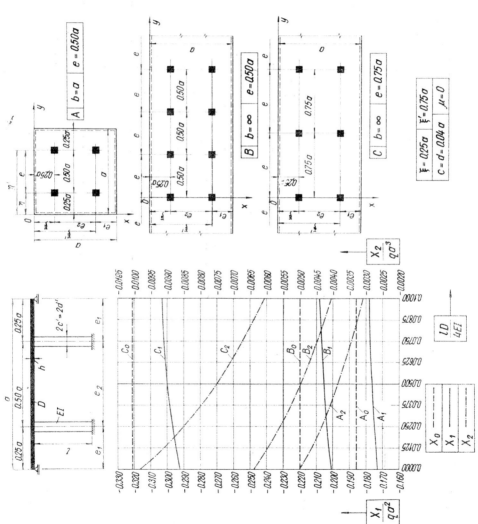

Figure 5.5 Slab structures under uniformly distributed gravity loads. Influence of relative slab-column stiffness and of parameters b/a and e/a on magnitude of associated unknowns X_1 (force-reaction) and X_2 (couple-reaction) at slab-column nodes. Variants V_0 and V_1 ($0 \leq 4EI/lD \leq \infty$).

Table 5.3

Slab structures subjected to uniformly distributed load q. Finite-boundary plates and infinitely long plates. The general force method — Variant 1

$$a \leqslant b \leqslant \infty; \quad \mu = 0$$

Position of columns: $\xi = 0.25a$; $\xi' = 0.75a$; $\eta = 0.25b$; $\eta' = 0.75b$ ($a \leqslant b \leqslant 3a$)
Cross-section of columns: $c = d = 0.04a$
Relative column-slab stiffness: $10 \leqslant 4EI/lD \leqslant \infty$

Coefficients of unknowns and free terms

Coefficients of unknowns and free terms	$\dfrac{4EI}{lD}$	$b = a$	$b = 1.5a$	$b = 2a$	$b = 3a$	$b = \infty$ $e = 0.75a$	$b = \infty$ $e = 0.50a$
$D\delta_{11}/a^2 = K_w^P$	—	0.01144	0.01517	0.01636	0.01676	0.02888	0.04212
$D\delta_{12}/a = K_w^M$	—	0.02946	0.04116	0.04490	0.04619	0.08414	0.12500
$D\delta_{21}/a = K_\varphi^{xP}$	—	0.02946	0.04116	0.04490	0.04619	0.08414	0.12500
$D\delta_{22} = K_\varphi^{xM}$	∞	0.23740	0.27322	0.28492	0.28907	0.40927	0.54415
$D\delta_{22} = K_\varphi^{xM} + K_\varphi^{st}$	80	0.24990	0.28572	0.29742	0.30157	0.42177	0.55665
	10	0.33740	0.37322	0.38492	0.38907	0.50927	0.64415
$D\Delta_{1s}^q/q\,a^4 = k_w^q$	—	0.00213	0.00412	0.00559	0.00733	0.00928	0.00928
$D\Delta_{2s}^q/q\,a^3 = k_\varphi^{xq}$	—	0.00630	0.01248	0.01706	0.02254	0.02865	0.02865

Comparative values of associated unknowns

$\dfrac{4EI}{lD}$	$b = a$ $(0.25 + 0.50 + 0.25)a$	$b = \infty$ $e = 0.50a$	$b = 1.5a$ $(0.375 + 0.75 + 0.375)a$	$b = \infty$ $e = 0.75a$	$b = 2a$ $(0.50 + 1.0 + 0.50)a$	$b = 3a$ $(0.75 + 1.50 + 0.75)a$
\multicolumn{7}{c}{X_1/qa^2}
∞	−0.1732	−0.2013	−0.2497	−0.2927	−0.3125	−0.3975
80	−0.1741	−0.2026	−0.2513	−0.2947	−0.3146	−0.4004
40	−0.1749	−0.2037	−0.2527	−0.2965	−0.3164	−0.4028
20	−0.1763	−0.2056	−0.2549	−0.2994	−0.3194	−0.4069
10	−0.1782	−0.2083	−0.2581	−0.3035	−0.3237	−0.4127
0*	−0.1862	−0.2203	−0.2715	−0.3213	−0.3417	−0.4374
\multicolumn{7}{c}{X_2/qa^3}
∞	−0.0050	−0.0064	−0.0081	−0.0098	−0.0106	−0.0145
80	−0.0047	−0.0060	−0.0075	−0.0091	−0.0099	−0.0134
40	−0.0044	−0.0056	−0.0070	−0.0085	−0.0092	−0.0125
20	−0.0039	−0.0050	−0.0062	−0.0075	−0.0081	−0.0110
10	−0.0031	−0.0041	−0.0050	−0.0061	−0.0066	−0.0089

* Comparative values calculated by use of the approximate method ($X_1 = X_0$; $X_2 = 0$).

Note. The accuracy condition imposed by the program is $ER = 10^{-6}$.

5.1 General force method. Transverse and in-plane loads

that the infinitely long plate and the finite-boundary plate (including the square plate as a limiting case) behave in a similar manner (see also Chaps. 3 and 4).

The influence of the parameter b/a is marked in the range $a \leqslant b \leqslant 3a$, for which the value of the unknowns X_1 increases 2.3 times and that of X_2 is 2.9 times larger (Table 5.3). For $b = \infty$, the influence of the parameter e is of the same order of magnitude. Thus, as the equidistance between columns increases from $e = 0.50a$ to $e = 0.75a$, the value of the force-unknown X_1 increases by 45—46 per cent and that of the couple unknown X_2 grows by 53—49 per cent.

If the span between columns is maintained equal ($e = 0.50a$), then the influence of the parameter b/a is more moderate: in the range $a \leqslant b \leqslant \infty$, the values of X_1 increase by only 16 per cent and those of X_2 grow by 28—32 per cent, greater values being obtained for small values of the relative stiffness $4EI/lD$. For the same value of the parameter b/a, the value of the associated force reactions X_1 becomes smaller and the value of the associated couple reactions X_2 increases substantially as the relative stiffness of columns becomes larger (Fig. 5.5 and Table 5.3).

The paramount influence of the relative column-slab stiffness ratio is made evident if we compare the values assumed by the couple unknown X_2 (curves A_2, B_2 and C_2) when $1/K_\varphi^{col}$ (5.23) varies within the ordinary range. Thus, in the range $10 \leqslant 4EI/lD \leqslant \infty$, X_2 increases by 61 per cent (for $b = a$) and by 62 per cent (for $b = 3a$), respectively. In the case of the infinitely long plate ($b = \infty$), the values of the couple reaction rises by 49 per cent (for $e = 0.50a$) and by 61 per cent (for $e = 0.75a$), respectively. As was predicted, X_2 assumes maximum values under the assumption of perfectly stiff columns ($4EI/lD = \infty$).

The maximum value of the force unknown X_1 is X_0, which represents the value of the reaction calculated in the approximate variant V_0. The values of X_0 (curves A_0, B_0 and C_0) are no doubt, independent of the value of the stiffness ratio $1/K_\varphi^{col}$ (Fig. 5.5). Meanwhile, the value of the force unknown X_1 (curves A_1, B_1 and C_1) diminishes with respect to its highest value X_0 — in the ordinary range of the stiffness ratio $10 \leqslant 4EI/lD \leqslant \infty$ — by 5—8 per cent (for $b = a$) and by 6—10 per cent (for $3a \leqslant b \leqslant \infty$), respectively. The values of X_1 decrease mostly (by 8 per cent through to 10 per cent) also under the assumption of the perfectly rigid columns, which is in compliance with expectations (Table 5.3). It should be noted that the variation of the associated force unknown X_1 with respect to the relative stiffness ratio $1/K_\varphi^{col}$ is only slightly dependent on the value of the parameter b/a and is almost unaffected by the parameter e/a in the case of the infinitely long plate ($b = \infty$).

5 A new method of elastic analysis

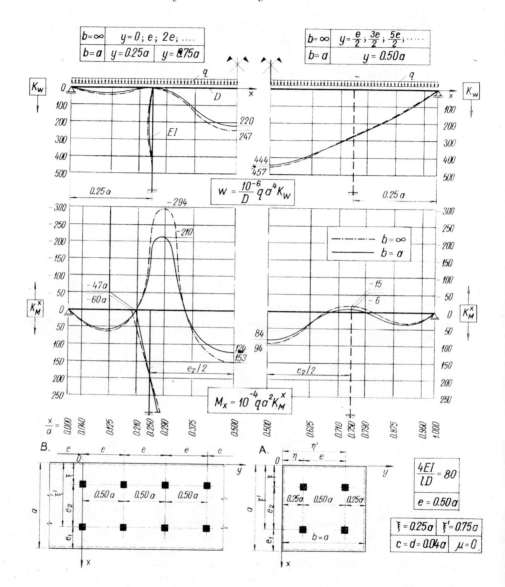

Figure 5.6 Slab structures under uniformly distributed gravity load q. Influence of parameter b/a on magnitude of transverse elastic displacements w and of sectional bending moments M_x. Variant V_1 ($4\,EI/lD = 80$).

5.1 General force method. Transverse and in-plane loads

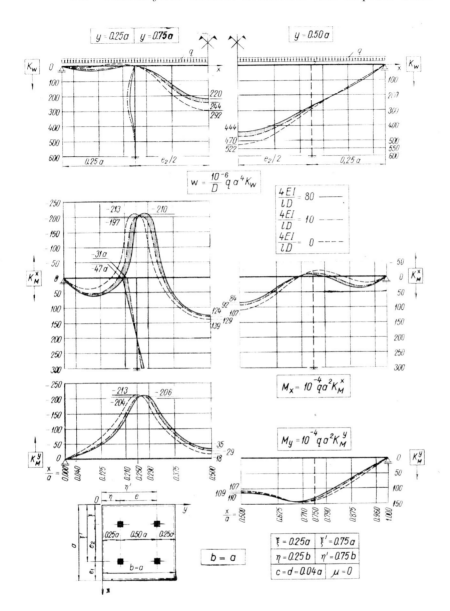

Figure 5.7 Slab structures under uniformly distributed gravity load q. Columns located the quarter of span. Influence of relative slab-column stiffness on magnitude of transverse elastic displacements w and of sectional bending moments M_x and M_y. Variants V_0 and V_1 ($0 \leqslant 4EI/lD \leqslant 80$).

5 A new method of elastic analysis

Further exploration of the structural effects was made by comparing schemes A (the square plate) and B (the infinitely long plate), where the columns are spaced at an equal distance ($e = 0.50\ a$) and the value of the relative slab-column stiffness $4EI/lD = 80$ is constant (Fig 5.6). An examination of the diagrams K_w and K_M^x shows that the static models employed reflect faithfully the behaviour of the real phenomenon. Thus, we notice that the two structures compared behave in a similar manner and that the values of the transverse displacements w and those of the bending moments M_x are only slightly different. Differences in values occur with no exception in compliance with expectations, i.e. larger absolute values in the case of the less rigid structure ($b = \infty$).

Particularly interesting is the existence of a region of negative moments M_x in the plane of transverse symmetry $y = 0.50\ a$ ($y = e/2$). This region extending in the free span along the column rows $x = 0.25\ a$ and $x = 0.75\ a$ justifies the use of the beam-effect in the equivalent frame method. This region also justifies the use of the negative coefficients for the distribution of the over-all moment on the supporting strips in the direct design method [155], [160], [162]. Meanwhile, the very low values of the negative moments in the free span, along the line of interior supports, seem to contradict the relatively high values of these empirical coefficients which are recommended by present specifications. However, this statement can hardly be generalized, as the two static diagrams under examination are outside the applied range encompassed by the approximate methods. Indeed, the value $e_2/e_1 = 2$ of the central-to-edge span ratio exceeds the prescribed limits.

In the static approach to the slab structure under consideration, a comparison was made of the deflected mid-surfaces w and of the diagrams of the moments M_x and M_y, obtained for the limit case of the square plate ($b = a$; $e = 0.50\ a$). Two extreme cases, for which account is taken in the analysis of the influence of the relative slab-column stiffness (Variant V_1) are referred to a third scheme which is obtained by an approximate calculation method where the effect of rigid slab-column interaction (variant V_0) is neglected (Fig. 5.7).

The hatched areas, which cover the range $10 \leqslant 4EI/lD \leqslant 80$, show that the values obtained by use of the simplified design procedure (where the effect of the slab-column interaction — dotted line — is neglected) are always outside the range covered by the ordinary stiffness values for slabs and columns. In compliance with expectations, the largest differences exist between the approximate values and those obtained for the case of very stiff columns ($4EI/lD = 80$).

The magnitude of the transverse elastic displacements w, calculated in the longitudinal axis of the slab ($x = 0.50\ a$), de-

5.1 General force method. Transverse and in-plane loads

creases by 33 per cent (along the column row $y = 0.25\,a$ and $y = 0.75\,a$), and by 18 per cent (in the transverse axis of symmetry of the slab $y = 0.50\,a$) when the relative stiffness ratio grows from zero to $4EI/lD = 80$. For the same range of $1/K_{\varphi}^{col}$, the maximum value of the positive moments M_x diminishes by 12 per cent at $y = 0.25\,a$ ($y = 0.75\,a$), and by 27 per cent at $y = 0.50\,a$, respectively. These lower values of the design moments M_x are themselves a strong reason for adopting the present calculation method.

The extreme values of the negative bending moments M_x about the interior columns are far less influenced by disregarding the column stiffnes. However, whereas in the case of a rigid column, the extreme negative value appears on its interior face, in the case of a zero stiffness column it is recorded in the column axis, which is in compliance with expectations (Fig. 5.7). In the simplified structural analysis, where the columns are assumed to have zero stiffness, the shape of the region of the negative bending moments M_x in the free span along the column rows shows clearly that this region and the absolute value of the bending moments are larger than those obtained by considering the relative slab-column stiffness. Here too, the beam-effect is more moderate than is assumed in current code requirements.

Neglecting the stiffness of columns in calculation barely influences the value of the bending moments M_y. Additionally, the diagrams K_M^y allow us to verify the precision of the numerical calculations. Indeed, in the simplified calculation method, the square slab assumes a two-fold topological symmetry if it is subjected to the load q uniformly distributed throughout the surface of the floor and to the four force reactions X_0 acting in the axis of the columns. We notice that the values of the bending moments M_x and M_y are equal at the centre of the panel ($M_x = M_y = 0.0107\,qa^2$) and in the axes of the four columns ($M_x = M_y = 0.0213\,qa^2$). However, when account is taken in calculations of the relative slab-column stiffness, then the values of the ending moments M_x and M_y become slightly different due to the unsymmetrical effect induced by the associated couple reactions X_2.

The accuracy of the numerical calculations is also made evident if we refer the over-all rotation φ_x of the slab to the rotation φ_{col} of the head of the column, at the design section (Table 5.4).

In the last example illustrating the validity of the author's method in variant 1, we consider the case of the square plate with columns of larger cross-sectional area ($c = d = 0.06\,a$) located at the plate corners ($\xi = c$; $\eta = d$), subjected to a force $P = 4\,pcd$, locally distributed at the center of the panel (Fig. 5.8).

5 *A new method of elastic analysis*

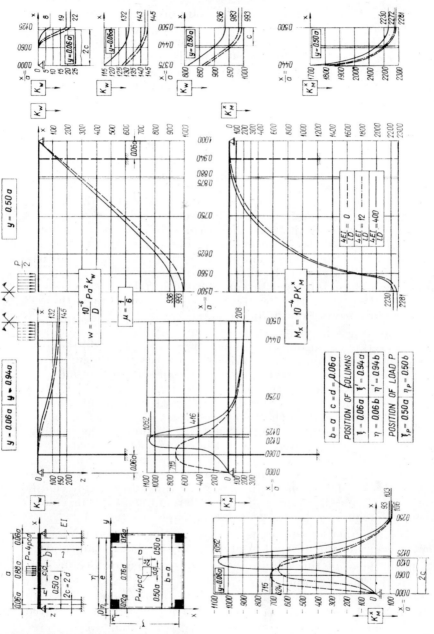

Figure 5.8 Slab structures under locally distributed gravity load $P = 4\ pcd$. Columns located at corners. Influence of relative slab-column stiffness on magnitude of transverse elastic displacements w and of sectional bending moments M_x. Variants V_0 and V_1 ($0 \leqslant 4EI/lD \leqslant 400$)

5.1 General force method. Transverse and in-plane loads

Table 5.4

Slab structures subjected to uniformly distributed load q

Comparative values of rotations φ_x of slab and φ_{col} calculated in column axis

Position of columns: $\xi = 0.25a$; $\xi' = 0.75a$; $e = 0.50a$; $c = d = 0.04a$; $\mu = 0$

Relative stiffness	$4EI/lD = 80$		$4EI/lD = 10$	
Plate	$b = a$	$b = \infty$	$b = a$	$b = \infty$
Rotation	slab/column	slab/column	slab/column	slab/column
$\varphi_x \left\vert \dfrac{qa^3}{D} \right.$	0.00005844	0.00007470	0.00031124	0.00040619
$\varphi_{col} \left\vert \dfrac{qa^3}{D} \right.$	0.00005843	0.00007470	0.00031124	0.00040618

The magnitude and distribution of the elastic displacements and sectional stresses were this time calculated for a different range of the relative column-slab stiffness ($0.75 \leq 4EI/lD \leq 12$) as was dictated by the different geometrical characteristics of the slab structure under examination. Additionally, the value of the bending moments was calculated assuming that Poisson's ratio is different from zero, i.e. $\mu = 1/6$, as is pertinent for reinforced concrete structures.

The comparative approach was extended to other two extreme cases: columns of zero-stiffness, in which case the slab supports are treated as hinges ($4EI/lD = 0$), and very rigid columns ($4EI/lD = 400$). It should be noticed that the limiting case $4EI/lD = \infty$ cannot be investigated here—nor could it be in the case of other slab structures studied by the author in which the columns (or shear-walls) were also located along the boundary of the floor — given the incompatibility with the geometrical and mechanical properties.

Variant 2

A comprehensive static approach to the effect of rigid connections in the design of slab structures was carried out by use of a second constraint for rotations in the direction of the Oy axis (5.18 c). Thus, resort was made to the system of three compatibility equations (5.18): one equation for the associated linear displacements w in the direction of the Oz axis and two equations for the associated angular displacements φ_x and φ_y in the two directions of the midplane xOy of the slab. In order to simplify the analysis and for a concise exposition of the results, only the sub-variants pertaining to the square plate ($b = a$) have been compared in this rigorous variant (Figs. 5.4, 5.9 and 5.10—Variant 2) [75], [78].

Thus, for instance, assuming that the floor is subjected to a uniformly distributed load q throughout its surface and inserting in (5.18) the expressions of the relative unit displacements δ_{ij} and of the relative displacements Δ_{is} (5.6)—(5.12) and (5.28), the condition equations (5.1) become:

$$X_1 \frac{a^2}{D} K_w^P + X_2 \frac{a}{D} K_w^{M(x)} + X_3 \frac{a}{D} K_w^{M(y)} + \frac{qa^4}{D} k_w^q = 0; \quad (5.26\text{a})$$

$$X_1 \frac{a}{D} K_\varphi^{xP} + X_2 \left(\frac{1}{D} K_\varphi^{xM(x)} + \frac{l}{4EI_x}\right) + X_3 \frac{1}{D} K_\varphi^{xM(y)} +$$

$$+ \frac{qa^3}{D} k_\varphi^{xq} = 0; \quad (5.26\text{b})$$

$$X_1 \frac{a}{D} K_\varphi^{yP} + X_2 \frac{1}{D} K_\varphi^{yM(x)} + X_3 \left(\frac{1}{D} K_\varphi^{yM(y)} + \frac{l}{4EI_y}\right) +$$

$$+ \frac{qa^3}{D} k_\varphi^{yq} = 0. \quad (5.26\text{c})$$

The expressions for the associated rotations induced at the heads of the columns of length l by the set of couple interacting forces X_2 (in the direction Ox) and X_3 (in the direction Oy) respectively,

$$\varphi_{col}^x = \frac{X_2 l}{4EI_x} \text{ (rad) } (5.27\text{a}); \quad \varphi_{col}^y = \frac{X_3 l}{4EI_y} \text{ (rad)} \quad (5.27\text{b})$$

are valid for non-displacing nodes when the far end of the column is clamped (symmetrical structures subjected to symmetrical loads).

The expressions for the associated unit displacements, induced by the couple interacting force $X_3=1$, are deduced from (5.6) recalling that $X_j = X_3$ is an associated reaction of type $M(y)$, the moment vector of which is parallel to the Ox axis:

(5.28) $\begin{cases} \delta_{13} = \dfrac{a}{D} K_w^{M(y)} & \text{— represents a linear displacement since } X_i=X_1 \text{ is an associated force reaction of type } P; \\[1em] \delta_{23} = \dfrac{1}{D} K_\varphi^{xM(y)} & \text{— is a rotation in the } Ox \text{ direction since } X_i = X_2 \text{ is an associated couple unknown of type } \overrightarrow{M} \| Oy; \\[1em] \delta_{33}^* = \dfrac{1}{D} K_\varphi^{yM(y)} & \text{— expresses a rotation in the } Oy \text{ direction since } X_i = X_3 \text{ denotes an associated couple reaction of type } \overrightarrow{M} \| Ox. \end{cases}$

5.1 General force method. Transverse and in-plane loads

δ_{33}^* above denotes the relative angular displacement of the slab in the direction X_3, due to the associated unknown $X_3 = 1$. This rotation must not be confused with

$$\delta_{33} = \delta_{33}^* + \frac{l}{4EI_y} \quad (\text{rad}/N),$$

which in (5.26c) represents the summed unit displacement of the slab and column, induced in the direction of the Oy axis by $X_3 = 1$.

Since all four columns have a square cross-section of equal size ($c = d = 0.04\ a$), $I_x = I_y = I$, the static diagram under study is symmetrical in both directions of the slab plane. This two-fold symmetry leads to the following equalities:
— the associated couple unknowns $X_2 = X_3$;
— the associated rotations of the column ends

$$\varphi_{col}^x = \varphi_{col}^y = \frac{X_2 l}{4EI};$$

— the unit linear displacements $\delta_{12} = \delta_{13} = X_2 \dfrac{a}{D} K_w^{M(x)}$;

— the unit angular displacements $\delta_{22}^* = \delta_{33}^* = X_2 \dfrac{1}{D} K_\varphi^{xM(x)}$;

— the relative rotations $\Delta_{2s} = \Delta_{3s} = \dfrac{qa^3}{D} k_\varphi^{xq}$.

Additionally, extending the reciprocity theorem of unit displacements to the case of plates,

$$\delta_{23} = \delta_{32} = \frac{1}{D} K_\varphi^{xM(y)} = \frac{1}{D} K_\varphi^{yM(x)}.$$

By virtue of the above statements, equations (5.26b) and (5.26c) become identical and system (5.26) reduces to only two equations with two unknowns X_1 and $X_2 = X_3$:

$$X_1 a K_w^P + 2 X_2 K_w^M + qa^3 k_w^q = 0; \quad (5.29a)$$
$$X_1 a K_\varphi^P + X_2 (K_\varphi^{xM(x)} + K_\varphi^{xM(y)} + K_\varphi^{col}) + qa^3 k_\varphi^q = 0. \quad (5.29b)$$

In the second variant of analysis, computations were performed by resorting to the system of equations (5.29).

★

We give below the values of the coefficients of the unknowns and of the free terms (Table 5.5), and those of the associated unknowns (Table 5.6), determined in both variants of the method of analysis. The difference in percentage between the values obtained in variants 1 and 2 ranges within 4.3 per cent — 6.5 per cent both for the force unknowns X_1 and the couple unknowns $X_2 = X_3$. Larger differences are obtained for the higher value of

Table 5.5

Slab structures subjected to uniformly distributed load q
General force method — Variant 1 and 2

Coefficients of unknowns and free terms

$b = a$; $\mu = 0$

Position of columns: $\xi = 0.25a$; $\xi' = 0.75a$; $\eta = 0.25b$; $\eta' = 0.75b$.

Column cross-section: $c = d = 0.04a$.

Relative column-slab stiffness: $10 \leqslant \dfrac{4EI}{lD} \leqslant 80$

Coefficients of the unknowns and free terms	$\dfrac{4EI}{lD}$	$b = a$ Variant 1 System of 2 equations	$b = a$ Variant 2 System of 3 equations	Coefficients of the unknowns and free terms	$\dfrac{4EI}{lD}$	$b = a$ Variant 1 System of 2 equations	$b = a$ Variant 2 System of 3 equations
$D\delta_{11}/a^2 = K_w^P$		—	0.01144	$D\delta_{23} = K_\varphi^{xM(y)}$		—	0.08679
$D\delta_{12}/a = K_w^{M(x)}$		0.02916	0.02946	$D\delta_{31}/a = K_\varphi^{yP}$		—	0.02946
$D\delta_{13}/a = K_w^{M(y)}$		—	0.02946	$D\delta_{32} = K_\varphi^{yM(x)}$		—	0.08679
$D\delta_{21}/a = K_\varphi^{xP}$		0.02946	0.02946	$D\delta_{33} = K_\varphi^{yM(y)} + K_\varphi^{col}$	80	—	0.24990
					10	—	0.33740
$D\delta_{22}^* = K_\varphi^{xM(x)}$	∞	0.23740	0.23740	$D\Delta_{1s}^q/qa^4 = k_w^q$		0.00213	0.00213
$D\delta_{22} = K_\varphi^{xM(x)} + K_\varphi^{col}$	80	0.24990	0.24990	$D\Delta_{2s}^q/qa^3 = k_\varphi^{xq}$		0.00630	0.00630
	10	0.33740	0.33740	$D\Delta_{3s}^q/qa^3 = k_\varphi^{yq}$		—	0.00630

the stiffness ratio ($4EI/lD = 80$). Major differences from 9 per cent through to 13.9 per cent (Table 5.6) exist only when X_0 (calculated by the approximate variant) is referred to X_i^2 (calculated with the author's method under variant 2).

The efficiency of the present method of static analysis is made evident by a comparative examination of the influence exerted by the effect of the rigid slab-column connection on the static quantities. Thus, for a range $10 \leqslant 4EI/lD \leqslant 80$ of the stiffness ratio (Table 5.6), the value of the associated couple interacting forces X_2 increases by 50 per cent (variant V_1), and that of the associated couple unknowns $X_2 = X_3$ by 47.5 per cent (variant V_2). These negative couple reactions which act at the slab-column nodes induce a decrease in the values of the positive bending moments max M_x and max M_y at the center panel of the floor and

5.1 General force method. Transverse and in-plane loads

in those of the transverse elastic displacements w (Figs. 5.9 and 5.10, Table 5.7).

Similarly, for the same range of variation of the relative slab-column stiffness ratio (Table 5.6), the value of the associated force reactions X_1 decreases but the difference in value is smaller: 2.3 per cent in variant V_1 and 4.5 per cent in variant V_2. However, if X_0, calculated in the approximate variant, is referred to X_1^{80}, calculated under the assumption of very stiff columns, then the differences come up to 6.9 per cent (V_1) and 13.9 per cent (V_2).

Table 5.6

Slab structures subjected to uniformly distributed load q
General force method — Variant 2

Comparative values of associated unknowns

Square plate ($b = a$); $\mu = 0$

Position of columns $\begin{cases} \xi = 0.25a \; ; \; \eta = 0.25b \\ \xi' = 0.75a \; ; \; \eta' = 0.75b \end{cases}$

Column cross-section: $c = d = 0.04a$

Relative column-slab stiffness $10 \leqslant \dfrac{4EI}{lD} \leqslant 80$

X_i, X_j	Approximate variant	New analysis method			
		Variant 1		Variant 2	
$4EI/lD =$	0	10	80	10	80
$X_0/qa^2 =$	-0.18619	—	—	—	—
$X_1/qa^2 =$	—	-0.17817	-0.17413	-0.17079	-0.16350
$\dfrac{X_0 - X_1}{X_1} =$	0%	4.5%	6.9%	9.0%	13.9%
$\dfrac{X_1^1 - X_1^2}{X_1^2} =$	—	0%	0%	4.3%	6.5%
$\dfrac{X_1^{10} - X_1^{80}}{X_1^{80}} =$	—	0%	2.3%	0%	4.5%
$X_2/qa^3 =$	—	-0.00312	-0.00468	-0.00299	-0.00441
$X_3/qa^3 =$	—	—	—	-0.00299	-0.00441
$\dfrac{X_2^1 - X_2^2}{X_2^2} =$	—	0%	0%	4.3%	6.1%
$\dfrac{X_2^{80} - X_2^{10}}{X_2^{10}} =$	—	50.0%	0%	47.5%	0%

Note. The accuracy condition imposed by the programs: $ER = 10^{-6}$.

5 A new method of elastic analysis

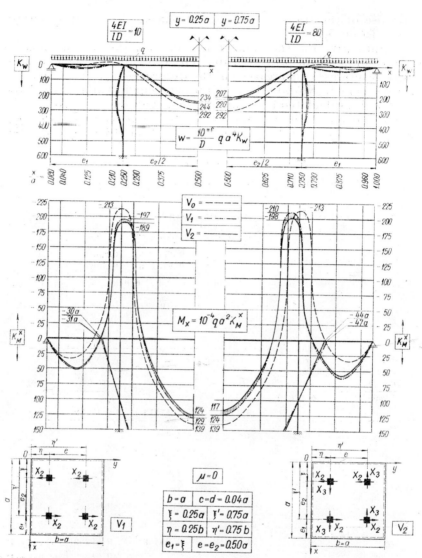

Figure 5.9 Slab structures under uniformly distributed gravity load q. Deflected mid-surface w and diagram of sectional bending moments M_x in column axis (vertical planes $y = 0,25\,b$ and $y = 0,75\,b$). Effect of additional condition for rotation. Variants V_0, V_1 and V_2 ($0 \leqslant 4EI/lD \leqslant 80$).

An examination of the deflected mid-surfaces w and of the diagrams of the bending moments M_x, drawn in the column axis in the vertical planes $y = 0.25b$ ($y = 0.75b$) (Fig. 5.9) and in the

5.1 General force method. Transverse and in-plane loads

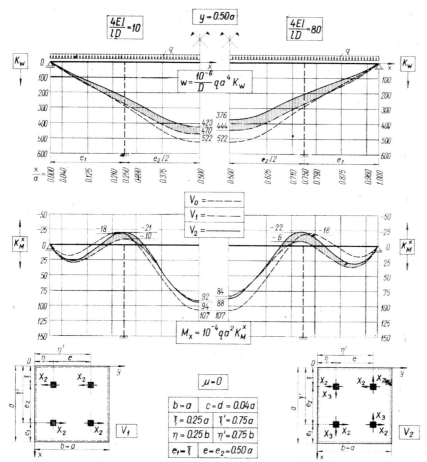

Figure 5.10 Slab structures under uniformly distributed gravity load q. Deflected mid-surface w and diagram of sectional bending moments M_x in symmetry axis of the floor (vertical plane $y = 0.50\,b$). Effect of additional condition for rotation. Variants V_0, V_1 and V_2 ($0 \leqslant 4EI/lD \leqslant 80$).

symmetry axis of the floor in the plane $y = 0.50\,b$ (Fig. 5.10) reinforces the earlier idea of the remarkable influence exerted by the second condition equation for rotations.

The magnitude of the transverse elastic displacements w and stresses M_x, calculated in the two variants of the present method, were refered to the corresponding values obtained in the approximate variant V_0 (Table 5.7). Thus, at the center of the floor ($x = 0.50\,a$; $y = 0.50\,b$), the values of w decrease in percentage down to two times their values in the second variant: for $4EI/lD = 10$,

Table 5.7

Slab structures subjected to uniformly distributed load q
General force method — Variant 2
Square plate

Position of columns : $\begin{cases} \xi = 0.25a\,; \ \eta = 0.25b \\ \xi' = 0.75a\,; \ \eta' = 0.75b \end{cases}$

Columns cross-section : $c = d = 0.04a$

Relative column-slab stiffness : $10 \leqslant \dfrac{4EI}{lD} \leqslant 80$

Comparative values of transverse elastic displacements

$$K_w = w \Big/ \dfrac{10^{-6}}{D} q a^4$$

Calculation point (x, y)		Approximate variant	New analysis method			
			Variant 1		Variant 2	
	$4EI/lD =$	0	10	80	10	80
$x = 0.50a$ $y = 0.50b$	$K_w =$	522	470	444	423	376
	$\dfrac{w^1 - w^0}{w^1} =$	0%	−11.1%	−17.6%	—	—
	$\dfrac{w^2 - w^0}{w^2} =$	0%	—	—	−23.4%	−38.8%
$x = 0.50a$ $y = 0.25b$ $(y = 0.75b)$	$K_w =$	292	244	220	234	207
	$\dfrac{w^1 - w^0}{w^1} =$	0%	−19.7%	−32.7%	—	—
	$\dfrac{w^2 - w^0}{w^2} =$	0%	—	—	−24.8%	−41.1%

from 11.1 per cent (V_1) to 23.4 per cent (V_2); for $4EI/lD = 80$, from 17.6 per cent (V_1) to 38.8 per cent (V_2). Meanwhile, at the same point and for the same range of variation of the relative stiffness ratio, the values of M_x decrease with respect to the approximate values by 16.3 per cent (13.8 per cent) up to 27.4 per cent (21.6 per cent). Mention should be made that the values obtained in variant V_2 (bracketed numbers) decrease in percentage more moderately than those calculated in variant V_1.

Considering now the points $(x = 0.50a\,;\ y = 0.25b)$ and $(x = 0.50a\,;\ y = 0.75b)$ along the column row, we notice that the values of the static quantities decrease mostly with respect to the values calculated in the approximate variant, as was predicted, in the case of very rigid columns $(4EI/lD = 80)$ and when the analysis is carried out in variant V_2. Thus, the value of the linear displacement w decreases by 41.1 per cent and that of the positive bending moment M_x by 18.8 per cent (Table 5.7).

5.1 General force method. Transverse and in-plane loads

Table 5.7 (continued)

Comparative values of sectional bending moments

$$K_M^x = M_x/10^{-4}qa^2$$

Calculation point (x, y)	$4EI/lD =$	Approximate variant	New analysis method			
			Variant 1		Variant 2	
		0	10	80	10	80
$x = 0.50a$ $y = 0.50b$	$K_M^x =$	107	92	84	94	88
	$\dfrac{M_x^1 - M_x^0}{M_x^1} =$	0%	-16.3%	-27.4%	—	—
	$\dfrac{M_x^2 - M_x^0}{M_x^2} =$	0%	—	—	-13.8%	-21.6%
$x = 0.50a$ $y = 0.25b$ $(y = 0.75b)$	$K_M^x =$	139	129	124	124	117
	$\dfrac{M_x^1 - M_x^0}{M_x^1} =$	0%	-7.8%	-12.1%	—	—
	$\dfrac{M_x^2 - M_x^0}{M_x^2} =$	0%	—	—	-12.1%	-18.8%

Note. The accuracy condition imposed by the programs is $ER = 10^{-6}$.

The hatched areas represent the range of variation of the static quantities w and M_x determined in variants V_1 and V_2 of the method of analysis suggested herein (Figs. 5.9 and 5.10). The values of w and M_x, calculated in the approximate variant V_0, where the effect of the rigid slab-column connection is neglected, fall constantly outside this range, being greater in absolute value, as was shown above (Table 5.7). Except for the bending moments M_x, calculated in the plane of symmetry $y = 0.50b$ (Fig. 5.10), all the values of w and M_x obtained in variant V_2 are smaller in absolute value than those calculated in variant V_1, where only one constraint is set for rotations.

However, in the vertical plane $y = 0.50b$, the values of M_x, calculated for the entire range $10 \leq 4EI/lD \leq 80$, are greater in variant V_2 than those obtained in variant V_1. This observation holds both for the positive moments M_x generated at the center of the floor ($x = 0.50a$; $y = 0.50b$) and for the negative moments M_x acting in the free span along the column row (Table 5.7).

As was predicted, the range of the negative bending moments M_x acting in the free span ($y = 0.50b$) along the column line

5 A new method of elastic analysis

($x = 0.25a$ and $x = 0.75a$) is more marked in variant 2 than it is in variant 1 (Fig. 5.10). The existence of this range constitutes a confirmation of the afore-mentioned beam-effect (Sub-sec. 5.1.2.1 — Variant 1). The difference between the values obtained in the two variants is greater in the case of very stiff columns ($4EI/lD = 80$). The absolute value of the negative moments falls constantly below the values of the positive design moments M_x acting at the free span. Indeed, in the plane $y = 0.50b$, the ratio $M_x(x = 0.50a)/M_x(x = 0.25a)$ varies between 4.4 and 4 within the range $10 \leqslant 4EI/lD \leqslant 80$.

5.1.2.2 In-plane loads.

We have already seen that most procedures of static analysis for slab structures, including Pfaffinger's method, do not permit the treatment of a system of horizontal loads acting in the plane of the slab (Sec. 1.2). Unlike these procedures, the author's method permits the analysis of such structures subjected to a seismic load S [75], [83], [86].

In order to illustrate the use of this method in the case of this type of load, resort was made to the static diagram considered in the analysis of slab structures subjected to gravity loads (Figs. 5.3 and 5.5): the square plate ($b = a$ — Scheme A) and the infinitely long plate ($b = \infty$) assuming two cases for the equidistance between columns ($e = 0.50 a$ — Scheme B; $e = 0.75 a$ — Scheme C). We refer to the free translation of the infinitely long plate in the direction of the forces S (along the Ox axis) and assume that the square plate undergoes a free lateral translation in the same direction (Figs. 5.11–5.13). Since the symmetrical structure is anti-symmetrically loaded, use is made of a simplified basic system by grouping the unknowns (Fig. 5.11). It is evident that a load of type S — i.e. for a system of horizontal forces acting in the plane of the slab — the second compatibility equation for rotations (5.18 c) will barely influence the value of the associated interacting forces. Thus, assuming that in the system of equations (5.18) $X_3 = 0$, the effect of the rigid slab-column connections can be described by setting the condition of zero relative rotation only

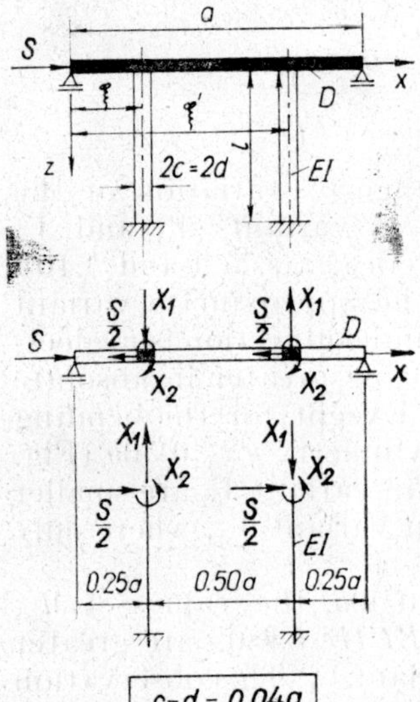

Figure 5.11 In-plane loads. Basic system associated to the comparatively analysed static diagram.

5.1 General force method. Transverse and in-plane loads

in the direction of the Ox axis. This can be achieved by use of the system (5.19) (Variant 1), where the terms of equation (5.19a) represent the associated linear displacements in the direction of the associated force unknowns X_1, such that the relative transverse displacement of the basic system due to the system of forces S is zero:

$$\Delta_{1s}^S = 0. \tag{5.30a}$$

The relative angular displacement of the unrestrained end of the columns, induced in the direction of the associated couple reaction X_2 by the system of forces S applied on the basic system (Fig. 5.11) is given by

$$\Delta_{2s}^S = -\frac{Sl^2}{4EI_x}, \tag{5.30b}$$

where the negative sign is due to the opposed directions in which the forces $S/2$ and the couple $X_2 = 1$ act at the free end of each column.

We substitute (5.8)–(5.9) in the calculation of the unit displacements δ_{ij} of the slab, corresponding to the associated unknowns X_1 and X_2. The expression for the associated relative rotation of the columns (of length l) at the slab-column nodes, induced by the set of couple reactions X_2 in the direction Ox

$$\varphi_{col}^x = \frac{X_2 l}{EI_x} \text{ (rad)}, \tag{5.31}$$

is valid in the case of displaceable nodes, assuming that the opposite end of the column is clamped. With these notations, the system of compatibility equations (5.19) becomes

$$X_1 \frac{a^2}{D} K_w^P + X_2 \frac{a}{D} K_w^M = 0 \tag{5.32a}$$

$$X_1 \frac{a}{D} K_\varphi^{xP} + X_2 \left(\frac{1}{D} K_\varphi^{xM} + \frac{l}{EI_x} \right) - \frac{Sl^2}{4EI_x} = 0. \tag{5.32b}$$

Since the associated unit displacements $\delta_{12} = \delta_{21}$ are zero for all the three static diagrams A, B and C under study,

$$K_w^M = K_\varphi^{xP} = 0; \Rightarrow X_1 = 0.$$

In order to define the relative column-slab stiffness in bending (Sub-sec. 5.1.2.1), we retain the notation

$$\frac{1}{K_\varphi^{col}} = \frac{4EI_x}{lD} \tag{5.23}$$

5 A new method of elastic analysis

and obtain the expression of the associated couple unknown X_2

$$X_2 = \frac{K_\varphi^{col}}{K_\varphi^{xM} + 4K_\varphi^{col}} Sl, \qquad (5.33)$$

which was used in the numerical calculations performed herein.

★

The effect of the relative column-slab stiffness was made evident also in the case of horizontal loads of type S, by the solution of Eqn. (5.33), for the five values considered in the approach to gravity loads: $4EI/lD = 10, 20, 40, 80$ and ∞. The sixth value ($4EI/lD = 0$) defining the columns of negligible stiffness is irrelevant to the loading diagram under consideration. Meanwhile, retaining the range of variation of the parameters b/a and e/a, the comparative approach carried out for the load of type S reinforces the earlier conclusions concerning the great influence exerted by the relative slab-column stiffness on the value of the associated couple unknown X_2 (Fig. 5.12 and Table 5.8). But much to the contrary, the influence of the parameters b/a and e/a is negligible in this case.

Table 5.8
Slab structures subjected to seismic load S
General force method

$a \leqslant b \leqslant \infty$; $\mu = 0$

Position of columns : $\xi = 0.25a$; $\xi' = 0.75a$; $\eta = 0.25b$; $\eta' = 0.75 b$ $(b = a)$

$\xi = 0.25a$; $\xi' = 0.75a$; $\eta = \lambda e$ $(b = \infty)$

Column cross-section : $c = d = 0.04\,a$

Coefficients of unknowns and free terms

Coefficients of the unknowns and free terms	$\dfrac{4\,EI}{lD}$	$b = \infty$ $e = 0.50a$	$b = a$ $e = 0.50\,a$	$b = \infty$ $e = 0.75\,a$
$D\delta_{11}/a = K_w^P$	—	0.00566	0.00399	0.00453
$D\delta_{12} = K_w^M$	—	—	0	0
$D\delta_{21}/a = K_\varphi^{xP}$	—	0	0	0
$D\Delta_{1s}^S = K_w^S$	—	0	0	0
$D\delta_{22}^* = K_\varphi^{xM}$	∞	0.12665	0.12530	0.12463
$D\delta_{22} = K_\varphi^{xM} + 4\,K_\varphi^{col}$	80	0.17665	0.17530	0.17463
	40	0.22665	0.22530	0.22463
	20	0.32665	0.32530	0.32463
	10	0.52665	0.52530	0.52463
$-D\Delta_{2s}^S/Sl = K_\varphi^{col}$	80	0.01250	0.01250	0.01250
	10	0.10000	0.10000	0.10000

5.1 General force method. Transverse and in-plane loads

Table 5.8 (continued)
Comparative values of associated unknowns X_2/Sl

$\dfrac{4EI}{lD}$	K_φ^{col}	$b = \infty$ $e = 0.50 a$	$b = a$ $e = 0.50 a$	$b = \infty$ $e = 0.75 a$
∞	0.00000	0.00000	0.00000	0.00000
80	0.01250	0.07076	0.07131	0.07158
40	0.02500	0.11030	0.11096	0.11129
20	0.05000	0.15307	0.15370	0.15402
10	0.10000	0.18988	0.19037	0.19061

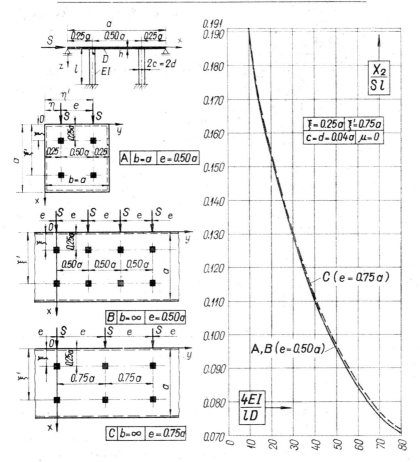

Figure 5.12 Slab structures under seismic load S. Influence of relative slab-column stiffness and of equidistance e spanned by columns on magnitude of associated unknown X_2 (couple-reaction) at the slab-column nodes ($10 \leqslant 4EI/lD \leqslant 80$).

5 A new method of elastic analysis

The diagrams of the bending moments induced in the slab (M_x) and columns (M_{col}), which was plotted for variants $A(b = a)$ and $B(b = \infty)$ assuming that the columns are spaced at an equal distance ($e = 0.50a$), shows the manner in which the parameter $4EI/lD$ influences the extreme values of the bending moments (Fig. 5.13). Thus, for an eight-fold increase of the relative column-

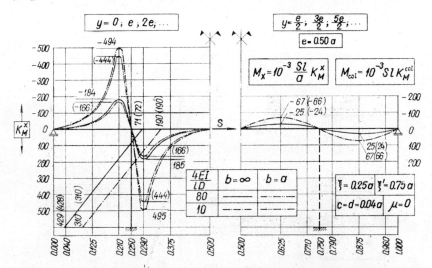

Figure 5.13 Slab structures under seismic load S. Influence of relative slab-column stiffness and of parameter b/a on extreme magnitude of sectional bending moments M_x in slab and M_{col} in columns ($10 \leqslant 4EI/lD \leqslant 80$).

slab stiffness, the extreme values of M_x decrease by 63 per cent, whereas the maximum values M_{col} are 38 per cent higher. The influence of the parameter b/a is moderate (bracketed values refer to the square plate $b = a$).

5.1.3 Effect of failure of supports on slab structures

The general applicability and the advantages of the author's method of static analysis for slab structures have been shown in the case of gravity-loads (Sub-sec. 5.1.2.1) and horizontal-loads, acting in the slab plane (Sub-sec. 5.1.2.2). Additionally, incorporation of the surface couple as a generalized interacting force permits the study of mixed structures, consisting of bars and plates, in the case of support failure of known magnitude: translation and rotation. Save for Pfaffinger's approach in which

5.1 General force method. Transverse and in-plane loads

the elastic longitudinal deformation of columns is solely incorporated [71], no other investigation into this topic is so far available.

In order to illustrate our method, we shall as always use simple statical schemes having at least one axis of geometrical symmetry, i.e. one-storied slab structures having the square plate simply supported on the entire boundary and one, two or four interior columns (Figs. 5.14 and 5.15). Of course, as is idealized in all the other cases examined in this book, free horizontal translation is assumed only upon the direction of the Ox axis. The simple design schemes will permit an easier understanding of the physical phenomenon [86].

The basic systems associated to the three hyperstatic systems are similar to those previously employed (Sub-secs. 5.1.1 and 5.1.2 — Figs. 5.2, 5.4 and 5.11). In order to simplify the continuity equations, the interacting force-couples of the type $\vec{M} \parallel Ox$ are neglected, only the reaction-couples with the vector moment $\vec{M} \parallel Oy$ being incorporated in the analysis (Variant 1). The difference from a rigorous calculation, with consideration of both equations for rotation (Variant 2), has been previously discussed (Sub-sec. 5.1.2.1 — Tables 5.6 and 5.7).

As the principle of effect supperposition permits the independent treatment of various types of given loads, in the sequel we shall examine only the support-failure effect on slab structures. The system of compatibility equations (5.1) in the most general form, will reduce here to

$$\mathbf{d\,X + C = 0}, \tag{5.1b}$$

where each term is as was defined earlier (Sub-sec. 5.1.1.1). The elements of the square matrix $\mathbf{d}$ of unit displacements δ_{ij} is here too calculated by use of formulae (5.2), (5.3) and (5.8), (5.9) provided that the unknowns X_i and X_j can be grouped (Sub-sec. 5.1.1.3). The unknowns of the type X_i ($i = 2r - 1$) will here be the locally distributed force-reactions of type P, and the unknowns of type $X_j (j = 2r)$ represent reaction-couples of type $\vec{M} \parallel Oy$.

Vector $\mathbf{C}$ contains the displacements Δ_{ic}, Δ_{jc} induced upon the direction of the unknowns X_i, X_j, respectively, by support failure of known magnitude. In the static schemes examined so far, both vertical traslations w_v and rotations φ_v upon the direction of the Ox axis have been incorporated in the analysis as support failure, i.e.

$$\Delta_{ic} = \mp w_v \text{ (m)} \quad (5.34); \quad \Delta_{jc} = \mp \varphi_v \text{ (rad)}. \tag{5.35}$$

Except for the one-column slab structure (Fig. 5.14), the geometrical and mechanical symmetry of the static schemes exa-

5 A new method of elastic analysis

mined permits the grouping of unknowns. In this context and adopting the assumptions stated above, the system of equations (5.1b) reduces to a very small number of continuity equations: only one equation for linear displacements (Fig. 5.14 A) and only one equation for rotations (Fig. 5.14 B), two equations (Fig. 5.15 A) or four equations (Fig. 5.15 B). In an explicit formulation, in the latter case the system (5.1b) becomes:

$$\begin{bmatrix} \delta_{11} & \cdots & \delta_{14} \\ \vdots & & \vdots \\ \delta_{41} & \cdots & \delta_{44} \end{bmatrix} \begin{bmatrix} X_1 \\ \vdots \\ X_4 \end{bmatrix} + \begin{bmatrix} \Delta_{1c} \\ \vdots \\ \Delta_{4c} \end{bmatrix} = 0, \qquad (5.36)$$

where X_1, X_3 are associated reaction-forces and X_2, X_4 are associated reaction-couples. Once the generalized interacting forces are determined, the static magnitudes — linear displacements, angular displacements and sectional stresses — on the slab of the real statically indeterminate system are calculated by use of formulae (5.13) — (5.16) or the like (Sub-sec. 5.1.1.4).

★

Numerical computations were performed for those values of the parameters defining the slab structures under examination that correspond to ordinary values (Figs. 5.14, 5.15 — Tables 5.9, 5.10). Our comparative approach is less concerned here with various numerical variants. Instead, we shall deal with the differences resulting from the different design schemes under comparison and the different types of support failure considered.

The slab structure having only one column ($c = d = 0.06a$) at the floor center ($\xi = 0.50 a$, $\eta = 0.50 b$) was examined under two loading assumptions:

— vertical tanslation of known magnitude of the column foundation

$$\Delta_{1c} = - w_\nu, \qquad (5.37)$$

in the positive sense of the Oz axis (Fig. 5.14 A);

— positive rotation of known magnitude of the column foundation

$$\Delta_{2c} = - \varphi_\nu, \qquad (5.38)$$

upon the direction of the Ox axis (Fig. 5.14 B).

In the first loading case $X_2 = 0$, and hence the system of equations (5.36) reduces to (Fig. 5.14 A):

$$X_1 \delta_{11} + \Delta_{1c} = 0, \qquad (5.39)$$

5.1 General force method. Transverse and in-plane loads

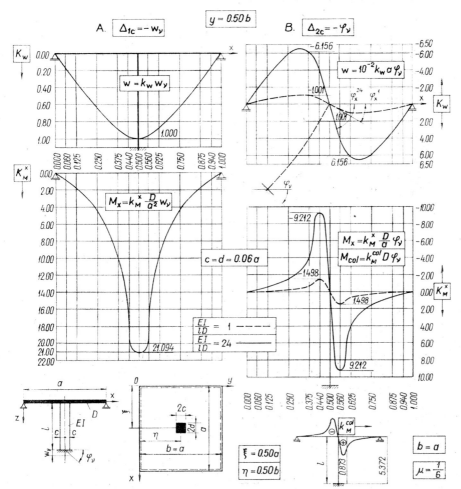

Figure 5.14 Slab structures under the effect of support failure. A. Vertical translation w_ν of column foundation; B. Rotation φ_ν of column foundation.

or to

$$X_1 \frac{a^2}{D} k_w^P = w_\nu, \tag{5.39a}$$

respectively.

Now, introducing the value $k_w^P = 1129 \cdot 10^{-5}$, corresponding to the computational point ($x = 0.50a$; $y = 0.50b$), we obtain

$$X_1 = 88.57396 \frac{D}{a^2} w_\nu,$$

a magnitude independent of the relative slab-column stiffness.

5 A new method of elastic analysis

In the second loading case $X_1 = 0$, and hence the system of equations (5.36) reduces to (Fig. 5.14 B):

$$X_2 \delta_{22} + \Delta_{2c} = 0, \tag{5.40}$$

or to

$$X_2 \left(\frac{1}{D} k_\varphi^{xM} + \frac{l}{EI} \right) = \varphi_v, \tag{5.40a}$$

respectively.

Introducing the value of the dimensionless coefficients $k_\varphi^{xM} = 14448 \cdot 10^{-5}$ and for the values of the relative stiffness factor that range between

$$1 \leqslant \frac{EI}{lD} \leqslant 24,$$

we obtain (see diagram k_M^{col}):

$$0.87376 \, D \, \varphi_v \leqslant X_2 \leqslant 5.37204 \, D \varphi_v.$$

The diagrams for the elastic displacement w and bending moments M_x in the slab of the structure were drawn for the values of interacting forces X_1 (Fig. 5.14 A) and X_2 (Fig. 5.14 B) (respectively) thus determined. These diagrams show the acute local perturbation induced by failure of supports at the slab-column contact area. The sectional stresses in the slab were calculated for the value $\mu = 1/6$ of Poisson's ratio, which is pertinent for reinforced concrete structures.

In the case of slab structure having two columns at the quarter of the span (in the vertical plane $\xi = 0.25\,a$), the failure of supports of known magnitude is likewise a translation which develops however in the negative sense of the Oz axis (Fig. 5.15 A)

$$\Delta_{1c} = w_v. \tag{5.41}$$

As the foundation of the two columns does not fail in rotation, $\Delta_{2c} = 0$.

Because of the geometrical non-symmetry with respect to any plane $x = $ constant, in the slab-column nodes both types of associated generalized interacting forces develop: $X_i = X_1$; $X_j = X_2$. Hence, the explicit form of the system of equations (5.36) is similar to that of system (5.22) (Sub-sec. 5.1.2.1 — Variant 1).

5.1 General force method. Transverse and in-plane loads

The only difference between these systems concerns the free terms and the expression for the rotation of the column free-end (see Sub-sec. 5.1.2.2):

$$\varphi_{col} = \frac{X_2 l}{EI}. \qquad (5.31)$$

We have

$$X_1 \frac{a^2}{D} k_w^P + X_2 \frac{a}{D} K_w^M + w_v = 0; \qquad (5.42a)$$

$$X_1 \frac{a}{D} K_\varphi^{xP} + X_2 \left(\frac{1}{D} K_\varphi^{xM} + \frac{l}{EI} \right) = 0. \qquad (5.42b)$$

The numerical variants under comparison were calculated for the range $10 \leqslant \frac{EI}{lD} \leqslant 80$ of values for the relative column-slab stiffness factor, to which added the extreme values $EI/lD = 0$, corresponding to zero-stiffness columns and $EI/lD = \infty$, corresponding to perfectly rigid columns (Table 5.9).

Table 5.9

Slab structures subjected to support failure of known magnitude
General force method — Variant 1

**Coefficients of unknowns and free terms
Comparative values of associated unknowns**

Square plate ($b = a$) — Two columns
Position of columns: $\xi = 0.25a$ $\begin{cases} \eta = 0.25b \\ \eta' = 0.75b \end{cases}$
Column cross-section: $c = d = 0.04 a$
Relative column-slab stiffness: $0 \leqslant \frac{EI}{lD} \leqslant \infty$

Coefficients of unknowns and free terms			$\frac{EI}{lD}$	Values of associated unknowns	
K_r^{st} ; Δ_{ic}		$D\delta_{22} = K_\varphi^{xM} + K_\varphi^{col}$		$X_1 \Big/ \frac{D}{a^2} w_v$	$X_2 \Big/ \frac{D}{a} w_v$
$D\delta_{11}/a^2 = K_w^P$	0.00772	0.18135	∞	−153.2903	12.4509
$D\delta_{12}/a = K_w^M$	0.01473	0.19385	80	−151.4987	11.5119
$D\delta_{21}/a = K_\varphi^{xP}$	0.01473	0.28135	10	−143.9094	7.5343
$\Delta_{1c} = w_v$		∞	0	−129.5337	0.0000

5 A new method of elastic analysis

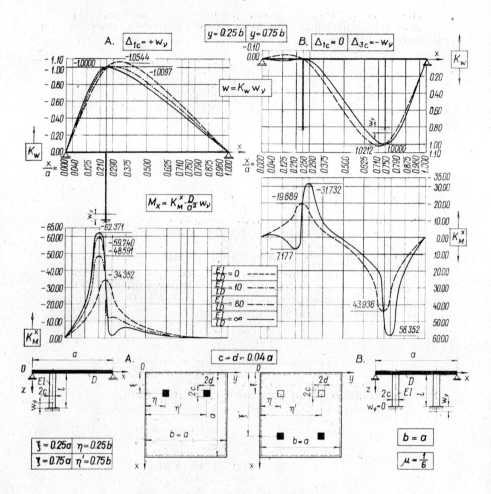

Figure 5.15 Slab structures under the effect of support failure w_y.
A. Structure with two columns; B. Structure with four columns.

In order to verify the accuracy of the statical analysis, the value of the slab rotation φ_x were compared to the value of the column-end rotation φ_{col} at the slab-column node:

$$\varphi_x^\infty = \varphi_{col}^\infty = 0;$$

$$\varphi_x^{10} = 0.7534337 \frac{w_y}{a}; \quad \varphi_{col}^{10} = 0.7534338 \frac{w_y}{a}.$$

5.1 General force method. Transverse and in-plane loads

The diagrams w (K_w) and M_x(K_M^x), drawn for the slab of the structure *, show that the physical reality was modelled with due accuracy (Fig. 5.15 A)**. The curves K_w, which represent the deflected mid-surface of the slab as a function of the parameter EI/lD are slightly influenced by the value of this parameter. Far larger differences are noted — in compliance with expectations— in the case of the bending moments M_x: the increase in percentage between the extreme values $|\min K_M^{x0}|$ and $|\min K_M^{x\infty}|$ is 81.6 percent. Finally, let us refer to the displacement of the peak of the curves K_M^x, from the column axis (for $EI/lD = 0$) to its face (for any $EI/lD > 0$). In distinction of the given transverse load case, when max (min) K_M^x developed on the inside face of the columns (Sub-sec.5.1.2 — Figs. 5.6—5.9), in the case of support failure w_v, these extreme values occur on the outside face of the columns.

The slab structure having a square plate and four columns at the quarter of the span is identical to that examined in detail under loads of type q, P and S (Sub-sec. 5.1.2). The behaviour of this structure under the effect of the failure of supports will be examined here assuming that the column foundations, located in the plane $\xi' = 0.75a$ (Fig. 5.15 B), undergo one vertical translation in the positive sense of the Oz axis:

$$\Delta_{3c} = -w_v. \tag{5.43}$$

The other two columns, located in the plane $\xi = 0.25\,a$, do not undergo failure of supports. Hence,

$$\Delta_{1c} = \Delta_{2c} = \Delta_{4c} = 0.$$

As the geometrical and mechanical symmetry with respect to the plane $y = 0.50\,b$ permits the grouping of unknowns, the basic system will be subjected to only four distinct generalized interacting forces: $X_1 - X_4$. In the system of four continuity equations (5.36), all the 16 coefficients of the unknowns are pairwise equal. Besides, equalities of the type $\delta_{ij} = \delta_{ji}$, between the coefficients outside the main diagonal there exist also the equalities

$$\delta_{11} = \delta_{33} \quad \text{and} \quad \delta_{22} = \delta_{44} \tag{5.44}$$

---

* The curves K_w^{col} and K_M^{col} for the columns are omitted so as to prevent overcrowding of the picture.
** The curve K_w^{80}, corresponding to the highly stiff column case ($EI/lD = 80$) could not be drawn as it almost borders the curve K_w^∞. The curve K_M^{x80} was only partially represented for the same reason.

5 A new method of elastic analysis

which result from the location of the four — all equally sized — collumns at the quarter of the slab span (Table 5.10).

Table 5.10
Slab structures subjected to support failure of known magnitude
General force method — Variant 1
Square plate — Four columns

Position of columns : $\begin{cases} \xi = 0.25\,a; & \eta = 0.25\,b \\ \xi' = 0.75\,a; & \eta' = 0.75\,b \end{cases}$

Column cross-section : $c = d = 0.04\,a$

Relative column-slab stiffness : $0 \leqslant \dfrac{EI}{lD} \leqslant \infty$

Coefficients of unknowns and free terms

$D\delta_{11}/a^2 = K_w^P$	0.00772	$K_w^P = D\delta_{33}/a^2$	$D\delta_{22} = K_\varphi^{xM} + K_\varphi^{col}$	0.18135 *	$D\delta_{44} = K_\varphi^{xM} + K_\varphi^{col}$		
$D\delta_{12}/a = K_w^M$	0.01473	$K_\varphi^{xP} = D\delta_{21}/a$	$D\delta_{23}/a = K_\varphi^{xP}$	0.01473	$D\delta_{32}/a = K_w^M$		
$D\delta_{13}/a = K_w^P$	0.00372	$K_w^P = D\delta_{31}/a$	$D\delta_{24} = K_\varphi^{xM}$	0.05605	$D\delta_{42} = K_\varphi^{xM}$		
$D\delta_{14}/a = K_w^M$	0.01473	$K_\varphi^{xP} = D\delta_{41}/a$	$D\delta_{34}/a = K_w^M$	0.01473	$D\delta_{43}/a = K_\varphi^{xP}$		

$\Delta_{1c} = \Delta_{2c} = 0\,;\qquad \Delta_{3c} = -w_v\,;\qquad \Delta_{4c} = 0$

* For $EI/lD = \infty$

Comparative values of associated unknowns

X_i, X_j	$EI/lD = \infty$	$EI/lD = 0$
$X_1 \Big/ \dfrac{w_v D}{a^2}$	189.2328	168.7063
$X_2 \Big/ \dfrac{w_v D}{a}$	-7.9709	0.0000
$X_3 \Big/ \dfrac{w_v D}{a^2}$	-60.7672	-81.2937
$X_4 \Big/ \dfrac{w_v D}{a}$	-7.9709	0.0000

The system of equations (5.36) was solved in two extreme variants: the perfectly-rigid column assumption ($EI/lD = \infty$) and the zero-stiffness columns ($EI/lD = 0$). In the first case the geometrical tangent to the deflected mid-surface of the slab is horizontal in the axis of all columns (Fig. 5.15B):

$$\varphi_x = \varphi_{col} = \frac{X_2 l}{EI} = \frac{X_4 l}{EI} = 0. \qquad (5.45)$$

5.1 General force method. Transverse and in-plane loads

In the second case, the slab-column nodes are treated as hinges and hence the reaction-couples are zero:

$$X_2 = X_4 = 0. \tag{5.46}$$

The accuracy of the numerical computations was checked in the axes of the columns that undergo failure of supports. Solving the system of four equations with four unknowns yielded:

$$w_3(x = \xi', y = \eta) = (1.0 + 10^{-9})\, w_\nu;$$

$$\varphi_{x3}(x = \xi', y = \eta) = 1.7 \cdot 10^{-9} \frac{w_\nu}{a}.$$

An examination of the diagrams K_w and K_M^x shows here too that the physical reality was modelled with due accuracy. As in the previous example, the shape of the deflected mid-surface w is slightly influenced by the increased stiffness of columns. However, the extreme values of the bending moments M_x in the slab increase, over the range $0 \leqslant EI/lD \leqslant \infty$, by 32.8 percent (max K_M^x) in the region of columns affected by failure of supports and by 59.5 percent ($|\min K_M^x|$) in the region of the columns that do not undergo settling (Fig. 5.15 B). It is worth noting that the extreme absolute values of the sectional stresses are smaller in the structure having a greater number of columns, hence, a higher degree of static indeterminacy. However, in compliance with expectations, the sums (max $M_x + |\min M_x|$) are larger in the latter case than the extreme values $|\min M_x|$ obtained from the analysis of the structure having a small degree of static indeterminacy (Fig. 5.15 A).

The foregoing conclusions on the applicability of the method of elastic analysis under examination and the faithful modeling of the physical reality are thus reinforced. We have already seen that the mathematical model of the rigid slab-column connection permits the use of the static analysis of slab structures under both transverse (Sub-sec. 5.1.2.1) and horizontal loads (Sub-sec.5.1.2.2) or under the effect of the failure of supports (Sub-sec. 5.1.3). Additionally, this method has proved to be efficient in the case of slab structures subjected to given loads of the type q, P or S as the additional constraints for rotation incorporated here permit a smaller value of the design stresses in slabs.

5.2 Use of the general displacement method
In-plane loads

Since slab structures can be treated in the static analysis as reticular structures (see Sec. 4.2), an eventual extension of the validity of the data obtained by use of the general displacement method has received special attention in the early investigations undertaken by the author. Whence the diversification of the approaches to the elastic analysis of this category of structures. In this respect, mention should be made that the results obtained in Chaps. 2, 3 and 4 can be adapted to deal with the plate subjected to displacements.

The systematizations required by the use of the general displacement method lead to a new procedure of elastic analysis for slab structures. We give below one of these new approaches to the static analysis, which is illustrated for a certain category of slab structures with which we are frequently confronted in the design practice.

5.2.1 Basic scheme

Let us consider the category of symmetrical multistoried reticular structures, subjected to a system of horizontal forces $S_1, S_2, \ldots, S_{n-1}, S_n$ which act in the planes of the floors $1, 2, \ldots, n-1, n$ (Fig. 5.16). The spans a_i between the columns, which are measured in the vertical plane of the system of forces S, are so arranged as to assume that the equidistance between the columns has an average value:

$$a = \frac{a_i + a_j}{2} \quad (i = 1, 2, \ldots, m-1; \; j = i + 1). \tag{5.47}$$

An examination of the distribution of the bending moments along the horizontal bars shows that for this particular category of multiply statically indeterminate structures the zero moment points are located in the transverse vertical mid-planes $1, 2, \ldots, i, j, \ldots, m$ (planes that are normal to the plane of forces S and pass through the axes of the spans a_i between columns).

For the design under a seismic load, we assume that the structure is indefinitely long in the direction of these mid-planes (i.e. the Oy direction) and that the value a of the equidistance between columns is given by (5.47). Hence, the calculation scheme becomes substantially simplified as the points where the bending moments are zero are assimilated with hinges that are located

5.2 General displacement method. In-plane loads

in the transverse mid-planes 1, 2, ..., i, j, ..., m (Fig. 5.16). This static diagram is frequently used in the analysis of similar reticular structures subjected to antisymmetrical loads.

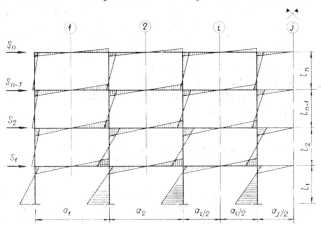

Figure 5.16 Symmetrical reticular structure under a system of horizontal forces S. Bending moment distribution in bars.

The results obtained from the study of the distribution of the bending moments induced in plates by the antisymmetrical sets of surface couples (Sub-sec. 4.1.1) allows us to extend the above static diagram to the analysis of symmetrical multi-storied slab structures subjected to a system of horizontal forces S acting in the plane of the slab. By virtue of this simplified diagram, the slabs at each storey are assumed to be simply supported (hinged) in the longitudinal direction Oy which is normal to the vertical plane of the forces S. Each of these slabs is subjected to a surface couple M, having the vector parallel to the line of supports and acting at the mid-span of the slabs (Fig. 5.17). This loading assumption was investigated here assuming that the infinitely long plate and the finite-boundary plates are subjected to only one surface couple (Secs. 2.2, 2.3, 3.2 and 3.3) and to a set of couples M associated in a simple row (Sub-sec. 4.1.1).

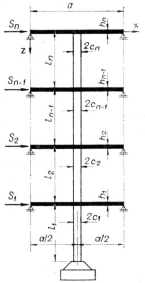

Figure 5.17 Symmetrical multistoried slab structure under seismic load S. Schematic diagram.

5   *A new method of elastic analysis*

The data obtained allow the calculation of the slab rotation φ_x induced by only one surface couple of vector $\vec{M} \parallel Oy$, at ($\xi = 0.50\,a$; $\eta = 0$), by use of the tabulated coefficients k_φ^x, and by some surface couples associated in a simple row at ($\xi = 0.50\,a$; $\eta = \lambda e$) by use of the tabulated values of K_φ^x (Chaps. 3 and 4). Thus, knowing the values of the slab rotation φ_x in the axes of the columns, we can estimate the effective plate width b_y assuming that the couples of the row act at $\xi = 0.50\,a$; $\eta = \lambda e$ (Sec. 4.2) and calculate the slab stiffness ρ_{si} (Sub-sec. 5,2.2). In this manner, the structural analysis can be carried out by use of the general displacement method.

5.2.2 *Use of the design procedure in the analysis of symmetrical multistoried slab structures*

Let us consider a four-storied symmetrical slab structure subjected to a system of horizontal forces $S_1 = S_2 = S_3 = S_4 = S$. The spans are assumed to be equal ($e = a$) in both directions of the slab plane and the length of the structure is assumed to be infinite in a direction normal to the vertical plane of the forces S. We shall adopt the previous simplified static diagram (Fig. 5.17), where :
  — the span of the slabs is $a = 5.0$ m
  — the thickness of the slabs $h = 0.20$ m
  — the equidistance between columns $e = a$
  — the height of each storey $l = 3.0$ m
  — the sizes of the cross-section of the columns
  $2c = 2d = 0.40$ m.

The effective plate width corresponding to the position $\xi = 0.50\,a$ of the row of surface couples is given by (Sub-sec. 4.2.2) :

$$b_y = \frac{a}{12(1-\mu^2)\,K_\varphi^x}. \tag{4.5}$$

For the value $c/a = 20/500 = 0.04$ (Table 4.8 or Fig. 4.33) and $e = a$, we obtain the mean $K_\varphi^x = 0.17240$. Assuming that the value of Poisson's ratio is negligible ($\mu = 0$) in the calculation of rotations, we find that $b_y = 0.483\,a = 2.415$ m.

In the equivalent frame method ([154], [155], [160], [162]) we assume only one value $b_y = 0.50\,a$ for the evaluation of the equivalent beam stiffness. In the procedure suggested herein for the analysis of slab structures subjected to lateral loads, the value of the effective plate width may be found to be much smaller in the case of relatively slender columns ($c/a = d/a \leqslant 0.04$) and

5.2 General displacement method. In-plane loads

much greater than $b_y = 0.50\, a$ in the case of relatively rigid columns, including the case of flat slabs with capitals ($c/a = d/a \geqslant \geqslant 0.06$).

For the structural analysis, we first evaluate the design stiffness, the distribution coefficients and the initial storey moments M^{sto} (Nm), i.e. perfect clamping moments induced by the lateral forces (Table 5.11). The iteration was strongly convergent so that calculations required only five cycles. The results of the iteration are given in Table 5.12.

Table 5.11

Symmetrical multistoried slab structure subjected to seismic load S
General displacement method

Computation scheme

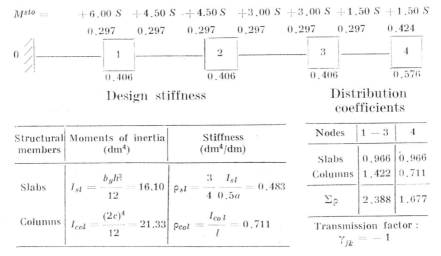

Design stiffness   Distribution coefficients

Structural members	Moments of inertia (dm⁴)	Stiffness (dm⁴/dm)
Slabs	$I_{sl} = \dfrac{b_y h^3}{12} = 16.10$	$\rho_{sl} = \dfrac{3}{4}\dfrac{I_{sl}}{0.5a} = 0.483$
Columns	$I_{col} = \dfrac{(2c)^4}{12} = 21.33$	$\rho_{col} = \dfrac{I_{col}}{l} = 0.711$

Nodes	1 − 3	4
Slabs	0.966	0.966
Columns	1.422	0.711
$\Sigma\rho$	2.388	1.677

Transmission factor: $\gamma_{jk} = -1$

The diagram of the bending moments M_c shows correctly the linear variation of these moments in the columns. For the sake of simplicity, the slabs of the structure were treated here as equivalent beams and so the bending moments M_{sl} vary also linearly (Fig. 5.18A).

The real distribution of the sectional bending moments M_x in the slabs is obtained assuming that the over-all end moments M_{sl} at each node (Table 5.12) act as a surface couple, i.e.

$$M_x = \frac{M_{sl}}{a} K_M^x. \tag{2.54}$$

5 A new method of elastic analysis

Table 5.12

Symmetrical multistoried slab structure subjected to seismic load $S(N)$
General displacement method

Values of end bending moments M_c and M_{sl} at column-slab nodes (in Nm)

Node	Moments M_c/S in columns	Moments M_{sl}/S in slabs		
		Over-all	Left	Right
4	$M_{4c}^i = +\ 2.76$	$M_{4sl} = -2.76$	-1.38	-1.38
3	$M_{3c}^s = +\ 0.24$ $M_{3c}^i = +\ 4.24$	$M_{3sl} = -4.48$	-2.24	-2.24
2	$M_{2c}^s = +\ 1.76$ $M_{2c}^i = +\ 4.44$	$M_{2sl} = -6.20$	-3.10	-3.10
1	$M_{1c}^s = +\ 4.56$ $M_{1c}^i = +\ 1.54$	$M_{1sl} = -6.10$	-3.05	-3.05
Foundation	$M_0 = +\ 10.46$	—	—	—

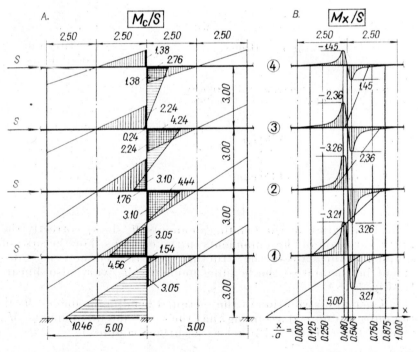

Figure 5.18 Symmetrical multistoried slab structure under seismic load S. A. Diagrams of bending moments M_c and M_{sl} in bars of equivalent reticular structure. B. Diagrams of sectional bending moments M_x in slabs of real structure.

5.2 General displacement method. In-plane loads

The values M_x (Fig. 5.18B) were computed for $a = 5.0$ m, taking into account the effect of sets of surface couples M_{sl} associated with the row of columns under study, for each of the four levels.

It should be noted that the extreme values of the sectional bending moments M_x (Nm/m) that are considered in sizing the slabs of the real structure, are 31 per cent greater than the design moments $M_{sl}/0.50\,a$ (Nm/m) considered in the equivalent frame method. We note however that the distribution of the sectional bending moments in slabs shows that the values max (min) M_x assume the characteristics of a local perturbation : these extreme values decrease rapidly along the slab sizes. Thus, at a distance of only $(0.625-0.540)\,5.0 = 0.425$ m from the maximum (minimum) point they become four times smaller (Fig. 5.18B). The linear decrease of the moments M_{sl} is obviously much slower.

6

Introduction to the dynamic analysis of slab structures

6.1 Calculation elements for the dynamic analysis of slab structures

Among the basic questions in the area of structural response to dynamic actions, much attention has been paid to: the magnitude of moving masses including that of unbalanced masses, the high value of the accelerations during the service period of a large category of structures and especially of industrial buildings, and the importance of spans and the height of ordinary structures. These studies, which were much accelerated with the advent of computers, have led to rigorous procedures of dynamic analysis for several categories of structures, e.g. reticular systems, tall chimneys and machine foundations. These procedures have already become established in design practice [13], [15], [17].

However, the dynamic analysis of slab structures is not well understood. Recalling the widespread use and the growing functional adaptability of this category of structures, the need arises to add design data derived from the dynamic analysis of structures consisting of bars and plates to those obtained in the static analysis.

In what follows, we shall use the method suggested for the elastic analysis of slab structures subjected to gravity loads as well as for the investigation of the response of this structural system to the forces acting in the planes of the slabs. In this manner, the difficulties encountered in the investigation of free and forced vibrations of slab structures are comparable to those inherent in the dynamic analysis of reticular structures [83], [86].

In order to do this, it is essential to adopt a correct dynamic representation for both categories of structures, where the masses

6 *Dynamic analysis of slab structures*

are discretized as faithfully as possible to their real distribution. The procedure suggested for the dynamic analysis of slab structures is rigorous within the range covered by the basic assumptions (Sec. 2.1) of the theory of plates and of the structural statics and dynamics.

6.1.1 *Elements for the study of free vibrations*

The vibration given by an initial pulse-excitation of an elastic structural system, which besides its own continuous but arbitrarily distributed mass, bears also other distributed or concentrated masses, represents a highly complex phenomenon. In order to construct a mathematical model for the applied study of these vibrations, several symplifying assumptions pertaining to the theory of structural dynamics must be adopted :

— the free vibrations of the elastic structure are assumed to belong to the range of small vibrations; this means one must adopt the principle of superposition of effects, with the general shape of the free vibration resulting from the composition of the natural shapes of vibrations of the structure under study ;

— the continuity effect of the mass of the structure is neglected; this means one must adopt a dynamic representation having a finite number of discrete masses and, hence, a finite number of n degrees of freedom ;

— the damping influence is neglected [13], [15], [17].

Let us consider a vibrating system having n degrees of freedom, represented by a rectangular plate (Fig. 6.1). For the sake of simplicity, we represent only the mass m_j having the co-ordinates (x_m, y_m) and the sizes $(2c_m, 2d_m)$ in the plane xOy. Due to an initial pulse-excitation, the system undergoes free transverse vibrations in the directions of the Oz axis. The angular displacements of the mass m_j can be assumed to be negligible in the first stage. Denoting the displacement of the mass m_j by

$$z_j = z_j(t) \quad (j = 1, 2, \ldots, n) \tag{6.1}$$

at an instant $t = t$, the following inertia forces arise in the direction of each of the n degrees of freedom of the system :

$$I_j = -m_j \ddot{z}_j(t) \quad (j = 1, 2, \ldots, n). \tag{6.2}$$

According to d'Alembert's principle, we apply the inertia forces I_j (6.2) to the dynamic scheme of the oscillating system and obtain

6.1 Calculation elements

the differental expressions of the dynamic displacements $z(t)$ of the moving masses m_j in the direction of the degrees of freedom:

$$z_j(t) = -\sum_{i=1}^{n} m_j z_{ji} \ddot{z}_j(t) \qquad (6.3)$$

$(i = 1, 2, \ldots, n; \quad j = 1, 2, \ldots, n),$

where
$\ddot{z}_j(t)$ denotes the acceleration of the mass m_j at an instant $t = t$ of the motion;
z_{ji} represents the unit displacement, i.e. the transverse elastic displacement of the centroid (x_m, y_m) of the mass m_j, induced by the force $P_i = 1$ (or the couple $M_i = 1$) applied at the point i (ξ_i, η_i).

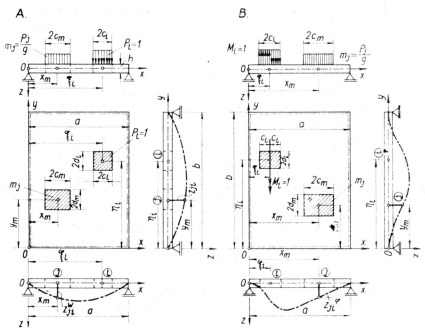

Figure 6.1 Plate as member of a vibrating system.
A. Unit transverse displacement z_{ji} upon the vibrating direction of mass $m_j(x_m, y_m)$ if a unit force $P_i = 1$ (ξ_i, η_i) is applied at i; B. Angular unit displacement z_{ji} upon the vibrating direction of mass m_j if a unit couple $M_i = 1$ is applied at i.

It is worth mentioning that in the method of elastic analysis for slab structures suggested herein, account can be taken in calculations of the in-plane sizes of the masses $m_j(2c_m, 2d_m)$,

321

6 Dynamic analysis of slab structures

and the force $P_i = 1$, which is assumed to be a locally distributed force, and of the couple $M_i = 1$, which is assumed to be a surface couple. The unit force P_i or the unit couple M_i are applied on an area of size $(2c_i, 2d_i)$ in the slab plane (Fig. 6.1).

Using the notation adopted in defining the dimensionless coefficients (Sec. 2.3), the expressions of the unit displacements z_{ji} to be employed in the dynamic analysis of slab structures are given by

(6.4) $\begin{cases} z_{ji} = \dfrac{a^2}{D} k_w^P, \text{ which represents the transverse unit displacement in the direction in which the mass } m_j \\ \quad\text{vibrates, if a force } P_i = 1 \text{ is applied at } i; \\ \\ z_{ji} = \dfrac{a}{D} k_w^M, \text{ which expresses the transverse unit displacement in the direction in which the mass } m_j \\ \quad\text{vibrates, if a couple } M_i = 1 \text{ is applied at } i. \end{cases}$

In the case of slab structures having discrete masses of large sizes, where the rotatory inertia can no longer be neglected, the angular unit displacements z_{ji} are calculated by means of the following expressions:

(6.5) $\begin{cases} z_{ji} = \dfrac{a}{D} k_\varphi^{xP}, \text{ which represents the unit rotation in the direction of the } Ox \text{ axis, at the mass } m_j, \text{ if a force} \\ \quad P_i = 1 \text{ is applied at } i; \\ \\ z_{ji} = \dfrac{1}{D} k_\varphi^{xM}, \text{ which expresses the unit rotation in the direction of the } Ox \text{ axis, at the mass } m_j, \text{ if a couple} \\ \quad M_i = 1 \text{ is applied at } i. \end{cases}$

We have used the following notation (Fig. 6.1):

a is the span of the slab measured in the direction of the Ox axis;

$D = \dfrac{Eh^3}{12(1-\mu^2)}$ is the cylindrical rigidity of the plate (6.6),

where E denotes the modulus of elasticity of the plate material;

h is the plate thickness which is assumed to be constant; and

μ denotes Poisson's ratio.

6.1 Calculation elements

The system of second-order differential equations with respect to t

$$z_j(t) + \sum_{i=1}^{n} m_j z_{ji} \ddot{z}_j(t) = 0 \qquad (6.7)$$

$(i = 1, 2, \ldots n; \; j = 1, 2, \ldots n),$

is verified by particular harmonic solutions of the form

$$z_j(t) = A_j \sin(\omega t + \varphi), \qquad (6.8)$$

where

A_j is the amplitude of the free vibration of the mass m_j m $(j = 1, 2, \ldots, n)$;

ω denotes the natural circular frequency of the elastic system (rad/sec.);

and

φ represents the initial phase angle (rad).

In a matrix formulation, the system of equations (6.7) becomes

$$\mathbf{Z} + \mathbf{z}\,\mathbf{M}\ddot{\mathbf{Z}} = \mathbf{0}, \qquad (6.9)$$

where

- $\mathbf{Z}$ denotes the vector of dynamic displacements z_j of the masses m_j $(j = 1, 2, \ldots, n)$ (see also Table 6.1);
- $\mathbf{z}$ is the flexibility matrix, which represents the square matrix of the transverse unit displacements z_{ji} (6.4) $(i = 1, 2, \ldots, n; \; j = 1, 2, \ldots, n)$ (see also Table 6.1);
- $\mathbf{M} = [\mathbf{M}]$ is the inertia matrix which represents the diagonal matrix of the masses m_j $(j = 1, 2, \ldots, n)$ of the discretized structure;
- $\ddot{\mathbf{Z}}$ denotes the vector of the accelerations $\ddot{z}_j(t)$ of the masses m_j at an instant $t = t$ of the motion.

Inserting the particular solutions (6.8) into the system of differential equations (6.9), where

$$\ddot{z}_j(t) = -\omega^2 A_j \sin(\omega t + \varphi) \qquad (6.10)$$

and letting

$$\lambda = \frac{1}{\omega^2}, \qquad (6.11)$$

we obtain a linear homogeneous system of algebraic equations of the form

$$\mathbf{SA} = \mathbf{0}, \qquad (6.12)$$

where

$$S = \begin{bmatrix} m_1 z_{11} - \lambda & \ldots & m_j z_{1j} & \ldots & m_n z_{1n} \\ \vdots & & \vdots & & \vdots \\ m_1 z_{j1} & \ldots & m_j z_{jj} - \lambda & \ldots & m_n z_{jn} \\ \vdots & & \vdots & & \vdots \\ m_1 z_{n1} & \ldots & m_j z_{nj} & \ldots & m_n z_{nn} - \lambda \end{bmatrix} ; \quad A = \begin{bmatrix} A_1 \\ \vdots \\ A_j \\ \vdots \\ A_n \end{bmatrix}.$$

The column matrix **A** expresses the vector of the amplitudes A_j of the masses m_j in the directions of their free vibration, corresponding to the n degrees of freedom of the dynamic diagram under study. These amplitudes represent the unknowns of the system of equations (6.12).

The equation of the natural circular frequency (the secular equation), which is obtained by use of the flexibility matrix method

$$D[S] = 0, \qquad (6.13)$$

describes formally the zeroing condition of the main determinant D of the matrix **S**. In a physical interpretation, this equation represents the necessary and sufficient condition for the elastic system to vibrate, i.e. the condition that at least one of the elements A_j of the amplitude vector is other than zero.

The algebraic equation of degree n with respect to λ (i.e. ω^2), which is obtained by expanding the determinant (6.13), has n real positive roots,

$$\lambda_1 > \lambda_2 \ldots > \lambda_r > \ldots \lambda_n,$$

which permits the determination of the value of natural circular frequencies of the structure under study:

$$\omega_1 < \omega_2 < \ldots < \omega_r < \ldots < \omega_n \text{ (rad/sec)}. \qquad (6.14)$$

ω_1 above denotes the lowest natural circular frequency, i.e. the fundamental frequency and ω_r represents the natural circular frequency of order r* to which there corresponds the natural frequency

$$f_r = \frac{\omega_r}{2\pi}, \text{ (Hz)} \qquad (6.15)$$

and the period of vibration

$$T_r = \frac{2\pi}{\omega_r} \text{ (sec)}. \qquad (6.16)$$

---

* The notation adopted for the eigenvalues are those established in the Romanian specifications for the earthquake-resistant design of structures P. 100−81[163].

6.1 Calculation elements

The number n of the eigenvalues ω_r (f_r, T_r) ($r = 1, 2, \ldots, n$) of a vibrating system is equal to the number of its degrees of dynamic freedom. The eigenvalue spectrum represents physical characteristics of the structure, which depend only on the distribution of masses and the elastic properties of the system.

★

Knowledge of the eigenvalues is essential both for the approach of the dynamic behaviour of a given structure subjected to a system of disturbing forces (Sub-sec. 6.1.2) and to the earthquake-resistant design of structures. The analysis of the dynamic response of a structural system to a seismic action requires the determination of the vibration shapes of this system for each eigenvalue $\omega_r(f_r, T_r)$ ($r = 1, 2, \ldots, n$), i.e. knowledge of the natural shapes of the elastic system. The natural vibration mode r, which consists of the eigenvalue ω_r and the corresponding natural shape, is directly involved in the evaluation of seismic lateral loads S_r [13], [15], [17], [163].

In order to have a geometrical configuration of the natural shapes of the system, in the system of motion equations (6.12) we introduce successively the values of the natural frequencies ω_r (or λ_r). These values represent n roots of the equation of the natural circular frequency (6.13). The new system of equations given by

$$\mathbf{S}_r \mathbf{A}_r = \mathbf{0}, \tag{6.17}$$

where the elements of the vector $\mathbf{A}_r$ are the amplitudes $A_{j,r}$ of the masses m_j ($j = 1, 2, 3, \ldots, n$) corresponding to the natural mode of vibration r, is still an algebraic homogeneous system with respect to $A_{j,r}$, such that the values of the unknowns $A_{j,r}$ cannot be determined. The matrix $\mathbf{S}_r$ is the matrix $\mathbf{S}$ (6.12) including the values $\lambda = \lambda_r$ corresponding to the natural vibration mode r. However, in the design of structures sited in seismic areas, it is not the value of the amplitudes $A_{j,r}$ which occur but that of their ratios $u_{j,r}$ (6.18):

$$u_{j,r} = \frac{A_{j,r}}{A_{1,r}} \quad (j = 1, 2, \ldots, n). \tag{6.18}$$

This holds also in the evaluation of seismic lateral loads $S_{j,r}$ (6.19) at each level of the structure. Indeed, in compliance with Romanian specifications [163]:

$$S_{j,r} = K_s \beta_r \psi \, \eta_{j,r} G_j, \tag{6.19}$$

where

6 Dynamic analysis of slab structures

K_s denotes the coefficient of seismic intensity;

β_r is the spectral (dynamic) coefficient whose value is inversely proportional to the natural vibration period T_r (6.16) of the structure, corresponding to the natural vibration mode r;

ψ denotes the dissipation coefficient whereby account can be taken of the damping capacity of the structure and its ductility;

G_j is the resultant of the gravity loads at the level j.

The ratio $\eta_{j,r}$ of the distribution of seismic lateral loads at the level j, which depends on the natural vibration mode r and on the magnitude and distribution of the gravity loads G_j, is calculated by means of the expression:

$$\eta_{j,r} = \frac{\sum_{j=1}^{n} G_j u_{j,r}}{\sum_{j=1}^{n} G_j u_{j,r}^2} u_{j,r}. \tag{6.20}$$

It is assumed that the values of $u_{j,r}$ (6.18), which are the ratios of the amplitudes $A_{j,r}$ with respect to one of these amplitudes, say $A_{1,r}$, represent the ordinates of the natural shape of vibration r.

By dividing each term in the system of equations (6.17) by $A_{1,r}$, the first ordinate of any natural shape r is

$$u_{1,r} = 1.$$

The system of equations thus obtained is:

$$S_r U_r = 0, \tag{6.21}$$

where

$$S_r = \begin{bmatrix} m_1 z_{11} - \lambda_r & \dots & m_j z_{1j} & \dots & m_n z_{1n} \\ \vdots & & \vdots & & \vdots \\ m_1 z_{j1} & \dots & m_j z_{jj} - \lambda_r & \dots & m_n z_{jn} \\ \vdots & & \vdots & & \vdots \\ m_1 z_{n1} & \dots & m_j z_{nj} & \dots & m_n z_{nn} - \lambda_r \end{bmatrix} ; \quad U_r = \begin{bmatrix} 1 \\ \vdots \\ u_{j,r} \\ \vdots \\ u_{n,r} \end{bmatrix}.$$

Since the system of n algebraic equations (6.21) has free terms and only $(n-1)$ unknowns $u_{j,r}$ $(j = 2, 3, \ldots, n)$, only $(n-1)$ equations will be used to determine these unknowns. The excess equation is used to verify numerical results. Since the value of the $(n-1)$ ordinates $u_{j,r}$ of the natural shape defining the natural vibration mode r is referred to the value $u_{1,r} = 1$ of the first

6.1 Calculation elements

ordinate, the geometrical configuration of the vibration shape r is known. By a successive introduction of the values $\lambda_r (r = 1, 2, \ldots, n)$ corresponding to the n eigenvalues into the system of equations (6.21) we can determine all the n shapes of vibration of the elastic system under study.

In order to verify the values obtained for the natural shapes, use is made of the orthogonality property of the natural shapes, which is expressed by the condition

$$\sum_{j=1}^{n} u_{j,r} m_j u_{j,s} = 0 \quad (r \neq s), \; (j = 1, 2, \ldots, n), \tag{6.22}$$

where $u_{j,r}$ and $u_{j,s}$ are the ordinates of the natural shapes corresponding to the two distinct natural vibration modes r and s, respectively.

★

In the particular case of slab structures, the topological symmetry and even the regular distribution of concentrated masses and of the supports allow us to group the unit displacements z_{ji} (6.4) and (6.5), which are included in the flexibility matrix (6.9), on condition that the associated sets are linearly independent.

In order to determine the values of the natural circular frequency ω_r and of the ordinates $u_{j,r}$ of the natural modes of vibration, in the secular equation (6.13) and the system of equations (6.21) the sets of parameters represent sets of unit displacements of type (Secs. 2.3, 4.1 and 5.1)

$$K_w^P; \; K_w^M; \; K_\varphi^{xP}; \; K_\varphi^{xM}; \; K_\varphi^{yP}; \; K_\varphi^{yM}. \tag{2.50}$$

Besides the advantages inherent to the work with symmetrical and antisymmetrical sets of parameters which are conducive to the separation of the equations in the system of compatibility equations (Sec. 5.1), the dynamic analysis of slab structures can be substantially simplified if use is made of the dimensionless coefficients calculated by the author (Chaps. 3—5 and the Appendix).

6.1.2 Elements for the study of forced vibrations

A system of m disturbing forces $P_k(t)$, $(k = 1, 2, \ldots m)$ acting on a given elastic structure with n finite degrees of dynamic freedom ($\forall m, n$) induces in the masses of this structure oscillations other than free. By superposing the effects of the two types (free and forced) of vibrations, the real phenomenon becomes more complex than in the free vibration case, at least at the beginning of the motion. In applications, the study of forced vibrations is carried out assuming that after a certain time period, the natural vibrations are damped, whereupon at instant $t = t$ of the motion the elastic system reaches a steady state of vibration defined by the frequency of the disturbances $P_k(t)$ [13], [15], [17].

For convenience, we also assume that the disturbing forces can be expressed by harmonic functions of the form (Fig. 6.2A)

$$P_k(t) = P_k^0 \sin pt \quad (k = 1, 2, \ldots, m), \tag{6.23}$$

where

P_k^0 denotes the amplitude value of the disturbing forces $P_k(t)$

and

p represents the frequency of the harmonic disturbing forces, which is assumed to be equal for all the disturbances.

Since the disturbing forces exciting the structure are assumed to act in phase, in the expression (6.23) the initial phase angle φ was assumed to be equal to zero.

The proposed method for the elastic analysis of slab structures [75], [78], [86] also allows the mathematical modelling of the effects of the disturbing couples, e.g. a *surface couple* $M_k(t)$ acting on the slab of the floor (Fig. 6.2B)

$$M_k(t) = M_k^0 \sin pt, \quad (k = 1, 2, \ldots, m) \tag{6.24}$$

where

M_k^0 denotes the amplitude value of the disturbance $M_k(t)$.

For the sake of simplicity, the dynamic scheme of the oscillating system consisting of a *rectangular plate* was reduced to only one mass m_j at j and only one disturbance $P_k(t)$ (Fig. 6.2A) or $M_k(t)$ (Fig. 6.2B) at k, respectively.

In the case of a given oscillating system under the above-mentioned assumptions, the motion of the masses m_j characterizing the dynamic scheme of the system will be given by the forced vibrations induced by the disturbing forces $P_k(t)$ (6.23), and by the disturbing couples $M_k(t)$ (6.24), respectively, such that in the steady-state vibrations, the law of motion for the mass m_j is of the form

$$z_j(t) \cong z_{jP,M}(t).$$

Since the steady-state response of the elastic structure to harmonic distrurbances can be represented by harmonic func-

6.1 Calculation elements

tions of frequency p, the dynamic displacement of the mass m_j in the direction of vibration will be expressed by

$$z_j(t) = B_j \sin pt, \tag{6.25}$$

where

B_j denotes the amplitude of the forced vibration of the mass m_j ($j = 1, 2, \ldots, n$) in the direction of the degree of freedom j attained at the instant when the disturbing forces reach their highest value P_k^0 (M_k^0, respectively).

The law of variation of the inertia forces I_j which develop in the direction of each of the n degrees of freedom of the oscillating system under the forced vibration, is given by

$$I_j = -m_j \ddot{z}_j(t) = I_j^0 \sin pt \quad (j = 1, 2, \ldots, n). \tag{6.26}$$

By

$$I_j^0 = m_j p^2 B_j \tag{6.27}$$

we denote the extreme (highest or lowest) value of the inertia forces I_j applied on the masses m_j. This value is simultaneously attained by all the inertia forces when the disturbances reach their amplitude values.

Hence, by virtue of the simplifying assumptions used in the design practice of forced vibrations in the case of elastic structures, we assume that under forced vibrations both the dynamic displacements $z_j(t)$ of the masses m_j and their corresponding inertia forces I_j simultaneously reach the extreme values when the amplitudes P_k^0 (M_k^0, respectively) of these disturbances are attained.

If in compliance with d'Alembert's principle, the dynamic scheme is statically loaded with the inertia forces I_j in the direction of the degrees of freedom j of the masses of the system and with the disturbing forces $P_k(t)$, ($k = 1, 2, \ldots, m$), we obtain the elastic displacements $z_j(t)$ of these masses in the steady-state forced vibrations:

$$z_j(t) = \sum_{i=1}^{n} z_{ji} I_j + \sum_{k=1}^{m} z_{jk} P_k(t) \quad (j = 1, 2, \ldots, n), \tag{6.28}$$
$$(k = 1, 2, \ldots, m)$$

where, assuming that the system is assimilated with a plate, z_{jk} denotes the unit displacement, or the elastic, linear or angular, displacement of the mass $m_j(x_m, y_m)$ produced in the direction of the degree of freedom j by the force $P_k^0 = 1$ acting at point $k(\xi_k, \eta_k)$. Mention should be made of the fact that some of the disturbing forces $P_k(t)$ act directly on the elastic system, hence they may not be applied on the masses m_j.

The unit displacements z_{ji} have been also defined above for the case of slab structures (Fig. 6.1; formulae (6.4) and (6.5)).

6 Dynamic analysis of slab structures

When the slab structures have geometrical or mechanical symmetry and the grouping of the parameters requires calculation of the associated unit displacements, the dimensionless coefficients k_w^P, k_w^M, k_φ^{xP}, k_φ^{xM}, k_φ^{yP}, k_φ^{yM} are replaced in formulae (6.4) and (6.5)

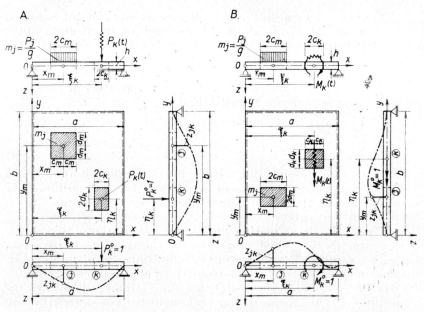

Figure 6.2 Plate as member of a vibrating system.
A. Transverse unit displacement z_{jk} upon the vibrating direction of mass m_j (x_m, y_m) if a unit force $P_k^0 = 1$ (ξ_k, η_k) is applied at k. B. Transverse unit displacement z_{jk} upon the vibrating direction of mass m_j if a unit couple $M_k^0 = 1$ is applied at k.

by the coefficients corresponding to the sets of parameters K_w^P, K_w^M, K_φ^{xP}, K_φ^{xM}, K_φ^{yP}, K_φ^{yM} (2.50) (see Fig. 6.3). For instance, consider only the effect of rotations in the direction of the Ox axis. Hence, we have:

$$(6.29) \begin{cases} z_{jk} = \dfrac{a^2}{D} K_w^P, \text{ which represents the associated linear (transverse) displacement induced in the direction of the degree of freedom } j \text{ of the sets of masses } m_j, \text{ by a force } P_k=1 \text{ applied at } k; \\ \\ z_{jk} = \dfrac{a}{D} K_w^M, \text{ which is the associated linear (transverse) unit displacement induced in the direction of the degree of freedom } j \text{ of the set of masses } m_j \text{ by a couple } M_k^0=1 \text{ applied at } k. \end{cases}$$

6.1 *Calculation elements*

(6.30) $\begin{cases} z_{jk} = \dfrac{a}{D} K_{\varphi}^{xP}, \text{ which denotes the associated unit angular displacement (unit rotation) in the direction of the } Ox \text{ axis induced at the associated masses } m_j \text{ by a unit force } P_k^0 = 1 \text{ applied at } k; \\ \\ z_{jk} = \dfrac{1}{D} K_{\varphi}^{xM}, \text{ which represents the associated unit angular displacements (unit rotation) in the direction of the } Ox \text{ axis, induced at the associated masses } m_j \text{ by a couple } M_k^0 = 1 \text{ applied at } k. \end{cases}$

The expresions for the unit angular displacements z_{jk} of type (6.30) are required, together with the expressions (6.5), in the dynamic analysis of slab structures with discrete masses of large sizes, where the rotatory inertia cannot be neglected. To these as-

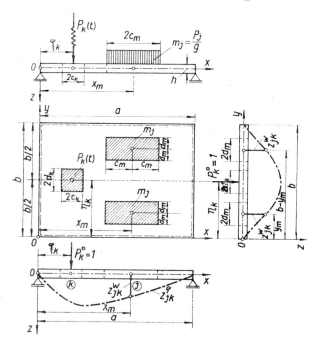

Figure 6.3 Plate as member of a symmetric vibrating system. Associated unit tranverse displacement z_{jk} upon the vibrating direction of the associated mass m_j if a unit force $P_k^0 = 1$ is applied at k.

6 *Dynamic analysis of slab structures*

sociated rotations in the direction of the Ox axis, there correspond sets of rotatory inertia forces I_j. Similar expressions can be obtained to calculate the associated rotations z_{jk} along the Oy axis when there exist corresponding degrees of freedom j for rotations (Table 6.1).

In the expressions for the dynamic displacements $z_j(t)$, given by Eqn. (6.28), of the masses m_j under steady-state forced

Table 6.1

The elastic plate as a member of an oscillating system

Expressions for unit displacements z_{ji} of mass m_j with co-ordinates (x_m, y_m)

No.	Unit force or unit couple applied at $i(\xi_i, \eta_i)$	z_{ji}	Nature of unit displacement in the direction of degree of freedom j
1	$P_i = 1$	$\dfrac{a^2}{D} k_w^P$	Transverse displacement in the direction of Oz axis
2	$M_i = 1$	$\dfrac{a}{D} k_w^M$	Transverse displacement in the direction of Oz axis
3	$P_i = 1$	$\dfrac{a}{D} k_\varphi^{xP}$	Unit rotation in the direction of Ox axis
4	$P_i = 1$	$\dfrac{a}{D} k_\varphi^{yP}$	Unit rotation in the direction of Oy axis
5	$M_i = 1$	$\dfrac{1}{D} k_\varphi^{xM}$	Unit rotation in the direction of Ox axis
6	$M_i = 1$	$\dfrac{1}{D} k_\varphi^{yM}$	Unit rotation in the direction of Oy axis.

Observations. 1. The expressions of the unit displacements z_{ji} at items 1 and 2 represent the linear elastic displacements which can be used in design, when account is taken of the translatory inertia of the discrete masses m_j alone.

2. The expressions of the unit displacements z_{ji} at the points 3, 4, 5 and 6 represent the elastic rotations appearing only when the masses m_j are of large sizes, such that the rotatory inertia cannot be neglected.

3. The values of the unit displacements z_{ji} are obtained for the real statically indeterminate system (Chap. 5).

4. The design expressions of the dimensionless coefficients $k_w^P, \ldots k_\varphi^{yM}$ and of those of type $K_w^P, \ldots K_\varphi^{yM}$, which represent associated unit displacements, are those used in the calculations required by the static analysis of slab structures (Secs. 2.3, 4.1 and 5.1).

6.1 Calculation elements

vibration, the unit displacements z_{ji} and z_{jk} are static characteristics of the elastic structure. Inserting in (6.28) the expressions for the disturbing forces $P_k(t)$ (6.23), the disturbing couples $M_k(t)$ (6.24) (respectively), and the corresponding inertia forces I_j (6.26), and by simplification, we obtain the system of linear algebraic equations in I_j^0 which describe the motion of the masses m_j of the oscillating system under steadystate vibration when the amplitude values are attained:

$$\mathbf{z^* I^0} + \mathbf{Z}_P^0 = \mathbf{0}, \qquad (6.31)$$

where

$$\mathbf{z^*} = \begin{bmatrix} z_{11}^* & \cdots & z_{1j} & \cdots & z_{1n} \\ \vdots & & \vdots & & \vdots \\ z_{j1} & \cdots & z_{jj}^* & \cdots & z_{jn} \\ \vdots & & \vdots & & \vdots \\ z_{n1} & \cdots & z_{nj} & \cdots & z_{nn}^* \end{bmatrix} ; \quad \mathbf{I}^0 = \begin{bmatrix} I_1^0 \\ \vdots \\ I_j^0 \\ \vdots \\ I_n^0 \end{bmatrix} ; \quad \mathbf{Z}_P^0 = \begin{bmatrix} z_{1P}^0 \\ \vdots \\ z_{jP}^0 \\ \vdots \\ z_{nP}^0 \end{bmatrix},$$

and where

$\mathbf{z^*}$ represents the corrected flexibility matrix derived from the square matrix $\mathbf{z}$ (6.9) of the unit displacement z_{ji} given by (6.3), (6.4) and (6.5) by inserting on the main diagonal the corrected unit displacements

$$z_{jj}^* = z_{jj} - \frac{1}{m_j p^2} \quad (j = 1, 2, \ldots n); \qquad (6.32)$$

$\mathbf{I}^0$ denotes the vector of the unknowns I_j^0 (6.27) which represent the inertia forces in the direction of the degrees of freedom j, at their highest value;

$\mathbf{Z}_P^0$ is the vector of the free terms z_{jP}^0 (6.33) which represent the static displacements produced in the direction of the degrees of freedom of the masses m_j only by the system of disturbing forces for the amplitude value P_k^0 and M_k^0 ($k = 1, 2, \ldots, m$), respectively.

For a system of disturbing forces $P_k(t)$,

$$z_{jP}^0 = \sum_{k=1}^{m} z_{jk} P_k^0 \quad (j = 1, 2, \ldots n), \qquad (6.33)$$

and for a system of disturbances that can be assimilated with surface couples

$$z_{jP}^0 = \sum_{k=1}^{m} z_{jk} M_k^0 \quad (j = 1, 2, \ldots n). \qquad (6.34)$$

As the extreme values $\pm I_j^0$ of the inertia forces I_j are determined by solving the system of equations (6.31), the highest or lowest values of the dynamic forces acting on the masses m_j of the elastic structure are obtained by superposing the effects under assumptions most inappropriate to the evaluation of design stresses:

$$P^0_{j\mathrm{dyn}} = P^0_j \pm I^0_j \quad (j = 1, 2, \ldots n). \tag{6.35}$$

P^0_j above represents the amplitude value of the disturbing force $P_j(t)$ acting on the mass m_j of the oscillating system.

The calculation of design stresses may require, if necessary, beside the dynamic forces $P^0_{j\mathrm{dyn}}$ from (6.35) also the disturbances P^0_k or M^0_k ($k \neq j$) which act on the structural members, in our case on the slabs.

★

In the dynamic analysis of slab structures, use is made of the flexibility matrix method. The plates forming the floors of the structure are assumed to be the members of an oscilating system. The dynamic diagram of this system can be obtained by a treatment similar to that employed in the case of reticular structures, i.e. by discretization of the masses at the level of each floor. However, unlike plane systems consisting of bars alone, slab structures are treated in calculations as three-dimensional systems. Indeed, the discrete masses $m_j (j = 1, 2, \ldots n)$ may have any position in the horizontal plane xOy of the floors, whereas the linear or angular displacements of these masses in the direction of the degrees of freedom may develop in the vertical planes xOz or yOz, which are normal to the plane of the floors. A more complex dynamic diagram may obviously include concentrated masses along the columns, if required.

An extension of the author's analysis for the static analysis of slab structures (Chaps. 4 and 5) to the dynamic analysis of this type of structure allows (Figs. 6.1, 6.2 and 6.3) us [83], [86]:

1. to express the unit displacements and the unit rotations of type z_{ji} occurring in the study of free (formulae 6.3, 6.7, 6.9, 6.12, 6.13 and 6.21) and forced vibrations (formulae 6.28, 6.31 and 6.32) of slab-column structures, using the dimensionless coefficients $k^P_w, \ldots k^{yM}_\varphi$ (Table 6.1);

2. to express in a similar manner the unit transverse displacements and the unit rotations of type z_{jk} occurring in the investigation of forced vibrations (formulae 6.28, 6.31, 6.33 and 6.34) of slab structures;

3. to express the associated unit transverse displacements and the associated unit rotations of type z_{ji} or z_{jk}, with the aid of the dimensionless coefficients $K^P_w, \ldots K^{yM}_\varphi$, which represent the sets of unit displacements (Secs. 2.3, 4.1 and 5.1);

4. to consider in the analysis the dimensions $(2c_m, 2d_m)$ of the discrete masses m_j in the plane of the slabs;

6.2 Comparative numerical studies

5. to take account of the sizes ($2c_i$, $2d_i$) of the area subjected to the force $P_i = 1$ as a locally distributed force, and to the couple $M_i = 1$, as a surface couple,
and

6. to include in calculations the sizes ($2c_k$, $2d_k$) of the area subjected to the disturbing force $P_k(t)$, and a disturbance that can be assimilated with a surface couple $M_k(t)$, respectively.

6.2 Behaviour of slab structures subjected to dynamic action. Comparative numerical studies

6.2.1 Determination of natural circular frequencies

In order to illustrate the use of the present analysis procedure of free vibrations, we revert to the simple diagram (Sec. 5.1) used for the static analysis of a slab structure (Fig. 5.3), assuming the same values for the relative slab-column stiffness. As was shown earlier, at present a structure of continuous mass $\overline{m}$ the slab can be schematically represented by the discretization of masses. A model where mass concentration would occur at each node of the computation diagram employed in the programs (Fig. 6.4A) would lead to an excessively large number of independent parameters ($9 \times 9 = 81$ degrees of dynamic freedom for the vertical displacements of the masses alone). A substantially simplified solution (Fig. 6.4B) leads to only two degrees of dynamic freedom, the masses whose motion is introduced in the analysis being given by $m_1 = m_1'$ and $m_2 = m_1'' + m_2' + m_2'' + m_3' + m_3''$. A free horizontal translation is assumed here to develop only in the direction of the Ox axis.

In order to have a clear image of the proposed procedure, we shall consider the most simple dynamic diagram of the structure under study (Fig. 6.4C), where a single discrete mass m at the centre of the floor ($x_m = 0.50a$; $y_m = 0.50b$) shapes the mass of the slab. The mass of the columns was neglected in all the calculation variants examined.

If, for simplifying reasons, we assume, here too, that a free horizontal translation may develop in the direction of the Ox axis, then the structure has also two degrees of dynamic freedom(Figs. 6.4 and 6.5). Mention should be made that to these degrees of freedom there correspond the fundamental frequency ω_1 and the second frequency ω_2 of natural vibrations, i.e. the low frequencies which are of much interest in engineering problems. Notwithstanding the excessively simplified representation of the real structure, the values ω_1 and ω_2 are slightly lower than those obtained in the continuous-mass assumption.

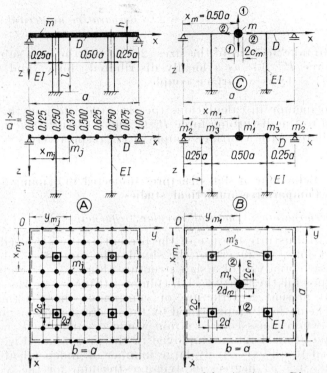

Figure 6.4 Dynamic analysis of slab structures. Discretization schemes for mass.
A. Dynamic diagram idealizing the mass $\overline{m}$ of plate as continuous; B, C. Intermediate dynamic diagrams; D. Dynamic diagram idealizing the whole mass m of plate as being concentrated at symmetry centre (x_m, y_m) on a limited area ($2c_m \times 2d_m$).

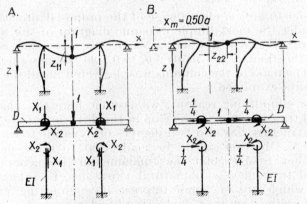

Figure 6.5 Free vibrations of slab structures. Basic system and schematic diagram for unit displacements z_{11} and z_{22} associated to the two degrees of freedom of mass m.

336

6.2 Comparative numerical studies

Since the mass is constant ($m_1 = m_2 = m$) for both degrees of freedom, the equation of the natural circular frequency (6.13) reduces to

$$\begin{vmatrix} mz_{11} - \lambda & mz_{12} \\ mz_{21} & mz_{22} - \lambda \end{vmatrix} = 0, \tag{6.36}$$

with two positive real roots λ_1 and λ_2.

As the unit displacements of type z_{ji} ($i \neq j$) are equal to zero ($z_{12} = z_{21} = 0$), these roots can be obtained by writing two equations involving λ, each with only one unknown:

$$\lambda_1 = mz_{11}; \tag{6.36a}$$
$$\lambda_2 = mz_{22}. \tag{6.36b}$$

In order to determine the values corresponding to the frequencies ω_1 and ω_2 of the natural vibrations (6.11), the unit displacements z_{11} and z_{22} are calculated for the dynamic diagram under consideration, which is assumed to be a statically indeterminate system (Fig. 6.5). Thus, z_{11} corresponds to the load due to the vertical force $P = 1$ and z_{22} corresponds to the load due to the horizontal force $P = 1$. Calculations are performed by grouping the unknowns in compliance with the procedure suggested for the static analysis of slab structures (Sec. 5.1).

For the unit component $P = 1$ acting in the vertical direction (Fig. 6.5 A):

$$X_1 a K_w^P + X_2 K_w^M + a k_w^P = 0 \tag{6.37a}$$
$$X_1 a K_\varphi^{xP} + X_2 (K_\varphi^{xM} + K_\varphi^{col}) + a k_\varphi^{xP} = 0. \tag{6.37b}$$

Knowing the value of the sets of unknowns X_1 and X_2 from the solution of (6.37), the unit displacement z_{11} in Eqn. (6.36a) is given by

$$z_{11} = \frac{a^2}{D} X_1 K_w^P + \frac{a}{D} X_2 K_w^M + \frac{a^2}{D} k_w^P. \tag{6.38}$$

For the horizontal component $P = 1$, and assuming that the columns are located at the quarter of the span ($\xi = 0.25 a$; $\xi' = 0.75 a$; $\eta = 0.25b$; $\eta' = 0.75 b$), it follows that $X_1 = 0$ (Sec. 5.1). In this case, the system of compatibility equations reduces to (Fig. 6.5):

$$X_2 \left(\frac{1}{D} K_\varphi^{xM} + \frac{l}{EI} \right) - \frac{l^2}{8EI} = 0 \tag{6.39}$$

or

$$X_2 = \frac{K_\varphi^{col}}{2(K_\varphi^{xM} + 4 K_\varphi^{col})} l. \tag{6.39a}$$

6 Dynamic analysis of slab structures

Knowing the value of the associated unknown X_2, the unit displacement z_{22} in Eqn. (6.36b) is given by

$$z_{22} = \frac{3}{4EI}\left(\frac{1}{3} - 2\frac{X_2}{l}\right). \tag{6.40}$$

★

Numerical calculations were performed for variants $a \leqslant b \leqslant 3a$ and

$$10 \leqslant \frac{4EI}{lD} \leqslant 80, \tag{5.17a}$$

which cover the range of possible values of the parameters defining the slab structure under consideration (Fig. 6.6 and Tables 6.2, 6.3 and 6.4).

Table 6.2

Slab structures — Free vibration study
Finite boundary plates ($a \leqslant b \leqslant 3a$)

Comparative values of natural vibration frequencies

$EI/l =$ constant
Position of columns: $\xi = 0.25a$; $\xi' = 0.75a$;
$\eta = 0.25b$; $\eta' = 0.75b$
Column cross-section: $c = d = 0.04a$;
$10 \leqslant 4EI/lD \leqslant 80$; $1 < a/2$
Position of mass m: $x_m = 0.50a$; $y_m = 0.50b$
Area subjected to mass: $0.02a \leqslant (c_m; d_m) \leqslant 0.06a$

		$b = a$		$b = 3a$
$c_m = d_m =$	0.02 a	0.04 a	0.06 a	0.04 a
$4EI/lD$		$\omega_1 \sqrt{\dfrac{1}{a}}\sqrt{\dfrac{EI}{ml}}$		
80	3.933	3.984	4.060	1.909
40	5.539	5.610	5.717	2.699
20	7.780	7.880	8.027	3.816
10	10.898	11.035	11.240	5.395
$4EI/lD$		$\omega_2 \sqrt{\dfrac{1}{l}}\sqrt{\dfrac{EI}{ml}}$		
∞*	3.464	3.464	3.464	3.464
80	3.907	3.907	3.907	3.910
40	4.242	4.242	4.242	4.246
20	4.719	4.719	4.719	4.725
10	5.291	5.291	5.291	5.296
0*	6.930	6.930	6.930	6.930

* Limit values of theoretical interest in the comparative approach.

6.2 Comparative numerical studies

If the value of the column stiffness EI/l is assumed to be constant, the value of the fundamental frequency mode ω_1 increases substantially as the cylindrical rigidity D of the slab is

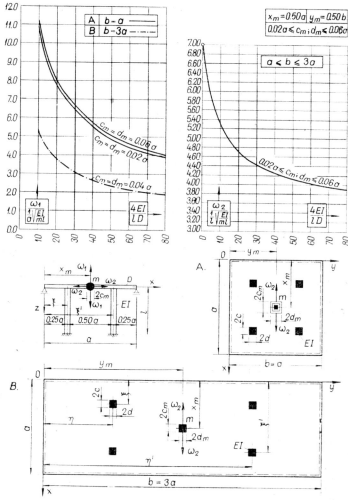

Figure 6.6 Free vibrations of slab structure. Influence of relative slab-column stiffness and of parameters b/a and c_m/a on magnitude of natural circular frequencies ω_1 and ω_2 ($10 \leqslant 4EI/lD \leqslant 80$).

higher, whereas the second frequency of the natural vibrations ω_2 is practically unaffected by the variation of this stiffness value (Table 6.2).

339

Table 6.3

Slab structures — Free vibration study

Finite boundary plates ($a \leqslant b \leqslant 3a$)

Comparative values of natural vibration frequencies

D = constant

$c_m = d_m =$	$b = a$			$b = 3a$
	$0.02\,a$	$0.04\,a$	$0.06\,a$	$0.04\,a$
$4EI/lD$		$\omega_1 \dfrac{1}{a}\sqrt{\dfrac{D}{m}}$		
80	17.590	17.817	18.158	8.535
40	17.517	17.742	18.078	8.534
20	17.397	17.620	17.949	8.533
10	17.231	17.447	17.772	8.530
0*	16.582	16.777	17.072	8.521
$4EI/lD$		$\omega_2 \dfrac{1}{l}\sqrt{\dfrac{D}{m}}$		
80	17.474	17.474	17.474	17.484
40	13.414	13.414	13.414	13.426
20	10.553	10.553	10.553	10.565
10	8.365	8.365	8.365	8.374

* Limit values of theoretical interest in the comparative approach.

Retaining now the same value for the plate stiffness D, we observe that the value of the fundamental frequency ω_1 is practically uninfluenced by the decreasing values of the column stiffness EI/l, whereas the frequency ω_2 lowers substantially as EI/l becomes smaller. This behaviour is in compliance with expectations (Table 6.3).

These observations are valid for the cases of the square plate ($b = a$) and the very long rectangular plate ($b = 3a$). As was predicted in both cases, the fundamental frequency ω_1 decreases substantially as the length b of the plate increases, whereas the second frequency remains constant (Tables 6.2 and 6.3).

Finally, in the comparative study of the natural vibration frequency, we consider a variant where the slab span a is expressed as a function of the column length l. In this case, we observe that as the values of $4EI/lD$ decrease to 20, and for ordinary values of column length, $\omega_1 < \omega_2$. It is only in the case of very

6.2 Comparative numerical studies

Table 6.4

Slab structures — Free vibration study
The square plate

Comparative values of natural vibration frequencies

$a/3 \leqslant l \leqslant a/2$
Area subjected to mass : $c_m = d_m = 0.04\ a$

$4EI/lD$	$\omega_1 \dfrac{1}{l}\sqrt{\dfrac{EI}{ml}}$		$\omega_2 \dfrac{1}{l}\sqrt{\dfrac{EI}{ml}}$
	$a = 3l$	$a = 2l$	
80	1.328	1.992	3.907
40	1.870	2.805	4.242
20	2.627	3.940	4.719
10	3.678	5.518	5.291

$4EI/lD$	$\omega_1 \dfrac{1}{l}\sqrt{\dfrac{D}{m}}$		$\omega_2 \dfrac{1}{l}\sqrt{\dfrac{D}{m}}$
	$a = 3l$	$a = 2l$	
80	5.939	9.909	17.474
40	5.914	8.871	13.414
20	5.873	8.810	10.553
10	5.816	8.724	8.365

flexible columns with respect to the slab stiffness ($l = a/2$ and for the limit value $4EI/lD = 10$) that the above inequality is reversed, i.e. the fundamental frequency ω_1 corresponds to the horizontal displacement of the mass m (Table 6.4).

All the conclusions derived from the comparative study reinforce the idea that the author's method of analysis provides a faithful image of the behaviour of the real phenomenon.

6.2.2 Dynamic deflected shape and stresses induced by disturbing forces

We consider the dynamic diagram of the slab structure having a two-fold geometrical and mechanical symmetry, used in the study of vibrations (Fig. 6.5). The single mass $m = m_1 = m_2$, concentrated at the centre of the plate, has two degrees of freedom : vertical translation in the direction of the Oz axis and horizontal translation in the direction of the Ox axis. The motion of the sys-

6 Dynamic analysis of slab structures

tem in the direction of the Oy axis is assumed to be obstructed and the dimensions of the mass m are taken sufficiently small, so that the effect of the degrees of freedom expressed in rotations (rotatory inertia) may be neglected.

We shall illustrate the approach to forced vibrations in the case of slab structures by considering that the mass m is excited by a harmonic disturbance of the type (Fig. 6.7A):

$$P(t) = P_0 \sin pt. \qquad (6.23)$$

By $P_0^V = P_0^H = P_0$, we have denoted the amplitude value of the disturbing force acting successively in the direction of the two degrees of freedom of the structure.

The system of linear algebraic equations for I_j (6.31), which describe the motion of the mass m of the oscillating system under steady-state forced vibrations, can be reduced to a system of two equations having two unknowns (Sub-sec. 6.1.2):

$$\begin{cases} \left(z_{11} - \dfrac{1}{m_1 p^2}\right) I_1^0 + z_{12} I_2^0 + z_{1P}^0 = 0\,; \\ z_{21} I_1^0 + \left(z_{22} - \dfrac{1}{m_2 p^2}\right) I_2^0 + z_{2P}^0 = 0, \end{cases} \qquad (6.41)$$

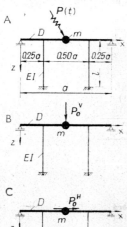

Figure 6.7 Forced vibrations of slab structures. Dynamic diagram assuming mass m as excited by a disturbing force $P(t)$.

where I_1^0 and I_2^0 represent the inertia forces developing in the direction of the two degrees of freedom for the amplitude value.

The free terms z_{jP}^0 (6.33), which express the displacements of the mass m produced in these two directions only by the disturbing force $P(t)$ (6.23), depend here on only one of the two components P_0^V and P_0^H, respectively, of the disturbance (Figs. 6.7B and 6.7C): $z_{1P}^0 = z_{11} P_0^V$; $z_{2P}^0 = z_{22} P_0^H$.

Owing to the topological symmetry of the slab structure under consideration, the unit displacements of type $z_{ji}(i \neq j)$ are zero in the direction of the two degrees of freedom of the mass m: $z_{12} = z_{21} = 0$. Thus, the system of two equations (6.41) splits into two equations with only one unknown each:

$$\left(z_{11} - \dfrac{1}{mp^2}\right) I_1^0 + z_{11} P_0^V = 0, \qquad (6.41a)$$

6.2 Comparative numerical studies

for symmetrical loading, and

$$\left(z_{22} - \frac{1}{mp^2}\right) I_2^0 + z_{22} P_0^H = 0, \tag{6.41b}$$

for antisymmetrical loading.

For the dynamic diagram under examination, the expressions of the unit displacements of the type $z_{ji}(i = j)$ about the principal diagonal of the matrix **S** (6.12) were established in the study of free vibrations (Sub-sec. 6.2.1): z_{11} given by (6.38) and z_{22} given by (6.40). Inserting the values of the natural vibration frequencies ω_1 and ω_2 in Eqns. (6.41a) and (6.41b), we obtain the highest values of the inertia forces:

$$I_1^0 = \frac{p^2}{\omega_1^2 - p^2} P_0^V; \quad I_2^0 = \frac{p^2}{\omega_2^2 - p^2} P_0^H, \tag{6.42}$$

where ω_1 denotes the fundamental frequency of the slab structure;

ω_2 represents the second natural frequency;

p is the frequency of the disturbing force $P(t)$ given by Eqn. (6.23);

and $P_0^V = P_0^H = P_0$ represents the amplitude value of this force.

Finally, we perform the superposition of effects and obtain the expressions for the dynamic forces for the amplitude value $P_{j\,\text{dyn}}^0$ (6.35), given by:

$$P_{1\,\text{dyn}}^0 = P_0^V + I_1^0 = \frac{1}{1 - \dfrac{p^2}{\omega_1^2}} P_0 = \psi_1 P_0 \tag{6.43a}$$

$$P_{2\,\text{dyn}}^0 = P_0^H + I_2^0 = \frac{1}{1 - \dfrac{p^2}{\omega_2^2}} P_0 = \psi_2 P_0, \tag{6.43b}$$

where

$$\psi_1 = \frac{1}{1 - \dfrac{p^2}{\omega_1^2}}; \quad \psi_2 = \frac{1}{1 - \dfrac{p_2}{\omega_2^2}} \tag{6.44}$$

represent the dynamic coefficients for the two loading diagrams with the components P_0^V and P_0^H, respectively, of the disturbing force max $P(t) = P_0$ acting directly on the mass m, upon the direction of the two degrees of freedom.

★

6 Dynamic analysis of slab structures

Knowing the values of the dynamic forces P_{dyn}^0, the dynamic displacements w_{dyn} and dynamic bending moments M_x^{dyn} at any point (x, y) of the slab (Figs. 6.8 and 6.9) are calculated using the

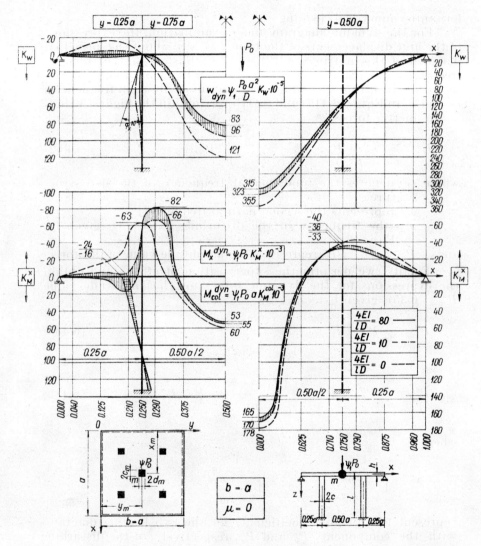

Figure 6.8 Forced vibration of slab structures. Influence of relative slab-column stiffness on dynamically deflected shape w_{dyn} and on magnitude of dynamic sectional moments M_x^{dyn}, associated to the first degree of freedom ($10 \leqslant 4EI/lD \leqslant 80$).

6.2 Comparative numerical studies

dimensionless coefficients K_w and K_M^x corresponding to the associated interacting forces X_1 and X_2 calculated for $P_0 = 1$ in the study of free vibrations (Sub-sec. 6.2.1).

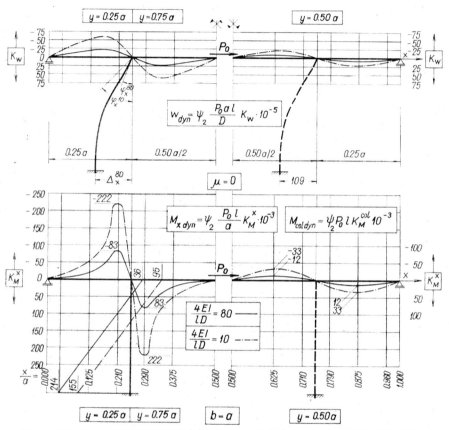

Figure 6.9 Forced vibration of slab structures. Influence of relative slab-column stiffness on dynamically deflected shape W_{dyn} and on magnitude of dynamic sectional moments M_x^{dyn} associated to the second degree of freedom ($10 \leqslant 4EI/lD \leqslant 80$).

For the vertical translation of the mass m in the direction of the Oz axis and assuming that the dynamic diagram is subjected to the force $P^0_{1\,\mathrm{dyn}}$ (6.43a), we obtain (Fig. 6.8):

$$w_{\mathrm{dyn}} = \frac{\psi_1 P_0 a^2}{D}\left(X_1 K_w^P + \frac{X_2}{a} K_w^M + k_w^P\right); \qquad (6.45)$$

$$M_x^{\mathrm{dyn}} = \psi_1 P_0 \left(X_1 K_M^{xP} + \frac{X_2}{a} K_M^{xM} + k_M^{xP}\right), \qquad (6.46)$$

where X_1 and X_2 are associated unknowns derived by use of equations (6.37) for $P_0^V = 1$.

For the horizontal translation of the mass m in the direction of the Ox axis, and assuming that the dynamic diagram is subjected to the force $P_{2\,dyn}^0$ (6.43b), we obtain (Fig. 6.9):

$$w_{dyn} = \frac{\psi_2 P_0 al}{D} \left(\frac{X^2}{l} K_w^M \right); \qquad (6.47)$$

$$M_x^{dyn} = \frac{\psi_2 P_0 l}{a} \left(\frac{X_2}{l} K_M^{xM} \right). \qquad (6.48)$$

X_2 above is the associated unknown obtained with the aid of Eqn. (6.39) for $P_0^H = 1$.

The dynamic bending moments M_{col}^{dyn} of columns and the displacement u_{dyn} of the upper ends of the columns along the Ox axis are obtained using the methods of the theory of reticular structures.

6.2.3 Comparative dynamic analysis of a multistoried slab structure

The method under discussion for the dynamic analysis of slab structures bears no restriction concerning the number of levels and, hence, the number of moving mass m_j. Also, no restrictions exist concerning the geometry of the slab structure (the disposition of columns in particular) or the in-plane distribution of the mass. However, in order to have a clear picture of the physical reality and of its pertinent aspects, we shall adopt a simple dynamic scheme consisting of two levels (Fig. 6.10). This schematic diagram, defined by the geometrical and mechanical parameters pertinent for the previously examined one-level structure (Sub-secs. 5.1.2, 6.2.1 and 6.2.2), permits also a comparison of the numerical results obtained [86].

In the dynamic analysis of the slab structure under study we proceed from the following simplifying assumptions:

a. The structure has a two-fold geometrical and mechanical symmetry; however, horizontal translation is assumedly free only upon the direction of the Ox axis.

b. The distribution of the mass reduces to only one concentrated mass m_j ($j = 1,2$) at the centre ($x_m = 0.50a$, $y_m = 0.50b$)

6.2 Comparative numerical studies

of each floor, discretizing the mass of the pertinent slab. The column mass is assumedly negligible.

c. The magnitude of the two moving masses is assumed to be equal ($m_1 = m_2 = m$).

d. The dimensions of the discrete mass are assumed to be sufficiently small ($c_m = d_m = 0.04\,a$) for their rotatory inertia to be negligible.

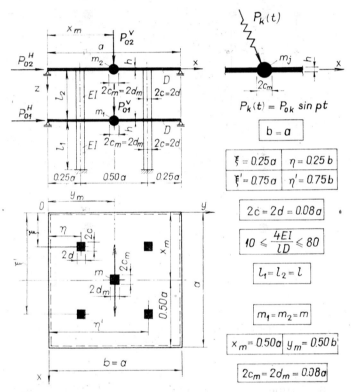

Figure 6.10 Dynamic analysis of multistoried slab structures. Design parameters.

It results that the dynamic scheme of our slab structure has only four degrees of freedom (Figs. 6.11 and 6.12), corresponding to the low frequency range (the fundamental plus the following three frequencies), that are particularly important in the design practice. Additionally, the value of the natural circular frequency of free vibrations $\omega_r (r = 1, \ldots 4)$ thus obtained will be close to and obviously smaller than those pertinent to the real slab struc-

6 Dynamic analysis of slab structures

ture (where the slab mass is continuous) and hence satisfy requirements of structural safety.

The natural circular frequencies are determined, as in the previous example (Sub-sec. 6.2.1), by use of the secular equation:

$$D\,[S] = 0. \tag{6.13}$$

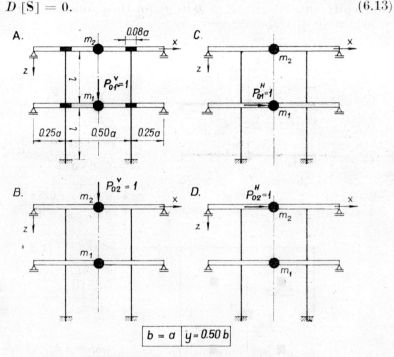

Figure 6.11 Two-level slab structures. Study of free vibrations. Dynamic scheme of the structure under study.

In the case under study only transverse unit displacements of the type z_{ji} (6.4) appear in the elements of the principal determinant of the matrix S defined in Sub-sec. 6.1.1. By neglecting the rotatory inertia, the angular unit displacements of type z_{ji} (6.5) are also disregarded. In order to define the subscripts j, i, we denote, for convenience, by $(j = 2r - 1;\ i = 2r - 1)$ the vertical motions and by $(j = 2r;\ i = 2r)$ the horizontal motions of the mass (Fig. 6.12). In the dynamic system under examination $(r = 1, \ldots 4)$, only the unit displacements whose subscripts represent displacements upon the same direction will differ from zero: $z_{11}, \ldots z_{44}$, $z_{13}, \ldots z_{42}$. Eight of the sixteen elements of the determinant

$D[S]$ will be zero: $z_{12} = z_{14} = \ldots = z_{41} = z_{43} = 0$. Hence, the secular equation (6.13) reduces to

$$\begin{vmatrix} z_{11} - \dfrac{\lambda}{m} & 0 & z_{13} & 0 \\ 0 & z_{22} - \dfrac{\lambda}{m} & 0 & z_{24} \\ z_{31} & 0 & z_{33} - \dfrac{\lambda}{m} & 0 \\ 0 & z_{42} & 0 & z_{44} - \dfrac{\lambda}{m} \end{vmatrix} = 0. \quad (6.49)$$

By expanding the determinant, Eqn. (6.49) of the fourth degree is split into two equations of second degree in λ, corresponding to the motion of the two mass upon the vertical

$$\left(z_{11} - \frac{\lambda}{m}\right)\left(z_{33} - \frac{\lambda}{m}\right) - z_{13}^2 = 0, \quad (6.49a)$$

and upon the horizontal

$$\left(z_{22} - \frac{\lambda}{m}\right)\left(z_{44} - \frac{\lambda}{m}\right) - z_{24}^2 = 0. \quad (6.49b)$$

The values of the unit displacements z_{ji} (6.4), which represent the coefficients of the unknowns $\lambda_1, \ldots \lambda_4$ and the free terms in Eqns. (6.49a) and (6.49b) are calculated by subjecting alternately the dynamic scheme with two mass m_1 and m_2 to the unit forces (Fig. 6.11 A—D)

$$P_{01}^V = 1, \quad P_{02}^V = 1, \quad P_{01}^H = 1, P_{02}^H = 1. \quad (6.50)$$

Considered as a static system, the slab structure under examination has four degrees of indeterminacy and a grouping of the unknowns $X_1, \ldots X_4$ can be performed. These associated unknowns are then determined by use of the method of static analysis defined earlier (see Sub-secs. 5.1.1—5.1.3), treating the unit forces P_0^V as locally distributed forces on the mass-slab contact area $(2c_m \times 2d_m)$ and the unit forces P_0^H as forces of the type S, acting in the plane of slabs.

The influence of the relative slab-column stiffness factor on the value of natural circular frequencies $\omega_r (r = 1, \ldots, 4)$ was examined for the following range

$$10 \leqslant \frac{4EI}{lD} \leqslant 80. \quad (5.17a)$$

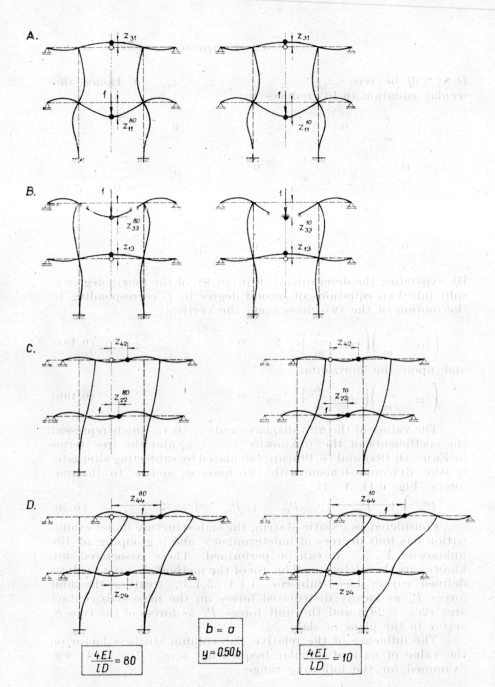

Figure 6.12 Two-level slab structures. Study of free vibrations. Dynamically deflected shapes. Influence of the relative column-slab stiffness (for magnitude of unit displacements z_{ij} and z_{ji}, see Table 6.5).

6.2 Comparative numerical studies

Hence, the values $X_1, \ldots, X_4$ and the unit displacement z_{ji} ($j = 1, \ldots 4; i = 1, \ldots 4$), induced by successive loads P_0 (6.50), were determined between these limits (Table 6.5).

The values thus obtained for the associated unknowns X_1, X_2 and for the unit displacements z_{11}, z_{22}, corresponding to the first level of the two-storied structure, are close to those obtained for the one-level structure (Sub-sec. 6.2.1), which is in compliance with expectations.

Once the magnitude of the unit displacements z_{ji} is determined, the root $\lambda_r (r = 1, \ldots, 4)$ of the split secular equations (6.49a) and (6.49b) permit the calculation of natural circular frequencies $\omega_r (r = 1, \ldots, 4)$ of the slab structure under study (Sub-secs. 6.1.1, 6.2.1 and Table 6.6).

The values ω_r associated to the horizontal unit displacements z_{22}, z_{44} decrease substantially as the relative stiffness factor becomes smaller (i.e. by 52.9 percent and 32.6 percent respectively over the range $80 \geqslant 4EI/lD \geqslant 10$), whereas the values of ω_r associated to the vertical displacements z_{11}, z_{33} are almost unaffected by this factor (Table 6.6 A).

For a comparison of the fundamental frequency ω_1 as a function of the relative slab-column stiffness, the values ω_r associated to the vertical components of motion were evaluated for two extreme variants of the column length : $l = a/4$ and $l = a/2$ (Table 6.6 B). After a suitable ordering of the values $\omega_r (\omega_1 < \omega_2 < \omega_3 < \omega_4)$, we observe that, except for the dynamic scheme having columns that are very short and rigid with respect to the slab ($4EI/lD = 80; l = a/4$), in all the other variants considered so far, the fundamental frequency ω_1 corresponds to the horizontal component of the motion of mass m_2 (level 2). However, the highest frequency ω_4 corresponds to the horizontal component of the motion of mass m_1 (level 1).

The values of the natural circular frequency $\omega_r (r = 1, \ldots, 4)$ are the most important dynamic properties of the slab structure under study. These can be used either straightforwardly in, say, the earthquake-resistant design of structures (see Sub-sec. 6.1.1) or in the study of forced vibrations, if this is required (see Sub-sec. 6.1.2). This analysis is not specific to slab structures as it can be performed as in the ordinary reticular-structure case.

★

Table 6.5

Multistoried slab structures — Free vibration study
The flexibility matrix method. Square plate — Two levels
Position of columns: $\xi = 0.25 a$; $\xi' = 0.75a$;
$\eta = 0.25b$; $\eta' = 0.75b$
Columns cross-section: $e = d = 0.04 a$

Relative column-slab stiffness: $10 \leq \dfrac{4EI}{lD} \leq 80$; $\dfrac{a}{4} \leq l \leq \dfrac{a}{2}$

Position of the mass m_1, m_2: $x_m = 0.50 a$; $y_m = 0.50b$. Mass-slab contact area: $e_m = d_m = 0.04 a$

Comparative values of associated unknowns

X_i, X_j	$4EI/lD$	Vertical unit forces P_0^V			X_i, X_j^*	$4EI/lD$	Horizontal unit forces P_0^H		
		1 Level	2 Levels				1 Level	2 Levels	
		$P_{01}^V = 1$	$P_{01}^V = 1$	$P_{02}^V = 1$			$P_{01}^H = 1$	$P_{01}^H = 1$	$P_{02}^H = 1$
X_1/P_0^V	80	−0.35324	−0.35124	0.00130	X_2/P_0^H	80	0.03565	0.01901	0.05373
	10	−0.37397	−0.36523	0.00538		10	0.09518	0.06359	0.15085
$X_2/P_0^V a$	80	−0.02406	−0.02484	−0.00050					
	10	−0.01601	−0.01941	−0.00209					
X_3/P_0^V	80	—	−0.35384	−0.35384	$X_4/P_0^H l$	80	—	0.01571	0.06613
	10	—	0.00518	−0.37558		10	—	0.02368	0.13463
$X_4/P_0^V a$	80	—	−0.00050	−0.02383					
	10	—	−0.00201	−0.01539	$*X_1 = X_3 = 0$				

Comparative values of unit displacements

$10^6 z_{ji}$	$4EI/lD$	Vertical unit forces P_0^V			$10^6 z_{ji} \left| P_0^{H^2}/D \right.$	$4EI/lD$	Horizontal unit forces P_0^H		
$\left. P_0^V a^2 \right/ D$		1 Level	2 Levels				1 Level	2 Levels	
		$P_{01}^V = 1$	$P_{01}^V = 1$	$P_{02}^V = 1$			$P_{01}^H = 1$	$P_{01}^H = 1$	$P_{02}^H = 1$
z_{11}	80	3150	3137	—	z_{22}	80	3275	3299	—
	10	3285	3228	—		10	14291	15879	—
z_{33}	80	—	—	3154	z_{44}	80	—	—	22691
	10	—	—	3295		10	—	—	68452
$z_{13} = z_{31}$	80	—	−8	−8	$z_{24} = z_{42}$	80	—	7420	7420
	10	—	−35	−35		10	—	26236	26236

Table 6.6

Multistoried slab structures — Free vibration study
The flexibility matrix method
Square plate — Two levels
Relative column-slab stiffness : $10 \leqslant 4EI/lD \leqslant 80$; $a/4 \leqslant l \leqslant a/2$

A. Comparative values of natural circular frequencies

z_{jj}	m_j^*	$\omega\sqrt{\frac{1}{a}\sqrt{\frac{D}{m}}}$		$\frac{4EI}{lD}$	$\omega\sqrt{\frac{1}{l}\sqrt{\frac{D}{a}}}$		m_j^*	z_{jj}
		2 Levels	1 Level		1 Level	2 Levels		
z_{11}	m_1^V	17.863	17.817	80	17.474	35.678	m_1^H	z_{22}
		17.640	17.447	10	8.365	14.105		
z_{33}	m_2^V	17.797	—	80	—	6.299	m_2^H	z_{44}
		17.384	—	10	—	3.551		

B. Comparative values of fundamental circular frequencies ω_1

m_j^*	1 Level	m_j^*	2 Levels	$\omega\frac{1}{a}\sqrt{\frac{D}{m}}$	2 Levels	m_j^*	1 Level	m_j^*
\multicolumn{4}{c}{$\frac{4EI}{lD}=80$}		\multicolumn{4}{c}{$\frac{4EI}{lD}=10$}
\multicolumn{4}{c}{$l = a/4$}		\multicolumn{4}{c}{$l = a/4$}
m_1^V	4.454	m_2^V	4.449	ω_1	3.551	m_2^H	4.362	m_1^V
		m_1^V	4.466	ω_2	4.346	m_2^V		
m_1^H	17.474	m_2^H	6.299	ω_3	4.410	m_1^V	8.365	m_1^H
		m_1^H	35.678	ω_4	14.105	m_1^H		
\multicolumn{4}{c}{$l = a/2$}		\multicolumn{4}{c}{$l = a/2$}
m_1^V	8.909	m_2^H	6.299	ω_1	3.551	m_2^H	8.365	m_1^H
		m_2^V	8.899	ω_2	8.692	m_2^V		
		m_1^V	8.932	ω_3	8.820	m_1^V		
m_1^H	17.474	m_1^H	35.678	ω_4	14.105	m_1^H	8.724	m_1^V

* The superscript of mass m_j indicates the direction in which mass moves (V, H).

The numerical results obtained from the analysis of the dynamic response of slab structures reinforce the earlier idea concerning the faithful reflection of the physical phenomenon by the author's method.

The simplicity of this method of dynamic analysis holds also in the case of more complex slab structures, where the dynamic diagram has a greater number of discrete masses and hence, of degrees of freedom, as well as more disturbances. The computational difficulty is comparable to that involved in the dynamic analysis of reticular structures.

Conclusions

As was shown in the preliminary section of this book, the category of slab structures includes flat slab structures, decks, slab bridges resting on columns, foundation rafts, certain shear-wall structures, foundation caissons, etc.

The main characteristic of all the types of mixed structures is the existence of rigid connections at the bar-plate nodes. Thorough knowledge of the study of the effects of these rigid connections was required for the investigation of the states of stress and deformation of slabs and throughout the structure.

With rare exceptions, the effect of interacting forces acting transverse to the plate has been uniquely approached in the available literature. Thus, the structure is schematically represented by a diagram having discrete interior supports that may or may not be point-like and the system of compatibility equations is reduced to the condition of zero relative displacement of the slab in the column axis. As a conclusion, we could state that a thorough elastic analysis is both necessary and possible. This may be done by inserting in the analysis the system of equations including conditions of zero relative rotation of the slab-column node in both directions of the slab plane.

The mathematical expressions derived from the analysis of the little studied effect of a surface couple on plates describes the real phenomenon with good accuracy. This is true both for the infinitely long plate case, to which several comparative data were given by Stiglat, and for the finite boundary plate case to which no data are available. From the comparative investigation undertaken by the author it could be establisehed that these two categories of plates behave in a similar manner with respect to elastic displacements and stresses. This result was reinforced in the case of the load due to only one surface couple, for a set of couples M as well as by the structural analysis performed.

The exploration of the effect of the rigid plate-bar connections on a structure, which was simulated with the aid of the surface couple, has led to several other results pertaining to the factors influencing the distribution of the elastic displacements and stresses of plates, the local perturbation characteristic of the load due to a surface couple M and the tendency towards symmetrization proper to the deflected mid-surface and the moment diagrams. Meanwhile, other main results concern an eventual analysis of unsymmetrical slab structures (structures that are severely unsymmetrical in the transverse direction, structures having rhombic or parallelogram-like panels) and the pronounced influence exerted by the relative slab-column stiffness in the structural analysis when the structure is subjected to loads due to both gravity and earthquakes.

The first advantage of the present approach to the elastic analysis of slab structures stems from the generality of the method, which permits an extension of the results thus obtained to the static and dynamic analysis of other types of slab structures. It is obvious that a procedure of analysis that tends to a faithful reflection of the real state of stress can only be implemented on a computer basis. The programs developed allow the determination of the elastic displacements w, φ_x, φ_y and the stresses M_x, M_y and M_{xy} of three categories of plates subjected to three types of loads, functions of six independent parameters, at any point (x, y) of the computation grid and adopting various conditions of precision, as required. The use of the numerical data tabulated in the form of auxiliary values, which are calculated for a wide range of ordinary values, are conductive to a shorter period and lower costs for design work. Additionally, the degree of complexity inherent to the analysis of the slab structure is similar to that noticed in the analysis of ordinary reticular structures.

An extension of the solutions obtained herein in the theory of plates allows an immediate diversification of the boundary conditions under consideration by modelling further edge conditions encountered in practice : the perfectly clamped edge, the cantilever free edge, etc. The modelling of the elastically supported edge or the elastically clamped edge requires however the development of further algorithms. The first attempt to deal with this problem is due to Pfaffinger, who resorted to the principle of the minimum potential strain energy. Other solutions might be obtained by means of the finite element methods.

Further exploration in the field might provide new mathematical models for the plates of finite length. Such models in which the expressions in double series are replaced by equivalent expressions in simple series will allow a faster computation rate. Likewise, this book gives several recommendations concerning the

Conclusions

use of the results obtained from the study of plates of infinite length, where the higher convergent expressions in simple series occur, in the analysis of finite boundary real structures.

In order to assure a simplified and wider use of the present method of analysis, the idea was suggested that several approximate solutions of the Lagrange-Sophie Germaine equation (of Ritz or Galerkin type) could be adopted in the study of plates subjected to surface couples. The reference values obtained herein by imposing severe conditions of accuracy on the programs will also allow a comparison of the results derived with the aid of variational methods, the establishment of the order of magnitude of the errors (with respect to the solutions that are assumed to be rigorous) and, hence, the specification of the conditions and applied range of the approximate solutions to the differential equation of plates.

Appendix

AUXILIARY VALUES

RECTANGULAR PLATES SUBJECTED TO A SURFACE COUPLE M

Values of dimensionless coefficients k_w, k_φ^x, k_φ^y, k_M^x and k_M^{xy} used to calculate transverse displacements w, rotations φ_x, φ_y, sectional bending moments M_x and torsional moments M_{xy}.

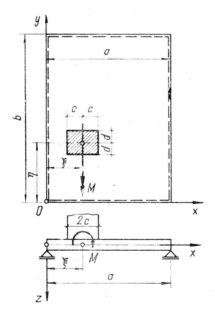

THE SQUARE PLATE
SUBJECTED TO A SURFACE COUPLE M

Transverse displacements w — Tables A.1—A.3
Rotations φ_x — Tables A.4—A.6
Rotations φ_y — Tables A.7—A.9
Sectional bending moments M_x — Tables A.10—A.15
Sectional torsional moments M_{xy} — Tables A.16—A.18

Formulae for the direct computation of
additional static magnitudes:

$$(M_y)_{\mu=0} = 3\left[(M_x)_{\mu=1/3} - (M_x)_{\mu=0}\right]$$
$$(M_y)_\mu = (M_y)_{\mu=0} + \mu(M_x)_{\mu=0}$$
$$(M_{xy})_\mu = (1-\mu)(M_{xy})_{\mu=0}$$

Note: Owing to the limited space of this book, the auxiliary values corresponding to the infinitely long plate have not been given here. However, the author will satisfy any question readers may have concerning the numerical data for the static magnitudes defining the behaviour of this category of plates subjected to a surface couple M or to a locally distributed force P. Likewise, any further question concerning the static magnitudes induced by a locally distributed force P upon the rectangular plates will be given due answer.

Appendix

Table A.1
TRANSVERSE DISPLACEMENTS w — COEFFICIENTS $10^5 k_w$

$$w = \frac{Ma^2}{D} k_w \qquad b = a \qquad c = d = 0.02\,a$$

$\xi = c$

$\eta = d$

y/b \ x/a	0.020	0.125	0.250	0.375	0.500	0.625	0.750	0.875	0.980
0.020	**87**	96	47	29	19	13	8	4	1
0.250	46	231	**277**	235	178	125	80	39	6
0.500	19	110	177	194	174	137	92	46	7
0.750	8	46	79	94	92	78	55	29	4
0.980	1	3	6	8	7	6	5	2	0

$\eta = 0.25\,b$

y/b \ x/a	0.020	0.125	0.250	0.375	0.500	0.625	0.750	0.875	0.980
0.020	46	231	277	235	178	125	80	39	6
0.250	**736**	2533	**2701**	2325	1820	1316	848	414	66
0.500	267	1492	2238	2313	2019	1562	1048	524	84
0.750	103	610	1031	1193	1133	930	648	330	53
0.980	8	46	79	94	92	78	55	29	4

$\eta = 0.50\,b$

y/b \ x/a	0.020	0.125	0.250	0.375	0.500	0.625	0.750	0.875	0.980
0.020	19	110	177	194	174	137	92	46	7
0.250	267	1492	2238	2313	2019	1562	1048	524	84
0.500	**840**	3142	**3732**	3518	2953	2246	1496	745	119
0.750	267	1492	2238	2313	2019	1562	1048	524	84
0.980	19	110	177	194	174	137	92	46	7

$\xi = 0.25\,a$

$\eta = d$

y/b \ x/a	0.020	0.125	0.250	0.375	0.500	0.625	0.750	0.875	0.980
0.020	−4	−32	**9**	53	28	16	9	5	1
0.250	−6	0	89	**180**	179	137	89	43	7
0.500	6	42	88	124	135	120	88	45	7
0.750	4	24	46	62	68	62	46	25	5
0.980	0	2	4	5	6	5	4	2	0

Table A.1 (continued)

η = 0.25 b

x/a \ y/b	0.020	0.125	0.250	0.375	0.500	0.625	0.750	0.875	0.980
0.020	−6	0	89	180	179	137	89	43	7
0.250	−35	−114	**910**	**1935**	1777	1370	910	451	72
0.500	−55	−407	1010	1533	1647	1421	1010	519	83
0.750	−47	−292	567	772	842	763	567	300	49
0.980	−4	−24	46	62	68	62	46	25	5

η = 0.50 b

x/a \ y/b	0.020	0.125	0.250	0.375	0.500	0.625	0.750	0.875	0.980
0.020	6	42	88	124	135	120	88	45	7
0.250	55	407	1010	1533	1647	1421	1010	519	83
0.500	11	178	**1477**	**2707**	2619	2133	1477	751	121
0.750	55	407	1010	1533	1647	1421	1010	519	83
0.980	6	42	88	124	135	120	88	45	7

ξ = 0.50 a

η = d

x/a \ y/b	0.020	0.125	0.250	0.375	0.500	0.625	0.750	0.875	0.980
0.020	−1	−8	−20	−44	0	44	20	8	1
0.250	−9	−55	**−100**	−98	0	98	**100**	55	9
0.500	−6	−29	−43	−32	0	32	43	29	6
0.750	−2	−8	−12	−9	0	9	12	8	2
0.980	0	−1	−1	−1	0	1	1	1	0

η = 0.25 b

x/a \ y/b	0.020	0.125	0.250	0.375	0.500	0.625	0.750	0.875	0.980
0.020	−9	−55	−100	−98	0	98	100	55	0
0.250	−82	−506	−930	**−1070**	0	**1070**	930	506	82
0.500	−64	−378	−598	−490	0	490	598	378	64
0.750	−24	−132	−193	−141	0	141	193	132	24
0.980	−2	−8	−12	−9	0	9	12	8	2

η = 0.50 b

x/a \ y/b	0.020	0.125	0.250	0.375	0.500	0.625	0.750	0.875	0.980
0.020	−6	−29	−43	−32	0	32	43	29	6
0.250	−64	−378	−598	−490	0	490	598	378	64
0.500	−106	−638	−1122	**−1211**	0	**1211**	1122	638	106
0.750	−64	−378	−598	−490	0	490	598	378	64
0.980	−6	−29	−43	−32	0	32	43	29	6

Appendix

Table A.2
TRANSVERSE DISPLACEMENTS w — COEFFICIENTS $10^5 k_w$

$$w = \frac{Ma}{D} k_w \qquad \boxed{\xi = e} \qquad \begin{array}{c} b = a \\ c = d = 0.04\,a \end{array}$$

x/a → y/b ↓	0.040	0.125	0.250	0.375	0.500	0.625	0.750	0.875	0.960
η = d									
0.040	194	320	181	114	76	50	31	15	5
0.250	170	440	**541**	464	353	250	159	77	24
0.500	74	216	349	382	345	271	184	92	29
0.750	30	90	157	187	183	153	108	56	18
0.960	5	14	24	29	29	25	18	10	3
η = 0.25 b									
0.040	170	440	541	464	353	250	159	77	24
0.250	**1045**	2414	**2661**	2310	1813	1313	847	414	132
0.500	512	1452	2203	2292	2007	1556	1046	523	167
0.750	202	600	1018	1182	1126	926	646	330	106
0.960	30	90	157	187	183	153	108	56	18
η = 0.50 b									
0.040	74	216	349	382	345	271	184	92	29
0.250	512	1452	2203	2292	2007	1556	1046	523	167
0.500	**1247**	3014	**3679**	3491	2940	2239	1493	744	238
0.750	512	1452	2203	2292	2007	1556	1046	523	167
0.960	74	216	349	382	345	271	184	92	29

$$\boxed{\xi = 0.25\,a}$$

x/a → y/b ↓	0.040	0.125	0.250	0.375	0.500	0.625	0.750	0.875	0.960
η = d									
0.040	−29	−100	**38**	181	104	64	38	18	6
0.250	−11	1	176	**356**	354	271	176	86	27
0.500	25	83	173	247	269	239	173	90	29
0.750	16	49	92	123	134	123	92	49	16
0.960	3	8	15	20	21	19	15	8	3

Table A.2 (continued)

x/a \ y/b	0.040	0.125	0.250	0.375	0.500	0.625	0.750	0.875	0.960
				η = 0.25 b					
0.040	−11	1	176	356	354	271	176	86	27
0.250	−63	−95	907	1909	1766	1364	907	449	143
0.500	115	408	1008	1527	1642	1416	1008	518	167
0.750	93	291	567	770	840	761	567	300	97
0.960	16	49	92	123	134	123	92	49	16
				η = 0.50 b					
0.040	25	83	173	247	269	239	173	90	29
0.250	115	408	1008	1527	1642	1416	1008	518	167
0.500	30	196	1473	2679	2606	2126	1473	749	241
0.750	115	408	1008	1527	1642	1416	1008	518	167
0.960	25	83	173	247	269	239	173	90	29

x/a \ y/b	0.040	0.125	0.250	0.375	0.500	0.625	0.750	0.875	0.960
				ξ = 0.50 a					
				η = d					
0.040	−10	−32	−74	−149	0	149	74	32	10
0.250	−35	−108	−196	−193	0	193	196	108	35
0.500	−20	−57	−86	−64	0	64	86	57	20
0.750	−6	−17	−25	−18	0	18	25	17	6
0.960	−1	−2	−3	−2	0	2	3	2	1
				η = 0.25 b					
0.040	−35	−108	−196	−193	0	193	196	108	35
0.250	−164	−502	−920	−1046	0	1046	920	502	164
0.500	−127	−376	−595	−486	0	486	595	376	127
0.750	−46	−132	−192	−140	0	140	192	132	46
0.960	−6	−17	−25	−18	0	18	25	17	6
				η = 0.50 b					
0.040	−20	−57	−86	−64	0	64	86	57	20
0.250	−127	−376	−595	−486	0	486	595	376	127
0.500	−210	−634	−1112	−1187	0	1187	1112	634	210
0.750	−127	−376	−595	−486	0	486	595	376	127
0.960	−20	−57	−86	−64	0	64	86	57	20

Appendix

Table A.3
TRANSVERSE DISPLACEMENTS w — COEFFICIENTS $10^5 k_w$

$$w = \frac{Ma}{D} k_w \qquad \xi = c \qquad \begin{array}{c} b = a \\ c = d = 0.06\, a \end{array}$$

$\eta = d$

x/a \ y/b	0.060	0.125	0.250	0.375	0.500	0.625	0.750	0.875	0.940
0.060	**282**	501	378	249	167	111	68	33	15
0.250	333	609	**783**	683	525	372	237	115	55
0.500	160	314	511	563	511	403	274	137	66
0.750	66	133	231	277	271	228	161	83	40
0.940	15	30	53	65	65	56	40	21	10

$\eta = 0.25\, b$

x/a \ y/b	0.060	0.125	0.250	0.375	0.500	0.625	0.750	0.875	0.940
0.060	333	609	783	683	525	372	237	115	55
0.250	**1194**	2183	**2592**	2284	1804	1310	846	414	198
0.500	717	1387	2144	2256	1988	1547	1042	521	250
0.750	293	585	996	1163	1113	918	642	328	158
0.940	66	133	231	277	271	228	161	83	40

$\eta = 0.50\, b$

x/a \ y/b	0.060	0.125	0.250	0.375	0.500	0.625	0.750	0.875	0.940
0.060	160	314	511	563	511	403	274	137	66
0.250	717	1387	2144	2256	1988	1547	1042	521	250
0.500	**1487**	2768	**3589**	3446	2916	2228	1488	742	356
0.750	717	1387	2144	2256	1988	1547	1042	521	250
0.940	160	314	511	563	511	403	274	137	66

$\xi = 0.25\, a$

$\eta = d$

x/a \ y/b	0.060	0.125	0.250	0.375	0.500	0.625	0.750	0.875	0.940
0.060	−78	−157	**82**	334	218	138	82	39	18
0.250	−20	3	261	**526**	522	401	261	128	61
0.500	57	123	258	368	402	356	258	134	65
0.750	35	73	138	184	200	183	138	73	36
0.940	9	18	33	44	48	44	33	18	9

364

Table A.3 (continued)

η = 0.25 b

x/a	0.060	0.125	0.250	0.375	0.500	0.625	0.750	0.875	0.940
0.060	−20	3	261	526	522	401	261	128	61
0.250	−76	−63	**902**	**1865**	1748	1355	902	447	214
0.500	177	409	1003	1516	1632	1409	1003	516	249
0.750	140	290	565	767	836	759	565	299	145
0.940	35	73	138	184	200	183	138	73	36

η = 0.50 b

x/a	0.060	0.125	0.250	0.375	0.500	0.625	0.750	0.875	0.940
0.060	57	123	258	368	402	356	258	134	65
0.250	177	409	1003	1516	1632	1409	1003	516	249
0.500	64	227	**1466**	**2632**	2584	2114	1466	746	359
0.750	177	409	1003	1516	1632	1409	1003	516	249
0.940	57	123	258	368	402	356	258	134	65

ξ = 0.50 a

η = d

x/a	0.060	0.125	0.250	0.375	0.500	0.625	0.750	0.875	0.940
0.060	−31	−67	−151	−263	0	263	151	67	31
0.250	−77	−159	**−287**	−284	0	284	**287**	159	77
0.500	−44	−85	−128	−96	0	96	128	85	44
0.750	−14	−26	−38	−27	0	27	38	26	14
0.940	−3	−5	−7	−5	0	5	7	5	3

η = 0.25 b

x/a	0.060	0.125	0.250	0.375	0.500	0.625	0.750	0.875	0.940
0.060	−77	−159	−287	−284	0	284	287	159	77
0.250	−242	−496	−905	**−1006**	0	**1006**	905	496	242
0.500	−188	−372	−590	−479	0	479	590	372	188
0.750	−68	−131	−191	−139	0	139	191	131	68
0.940	−14	−26	−38	−27	0	27	38	26	14

η = 0.50 b

x/a	0.060	0.125	0.250	0.375	0.500	0.625	0.750	0.875	0.940
0.060	−44	−85	−128	−96	0	96	128	85	44
0.250	−188	−372	−590	−479	0	479	590	372	188
0.500	−310	−627	−1096	**−1146**	0	**1146**	1096	627	310
0.750	−188	−372	−590	−479	0	479	590	372	188
0.940	−44	−85	−128	−96	0	96	128	85	44

Appendix

Table A.4
ROTATIONS φ_x — COEFFICIENTS $10^5 \, k_\varphi^x$

$$\varphi_x = \frac{M}{D} k_\varphi^x \qquad \begin{array}{c} b = a \\ c = d = 0.02\,a \end{array}$$

$\xi = c$

$\eta = d$

y/b \ x/a	0.020	0.125	0.250	0.375	0.500	0.625	0.750	0.875	0.980
0.020	**4286**	−671	−209	−101	−63	−45	−35	−30	−29
0.250	2288	1061	−148	−443	−446	−392	−344	−315	−305
0.500	948	746	319	−38	−243	−337	−364	−370	−370
0.750	383	331	197	47	−76	−159	−201	−219	−227
0.980	29	25	16	5	−5	−13	−16	−18	−9

$\eta = 0.25\,b$

y/b \ x/a	0.020	0.125	0.250	0.375	0.500	0.625	0.750	0.875	0.980
0.020	2288	1061	−148	−443	−446	−392	−344	−315	−305
0.250	**35219**	5989	−1719	−3810	−4129	−3896	−3590	−3368	−3291
0.500	13294	9313	2857	−1236	−3202	−3973	−4177	−4198	−4185
0.750	5150	4313	2338	317	−1146	−2018	−2441	−2612	−2643
0.980	384	331	197	47	−76	−159	−201	−219	−227

$\eta = 0.50\,b$

y/b \ x/a	0.020	0.125	0.250	0.375	0.500	0.625	0.750	0.875	0.980
0.020	948	746	319	−38	−243	−337	−364	−370	−370
0.250	13294	9313	2857	−1236	−3202	−3973	−4177	−4198	−4185
0.500	**40363**	10302	619	−3493	−5276	−5914	−6031	−5980	−5934
0.750	13294	9313	2857	−1236	−3202	−3973	−4177	−4198	−4185
0.980	948	746	319	−38	−243	−337	−364	−370	−370

$\xi = 0.25\,a$

$\eta = d$

y/b \ x/a	0.020	0.125	0.250	0.375	0.500	0.625	0.750	0.875	0.980
0.020	−209	−387	**4330**	−359	−120	−65	−46	−37	−35
0.250	−148	305	968	332	−185	−381	−376	−353	−344
0.500	319	354	330	206	−24	−206	−335	−353	−364
0.750	197	189	155	87	9	−87	−152	−189	−201
0.980	16	15	13	6	0	−7	−32	−17	−16

Table A.4 (continued)

η = 0.25 b

y/b \ x/a	0.020	0.125	0.250	0.375	0.500	0.625	0.750	0.875	0.980
0.020	−148	305	968	332	−185	−381	−376	−353	−344
0.250	−1719	963	21044	930	−2682	−3599	−3690	−3632	−3590
0.500	2857	4034	5168	2678	−675	−2732	−3702	−4087	−4177
0.750	2338	2318	1999	1153	−53	−1154	−1893	−2319	−2441
0.980	197	189	155	87	9	−87	−152	−189	−201

η = 0.50 b

y/b \ x/a	0.020	0.125	0.250	0.375	0.500	0.625	0.750	0.875	0.980
0.020	319	354	330	206	−24	−206	−335	−353	−364
0.250	2857	4034	5168	2678	−675	−2732	−3702	−4087	−4177
0.500	619	3282	23043	2084	−2735	−4753	−5583	−5951	−6031
0.750	2857	4034	5168	2678	−675	−2732	−3702	−4087	−4177
0.980	319	354	330	206	−24	−206	−335	−353	−364

ξ = 0.50 a

η = d

y/b \ x/a	0.020	0.125	0.250	0.375	0.500	0.625	0.750	0.875	0.980
0.020	−63	−72	−120	−352	4347	−352	−120	−72	−63
0.250	−446	−419	−185	371	1039	371	−185	−419	−446
0.500	−243	−189	−24	190	295	190	−24	−189	−243
0.750	−76	−56	9	56	79	56	9	−56	−76
0.980	−5	−4	0	3	49	3	0	−4	−5

η = 0.25 b

y/b \ x/a	0.020	0.125	0.250	0.375	0.500	0.625	0.750	0.875	0.980
0.020	−446	−419	−185	371	1039	371	−185	−419	−446
0.250	−4129	−3866	−2682	1197	21447	1197	−2682	−3866	−4129
0.500	−3202	−2622	−675	2569	4685	2569	675	−2622	−3202
0.750	−1146	−856	−53	856	1304	856	−53	−856	−1146
0.980	−76	−56	9	56	79	56	9	−56	−76

η = 0.50 b

y/b \ x/a	0.020	0.125	0.250	0.375	0.500	0.625	0.750	0.875	0.980
0.020	−243	−189	−24	190	295	190	−24	−189	−243
0.250	−3202	−2622	−675	2569	4685	2569	−675	−2622	−3202
0.500	−5276	−4721	−2735	2052	22751	2052	−2735	−4721	−5276
0.750	−3202	−2622	−675	2569	4685	2569	675	−2622	−3202
0.980	−243	−189	−24	190	295	190	−24	−189	−243

Appendix

Table A.5
ROTATIONS φ_x — COEFFICIENTS $10^5 \, k_\varphi^x$

$$\varphi_x = \frac{M}{D} k_\varphi^x \qquad \begin{array}{|c|} \hline b = a \\ c = d = 0.04\,a \\ \hline \end{array}$$

x/a \ y/b	0.040	0.125	0.250	0.375	0.500	0.625	0.750	0.875	0.960
$\xi = c$									
$\eta = d$									
0.040	**4660**	−1331	−746	−389	−244	−175	−139	−121	−116
0.250	4053	2161	−192	−845	−876	−778	−686	−628	−610
0.500	1824	1471	642	−74	−486	−667	−725	−737	−737
0.750	749	654	393	95	−150	−308	−397	−438	−448
0.960	113	100	64	20	−19	−47	−64	−67	−81
$\eta = 0.25\,b$									
0.040	4053	2161	−192	−845	−876	−778	−686	−628	−610
0.250	**24749**	7340	−1408	−3694	−4076	−3877	−3584	−3366	−3296
0.500	12502	9212	2980	−1135	−3141	−3941	−4168	−4190	−4180
0.750	4996	4257	2334	339	−1126	−1996	−2427	−2600	−2650
0.960	749	654	393	95	−150	−308	−397	−438	−448
$\eta = 0.50\,b$									
0.040	1824	1471	642	−74	−486	−667	−725	−737	−737
0.250	12502	9212	2980	−1135	−3141	−3941	−4168	−4190	−4180
0.500	**29745**	11597	926	−3355	−5202	−5874	−6010	−5966	−5946
0.750	12502	9212	2980	−1135	−3141	−3941	−4168	−4190	−4180
0.960	1824	1471	642	−74	−486	−667	−725	−737	−737

x/a \ y/b	0.040	0.125	0.250	0.375	0.500	0.625	0.750	0.875	0.960
$\xi = 0.25\,a$									
$\eta = d$									
0.040	−746	−971	**5211**	−867	−427	−250	−178	−147	−139
0.250	−192	599	1920	653	−504	−755	−743	−702	−686
0.500	642	703	716	414	−47	−412	−610	−701	−725
0.750	393	377	309	175	−5	−176	−303	−378	−397
0.960	64	60	48	27	0	−30	−57	−64	64

Table A.5 (continued)

y/b \ x/a	0.040	0.125	0.250	0.375	0.500	0.625	0.750	0.875	0.960
η = 0.25 b									
0.040	−192	599	1920	653	−504	−755	−743	−702	−686
0.250	−1408	1196	**16619**	1146	−2622	−3570	−3695	−3620	−3583
0.500	2980	4026	5130	2666	−670	−2717	−3688	−4077	−4168
0.750	2334	2314	2016	1149	−53	−1150	−1910	−2314	−2427
0.960	393	377	309	175	−5	−176	−303	−378	−397
η = 0.50 b									
0.040	642	703	716	414	−47	−412	−610	−701	−725
0.250	2980	4026	5130	2666	−670	−2717	−3688	−4077	−4168
0.500	926	3510	**18135**	2296	−2675	−4720	−5634	−5934	−6010
0.750	2980	4026	5130	2666	−670	−2717	−3688	−4077	−4168
0.960	642	703	716	414	−47	−412	−610	−701	−725

ξ = 0.50 a

y/b \ x/a	0.040	0.125	0.250	0.375	0.500	0.625	0.750	0.875	0.960
η = a									
0.040	−244	−277	−427	−841	**5275**	−841	−427	−277	−244
0.250	−876	−831	−504	728	2057	728	−504	−831	−876
0.500	−486	−377	−47	378	589	378	−47	−377	−486
0.750	−150	−110	−5	109	160	109	−5	−110	−150
0.960	−19	−14	0	10	11	10	0	−14	−19
η = 0.25 b									
0.040	−876	−831	−504	728	2057	728	−504	−831	−876
0.250	−4076	−3832	−2622	1408	**16534**	1408	−2622	−3832	−4076
0.500	−3141	−2605	−670	2554	4642	2554	−670	−2605	−3141
0.750	−1126	−852	−53	852	1255	852	−53	−852	−1126
0.960	−150	−110	−5	109	160	109	−5	−110	−150
η = 0.50 b									
0.040	−486	−377	−47	378	589	378	−47	−377	−486
0.250	−3141	−2605	−670	2554	4642	2554	−670	−2605	−3141
0.500	−5202	−4684	−2675	2260	**17789**	2260	−2675	−4684	−5202
0.750	−3141	−2605	−670	2554	4642	2554	−670	−2605	−3141
0.960	−486	−377	−47	378	589	378	−47	−377	−486

Appendix

Table A.6
ROTATIONS φ_x — COEFFICIENTS $10^5 \, k_\varphi^x$

$$\varphi_x = \frac{M}{D} k_\varphi^x$$

$b = a$
$c = d = 0.06\,a$

y/b \ x/a	0.060	0.125	0.250	0.375	0.500	0.625	0.750	0.875	0.940
ξ = c									
				η = d					
0.060	**4495**	1101	−1303	−804	−529	−385	308	−269	−259
0.250	5111	3255	−105	−1191	−1280	−1151	−1021	−937	−915
0.500	2568	2152	971	−73	−699	−980	−1075	−1097	−1099
0.750	1079	963	584	146	−216	−459	−590	−651	−666
0.940	246	222	143	45	−42	−105	−144	−163	−164
				η = 0.25 b					
0.060	5111	3255	−105	−1191	−1280	−1151	−1021	−937	−915
0.250	**18626**	10263	−849	−3497	−3998	−3849	−3572	−3363	−3305
0.500	11405	9020	3158	−972	−3039	−3882	−4138	−4172	−4168
0.750	4757	4165	2326	382	−1078	−1957	−2402	−2590	−2626
0.940	1079	963	584	146	−216	−459	−590	−651	−666
				η = 0.50 b					
0.060	2568	2152	971	−73	−699	−980	−1075	−1097	−1099
0.250	11405	9020	3158	−972	−3039	−3882	−4138	−4172	−4168
0.500	**23383**	14428	1477	−3116	−5077	−5806	−5975	−5953	−5931
0.750	11405	9020	3158	−972	−3039	−3882	−4138	−4172	−4168
0.940	2568	2152	971	−73	−699	−980	−1075	−1097	−1099

y/b \ x/a	0.060	0.125	0.250	0.375	0.500	0.625	0.750	0.875	0.940
ξ = 0.25 a									
				η = d					
0.060	−1303	−1040	4824	−841	−797	−520	−384	−322	−308
0.250	−105	887	2832	952	−714	−1105	−1098	−1040	−1021
0.500	971	1050	1062	615	−72	−614	−919	−1048	−1075
0.750	584	563	462	260	−4	−261	−452	−564	−590
0.940	143	136	108	60	−1	−62	−104	−137	−144

370

Appendix

Table A.6 (continued)

η = 0.25 b

y/b \ x/a	0.060	0.125	0.250	0.375	0.500	0.625	0.750	0.875	0.940
0.060	−105	887	2832	952	−714	−1105	−1098	−1040	−1021
0.250	−849	1618	**12798**	1538	−2514	−3523	−3663	−3603	−3572
0.500	3158	4011	5067	2649	−653	−2696	−3669	−4057	−4138
0.750	2326	2306	2001	1146	−52	−1146	−1897	−2306	−2402
0.940	584	563	462	260	−4	−261	−452	−564	−590

η = 0.50 b

y/b \ x/a	0.060	0.125	0.250	0.375	0.500	0.625	0.750	0.875	0.940
0.060	971	1050	1062	615	−72	−614	−919	−1048	−1075
0.250	3158	4011	5067	2649	−653	−2696	−3669	−4057	−4138
0.500	1477	3924	**14800**	2684	−2566	−4669	−5560	−5909	−5975
0.750	3158	4011	5067	2649	−653	−2696	−3669	−4057	−4138
0.940	971	1050	1062	615	−72	−614	−919	−1048	−1075

ξ = 0.50 a

η = d

y/b \ x/a	0.060	0.125	0.250	0.375	0.500	0.625	0.750	0.875	0.940
0.060	−529	−577	−797	−785	**4955**	−785	−797	−577	−529
0.250	−1280	−1215	−714	1061	3028	1061	−714	−1215	−1280
0.500	−699	−564	−72	565	882	565	−72	−564	−699
0.750	−216	−168	−4	167	241	167	−4	−168	−216
0.940	−42	−32	−1	31	49	31	−1	−32	−42

η = 0.25 b

y/b \ x/a	0.060	0.125	0.250	0.375	0.500	0.625	0.750	0.875	0.940
0.060	−1280	−1215	−714	1061	3028	1061	−714	−1215	−1280
0.250	−3998	−3779	−2514	1794	**13191**	1794	−2514	−3779	−3998
0.500	−3039	−2580	−653	2534	4572	2534	−653	−2580	−3039
0.750	−1078	−844	−52	844	1257	844	−52	−844	−1078
0.940	−216	−168	−4	167	241	167	−4	−168	−216

η = 0.50 b

y/b \ x/a	0.060	0.125	0.250	0.375	0.500	0.625	0.750	0.875	0.940
0.060	−699	−564	−72	565	882	565	−72	−564	−699
0.250	−3039	−2580	−653	2534	4572	2534	−653	−2580	−3039
0.500	−5077	−4624	−2566	2638	**14448**	2638	−2566	−4624	−5077
0.750	−3039	−2580	−653	2534	4572	2534	−653	−2580	−3039
0.940	−699	−564	−72	565	882	565	−72	−564	−699

Appendix

Table A.7
ROTATIONS φ_y — COEFFICIENTS $10^5 \, k_\varphi^y$

$$\varphi_y = \frac{M}{D} k_\varphi^y \qquad \xi = c \qquad \begin{array}{c} b = a \\ c = d = 0.02\,a \end{array}$$

y/b \ x/a	0.020	0.040	0.125	0.250	0.375	0.500	0.625	0.750	0.875	0.980
$\eta = d$										
0.000	5604	9498	4904	2375	1457	959	631	386	184	29
0.020	3620	6263	4519	2326	1442	953	628	385	183	29
0.040	659	1271	3576	2186	1400	935	620	380	181	29
0.250	−200	−379	−662	−148	225	317	281	198	101	16
0.500	−62	−122	−334	−444	−370	−251	−146	−77	−33	−5
0.750	−35	−69	−205	−344	−391	−365	−293	−205	−101	−16
0.980	−29	−58	−173	−304	−370	−369	−315	−226	−117	−19
$\eta = 0.25\,b$										
0.020	2326	4561	11560	13804	11696	8868	6256	3968	1922	304
0.230	3309	5757	5377	5843	5741	4945	3796	2536	1263	202
0.250	476	945	2754	4432	4830	4343	3413	2308	1157	185
0.270	−2356	−3863	140	3025	3914	3734	3019	2072	1046	168
0.500	−964	−1885	−4630	−4940	−3512	−2147	−1210	−626	−264	−40
0.750	−476	−945	−2754	−4432	−4830	−4343	−3413	−2308	−1157	−185
0.980	−385	−766	−2291	−3968	−4724	−4608	−3856	−2724	−1399	−226
$\eta = 0.50\,b$										
0.020	953	1895	5517	8868	9655	8679	6815	4608	2309	369
0.250	1348	2651	6918	8905	8235	6756	5068	3354	1664	266
0.480	2816	4778	2538	1339	913	664	473	306	150	21
0.500	0	0	0	0	0	0	0	0	0	0
0.520	−2816	−4778	−2538	−1339	−913	−664	−473	−306	−150	−21
0.750	−1348	−2651	−6918	−8905	−8235	−6756	−5068	−3354	−1664	−266
0.980	−953	−1895	−5517	−8868	−9655	−8679	−6815	−4608	−2309	−369

Table A.7 (continued)

$\xi = 0.25\ a$

x/a \ y/b	0.020	0.125	0.230	0.250	0.270	0.375	0.500	0.625	0.750	0.875	0.980

$\eta = a$

0.000	−211	−1674	−4597	479	5556	2715	1372	818	479	223	35
0.020	−200	−1514	−2300	476	3253	2546	1348	810	476	222	35
0.040	−168	−1108	864	467	72	2115	1278	789	467	219	34
0.250	69	454	264	159	54	−199	32	164	159	90	15
0.500	1	−17	−100	−119	−151	−241	−261	−202	−119	−56	9
0.750	−15	−93	−169	−183	−195	−230	−278	−247	−183	−96	−19
0.980	−16	−99	−173	−185	−197	−244	−266	−244	−185	−99	−16

$\eta = 0.25\ b$

0.020	148	0	3469	4432	5400	8977	8905	6811	4432	2166	344
0.230	53	204	264	2470	5208	4570	4195	3506	2470	1267	204
0.250	159	1036	1991	2178	2351	3092	3379	2999	2178	1129	183
0.270	265	1861	4249	1873	515	1613	2561	2487	1873	988	160
0.500	176	594	672	−1071	−1475	−2951	−2802	−1886	−1071	−471	−77
0.750	−159	−1036	−1991	−2178	−2351	−3092	−3379	−2999	−2178	−1129	−183
0.980	−198	−1218	−2150	−2308	−2458	−3080	−3354	−3068	−2308	−1230	−205

$\eta = 0.50\ b$

0.020	317	2065	3986	4343	4693	6183	6756	5993	4343	2255	365
0.250	22	625	2822	3379	3933	6031	6152	4954	3379	1702	268
0.480	−97	−788	−2220	332	2884	1480	819	531	332	161	26
0.500	0	0	0	0	0	0	0	0	0	0	0
0.520	97	788	2220	−332	−2884	−1480	−819	−531	−332	−161	−26
0.750	−22	−625	−2822	−3379	−3933	−6031	−6152	−4954	−3379	−1702	−268
0.980	−317	−2065	−3986	−4343	−4693	−6183	−6756	−5993	−4343	−2255	−365

Appendix

Table A.7 (continued)

					$\xi = 0.50\ a$				
x/a y/b	0.020	0.125	0.250	0.375	0.480	0.500	0.520	0.625	0.750	0.875	0.980

$\eta = d$

x/a \ y/b	0.020	0.125	0.250	0.375	0.480	0.500	0.520	0.625	0.750	0.875	0.980
0.000	—62	—411	—987	—2309	—5093	0	5093	2309	987	411	62
0.020	—62	—405	—964	—2141	—2793	0	2793	2141	964	405	62
0.040	—60	—389	—899	—1715	—380	0	—380	1715	899	389	60
0.250	1	26	166	390	114	0	—114	—390	—166	—26	—1
0.500	28	113	185	152	29	0	—29	—152	—185	—113	—28
0.750	9	50	67	52	9	0	—9	—52	—67	—50	—9
0.980	5	28	40	28	2	0	—2	—28	—40	—28	—5

$\eta = 0.25\ b$

x/a \ y/b	0.020	0.125	0.250	0.375	0.480	0.500	0.520	0.625	0.750	0.875	0.980
0.020	—444	—2725	—4940	—4891	—1035	0	1035	4891	4940	2725	444
0.230	—171	—976	—1659	—2040	—2708	0	2708	2040	1659	976	171
0.250	—119	—713	—1071	—806	—147	0	147	806	1071	713	119
0.270	—80	—453	—488	421	2413	0	—2413	—421	488	453	80
0.500	185	1152	2175	2217	474	0	—474	—2217	—2175	—1152	—185
0.750	119	713	1071	806	147	0	—147	—806	—1071	—713	—119
0.980	77	437	626	450	80	0	—80	—450	—626	—437	—77

$\eta = 0.50\ b$

x/a \ y/b	0.020	0.125	0.250	0.375	0.480	0.500	0.520	0.625	0.750	0.875	0.980
0.020	—251	—1429	—2147	—1619	—295	0	295	1619	2147	1429	251
0.250	—261	—1588	—2802	—2665	—553	0	553	2665	2802	1588	261
0.480	—34	—219	—513	—1168	—2553	0	2553	1168	513	219	34
0.500	0	0	0	0	0	0	0	0	0	0	0
0.520	34	219	513	1168	2553	0	—2553	—1168	—513	—219	—34
0.750	261	1588	2802	2665	553	0	—553	—2665	—2802	—1588	—261
0.980	251	1429	2147	1619	295	0	—295	—1619	—2147	—1429	—251

Table A.8
ROTATIONS φ_y — COEFFICIENTS $10^5 k_\varphi^y$

$\varphi_y = \dfrac{M}{D} k_\varphi^y$ $\xi = c$ $b = a$, $c = d = 0.04\,a$

y/b \ x/a	0.040	0.080	0.125	0.250	0.375	0.500	0.625	0.750	0.875	0.960
η = d										
0.000	5443	9768	8708	4662	2892	1910	1259	770	367	116
0.040	3613	6130	6541	4286	2778	1862	1235	759	362	114
0.080	687	1029	2664	3337	2461	1726	1168	724	347	110
0.250	−625	−1049	−1164	−286	440	627	557	393	200	64
0.500	−235	−449	−644	−866	−730	−491	−291	−154	−66	−20
0.750	−136	−268	−404	−678	−775	−723	−583	−400	−202	−65
0.960	−114	−225	−343	−604	−735	−733	−625	−448	−232	−75
η = 0.25 b										
0.040	4286	7998	11032	13411	11472	8745	6188	3932	1906	604
0.210	3770	6911	7390	7136	6581	5496	4152	2748	1362	434
0.250	924	1806	2699	4366	4783	4317	3399	2302	1154	369
0.290	−1908	−3275	−1962	1609	2960	3095	2603	1823	930	299
0.500	−1736	−3218	−4373	−4840	−3496	−2153	−1219	−633	−266	−81
0.750	−924	−1806	−2699	−4366	−4783	−4317	−3399	−2302	−1154	−369
0.960	−759	−1495	−2269	−3932	−4684	−4572	−3828	−2706	−1390	−448
η = 0.50 b										
0.040	1862	3636	5428	8745	9541	8591	6756	4572	2292	733
0.250	2491	4705	6632	8760	8175	6729	5055	3347	1662	531
0.460	2778	4981	4525	2632	1813	1322	942	610	300	95
0.500	0	0	0	0	0	0	0	0	0	0
0.540	−2778	−4981	−4525	−2632	−1813	−1322	−942	−610	−300	−95
0.750	−2491	−4705	−6632	−8760	−8175	−6729	−5055	−3347	−1662	−531
0.960	−1862	−3636	−5428	−8745	−9541	−8591	−6756	−4572	−2292	−733

Appendix

Table A.8 (continued)

	$\xi = 0.25\,a$									
x/a \ y/b	0.040	0.125	0.210	0.250	0.290	0.375	0.500	0.625	0.750	0.875	0.960

$\eta = d$

x/a \ y/b	0.040	0.125	0.210	0.250	0.290	0.375	0.500	0.625	0.750	0.875	0.960
0.000	−778	−2808	−4303	947	6213	4862	2663	1611	947	443	139
0.040	−625	−1984	−1673	924	3536	3972	2491	1555	924	434	136
0.080	−271	−373	1476	858	250	2179	2042	1398	858	408	129
0.250	339	885	701	320	−74	70	334	334	320	180	59
0.500	4	−32	−160	−244	−331	−479	−519	−400	−244	−111	−34
0.750	−59	−186	−309	−364	−414	−497	−544	−493	−364	−190	−61
0.960	−64	−197	−319	−369	−415	−488	−531	−488	−369	−198	−65

$\eta = 0.25\,b$

x/a \ y/b	0.040	0.125	0.210	0.250	0.290	0.375	0.500	0.625	0.750	0.875	0.960
0.040	−286	−49	−2475	4366	6270	8883	8760	6698	4366	2136	678
0.210	−77	−368	−278	2754	5774	5751	4955	3986	2754	1397	448
0.250	320	1034	1807	2170	2510	3083	3369	2991	2170	1127	364
0.290	705	2408	3864	1563	−769	410	1773	1975	1563	843	275
0.500	324	574	−311	−1065	−1842	−2917	−2771	−1874	−1065	−469	−144
0.750	−320	−1034	−1807	−2170	−2510	−3083	−3369	−2991	−2170	−1127	−364
0.960	−393	−1213	−1978	−2302	−2594	−3074	−3347	−3061	−2302	−1227	−400

$\eta = 0.50\,b$

x/a \ y/b	0.040	0.125	0.210	0.250	0.290	0.375	0.500	0.625	0.750	0.875	0.960
0.040	627	2040	3586	4317	5005	6161	6729	5962	4317	2240	723
0.250	70	642	2292	3369	4437	5991	6116	4936	3369	1698	544
0.460	−357	−1305	−1992	658	3314	2675	1597	1050	658	320	102
0.500	0	0	0	0	0	0	0	0	0	0	0
0.540	357	1305	1992	−658	−3314	−2675	−1597	−1050	−658	−320	−102
0.750	−70	−642	−2292	−3369	−4437	−5991	−6116	−4936	−3369	−1698	−544
0.960	−627	−2040	−3586	−4317	−5005	−6161	−6729	−5962	−4317	−2240	−723

Table A.8 (continued)

$\xi = 0.50\ a$

x/a \\ y/b	0.040	0.125	0.250	0.375	0.460	0.500	0.540	0.625	0.750	0.875	0.960
\multicolumn{12}{c}{$\eta = a$}
0.000	—246	—803	—1897	—4054	—5323	0	5323	4054	1897	803	246
0.040	—235	—761	—1736	—3178	—2665	0	2665	3178	1736	761	235
0.080	—203	—644	—1321	—1425	—566	0	566	1425	1321	644	203
0.250	4	—47	324	751	421	0	—421	—751	—324	—47	—4
0.500	75	224	367	303	113	0	—113	—303	—367	—224	—75
0.750	34	100	144	104	37	0	—37	—104	—144	—100	—34
0.960	20	57	81	57	20	0	—20	—57	—81	—57	—20
\multicolumn{12}{c}{$\eta = 0.25\ b$}
0.040	—866	—2662	—4840	—4847	—2030	0	2030	4847	4840	2662	866
0.210	—406	—1228	—2209	—2993	—3012	0	3012	2993	2209	1228	406
0.250	—244	—710	—1065	—802	—289	0	289	802	1065	710	244
0.290	—85	—201	55	1364	2424	0	—2424	—1364	—55	201	85
0.500	367	1143	2148	2186	909	0	—909	—2186	—2148	—1143	—367
0.750	244	710	1065	802	289	0	—289	—802	—1065	—710	—244
0.960	154	442	633	455	161	0	—161	—455	—633	—442	—154
\multicolumn{12}{c}{$\eta = 0.50\ b$}
0.040	—491	—1430	—2153	—1629	—588	0	588	1629	2153	1430	491
0.250	—519	—1577	—2771	—2632	—1067	0	1067	2632	2771	1577	519
0.460	—133	—429	—988	—2055	—2671	0	2671	2055	988	429	133
0.500	0	0	0	0	0	0	0	0	0	0	0
0.540	133	429	988	2055	2671	0	—2671	—2055	—988	—429	—133
0.750	519	1577	2771	2632	1067	0	—1067	—2632	—2771	—1577	—519
0.960	491	1430	2153	1629	588	0	—588	—1629	—2153	—1430	—491

Appendix

Table A.9
ROTATIONS φ_y — COEFFICIENTS $10^5 \, k_\varphi^y$

$$\varphi_y = \frac{M}{D} k_\varphi^y \qquad \xi = e \qquad \begin{array}{c} b = e \\ c = d = 0.06\,a \end{array}$$

x/a y/b	0.060	0.120	0.125	0.250	0.375	0.500	0.625	0.750	0.875	0.940
η = a										
0.000	5377	9675	9780	6697	4271	2842	1878	1151	548	260
0.060	3543	6001	6111	5565	3898	2683	1800	1112	531	252
0.120	625	871	954	3164	2954	2252	1581	999	483	230
0.250	−916	−1361	−1366	−379	640	924	826	585	298	144
0.500	−484	−882	−909	−1247	−1070	−729	−435	−232	−99	−45
0.750	−297	−569	−590	−996	−1143	−1071	−865	−595	−300	−144
0.940	−252	−490	−508	−895	−1089	−1087	−927	−664	−344	−167
η = 0.25 b										
0.060	5565	9861	10132	12775	11101	8538	6073	3870	1879	895
0.190	4272	7825	7982	8241	7348	6009	4485	2947	1455	696
0.250	1323	2518	2609	4257	4703	4271	3376	2292	1150	552
0.310	−1589	−2731	−2707	298	2006	2439	2172	1564	808	391
0.500	−2218	−3874	−3975	−4664	−3464	−2162	−1232	−643	−271	−124
0.750	−1323	−2518	−2609	−4257	−4703	−4271	−3376	−2292	−1150	−552
0.940	−1112	−2150	−2231	−3870	−4616	−4511	−3781	−2674	−1375	−664
η = 0.50 b										
0.060	2683	5094	5275	8538	9354	8448	6657	4511	2263	1087
0.250	3317	6003	6184	8508	8070	6682	5033	3337	1658	794
0.440	2813	5081	5143	3793	2678	1964	1404	909	448	214
0.500	0	0	0	0	0	0	0	0	0	0
0.560	−2813	−5081	−5143	−3793	−2678	−1964	−1404	−909	−448	−214
0.750	−3317	−6003	−6184	−8508	−8070	−6682	−5033	−3337	−1658	−794
0.940	−2683	−5094	−5275	−8538	−9354	−8448	−6657	−4511	−2263	−1087

Table A.9 (continued)

x/a \ y/b	0.060	0.125	0.190	0.250	0.310	0.375	0.500	0.625	0.750	0.875	0.940
\multicolumn{12}{c}{$\xi = 0.25\,a$}
\multicolumn{12}{c}{$\eta = d$}
0.000	−1498	−3201	−3717	1397	6559	6212	3804	2356	1397	655	309
0.060	−915	−1629	−1105	1323	3788	4437	3317	2182	1323	626	297
0.120	137	769	1868	1122	384	1506	2199	1729	1122	546	260
0.250	708	1269	1209	481	−284	−483	153	516	481	270	133
0.500	5	−44	−178	−360	−551	−710	−767	−589	−360	−164	−77
0.750	−133	−277	−419	−543	−652	−743	−813	−735	−543	−285	−139
0.940	−144	−295	−436	−552	−651	−731	−794	−730	−552	−296	−144
\multicolumn{12}{c}{$\eta = 0.25\,b$}
0.060	−379	−109	1544	4257	6994	8708	8508	6512	4257	2087	996
0.100	−290	−580	−186	3014	6195	6523	5626	4426	3014	1517	729
0.250	481	1032	1617	2161	2654	3067	3352	2977	2161	1122	543
0.310	1214	2578	3354	1257	−915	−399	1056	1481	1257	698	342
0.500	436	541	−20	−1056	−2143	−2861	−2730	−1854	−1056	−466	−216
0.750	−481	−1032	−1617	−2161	−2654	−3067	−3352	−2977	−2161	−1122	−543
0.940	−585	−1206	−1796	−2292	−2715	−3065	−3337	−3050	−2292	−1220	−595
\multicolumn{12}{c}{$\eta = 0.50\,b$}
0.060	924	1998	3168	4271	5281	6123	6682	5907	4271	2214	1071
0.250	153	671	1823	3352	4859	5924	6065	4906	3352	1690	813
0.440	−677	−1453	−1641	975	3605	3471	2298	1542	975	476	227
0.500	0	0	0	0	0	0	0	0	0	0	0
0.560	677	1453	1641	−975	−3605	−3471	−2298	−1542	−975	−476	−227
0.750	−153	−671	−1823	−3352	−4859	−5924	−6065	−4906	−3352	−1690	−813
0.940	−924	−1918	−3168	−4271	−5281	−6123	−6682	−5907	−4271	−2214	−1071

Appendix

Table A.9 (continued)

						$\xi = 0.50\ a$						
x/a y/b	0.060	0.125	0.250	0.375	0.440	0.500	0.560	0.625	0.750	0.875	0.940

$\eta = a$

y/b	0.060	0.125	0.250	0.375	0.440	0.500	0.560	0.625	0.750	0.875	0.940
0.000	−538	−1159	−2666	−5015	−5279	0	5279	5015	2666	1159	538
0.060	−484	−1030	−2218	−3285	−2563	0	2563	3285	2218	1030	484
0.120	−344	−704	−1208	−481	684	0	−684	481	1208	704	344
0.250	5	54	436	1052	821	0	−821	−1052	−436	−54	−5
0.500	163	330	543	451	247	0	−247	−451	−543	−330	−163
0.750	77	148	216	157	83	0	−83	−157	−216	−148	−77
0.940	45	87	124	88	46	0	−46	−88	−124	−87	−45

$\eta = 0.25\ b$

y/b	0.060	0.125	0.250	0.375	0.440	0.500	0.560	0.625	0.750	0.875	0.940
0.060	−1247	−2559	−4664	−4755	−2904	0	2904	4755	4664	2559	1247
0.190	−715	−1459	−2686	−3555	−3202	0	3202	3555	2686	1459	715
0.250	−360	−704	−1056	−795	−422	0	422	795	1056	704	360
0.310	−14	30	523	1910	2324	0	−2324	−1910	−523	−30	14
0.500	543	1128	2112	2135	1273	0	−1273	−2135	−2112	−1128	−543
0.750	360	704	1056	795	422	0	−422	−795	−1056	−704	−360
0.940	232	448	643	462	242	0	−242	−462	−643	−448	−232

$\eta = 0.50\ b$

y/b	0.060	0.125	0.250	0.375	0.440	0.500	0.560	0.625	0.750	0.875	0.940
0.060	−729	−1429	−2162	−1645	−878	0	878	1645	2162	1429	729
0.250	−767	−1560	−2730	−2579	−1505	0	1505	2579	2730	1560	767
0.440	−291	−621	−1393	−2550	−2662	0	2662	2550	1393	621	291
0.500	0	0	0	0	0	0	0	0	0	0	0
0.560	291	621	1393	2550	2662	0	−2662	−2550	−1393	−621	−291
0.750	767	1560	2730	2579	1505	0	−1505	−2579	−2730	−1560	−767
0.940	729	1429	2162	1645	878	0	−878	−1645	−2162	−1429	−729

BENDING AND TORSIONAL MOMENTS

Appendix

Table A.10
BENDING MOMENTS M_x — COEFFICIENTS k_M^x

$$M_x = \frac{M}{a} k_M^x \qquad \begin{array}{l} b = a \\ c = d = 0.02\,a \\ \mu = 0 \end{array}$$

$\xi = c$

y/b \\ x/a	0.020	0.040	0.125	0.250	0.375	0.500	0.625	0.750	0.875	0.960	0.980
						$\eta = d$					
0.020	0.593	**2.083**	−0.099	−0.015	−0.005	−0.002	−0.001	−0.001	0.000	0.000	0.000
0.250	0.044	0.086	0.140	0.043	0.006	−0.004	−0.004	−0.003	−0.002	−0.001	0.000
0.500	0.006	0.012	0.030	0.034	0.023	0.005	0.004	0.001	0.000	0.000	0.000
0.750	0.001	0.003	0.008	0.012	0.011	0.008	0.005	−0.003	−0.001	0.000	0.000
0.980	0.000	0.000	0.001	0.001	0.001	0.001	0.000	0.000	−0.001	−0.012	0.009
						$\eta = 0.25\,b$					
0.020	0.044	0.086	0.140	0.043	0.006	−0.004	−0.004	−0.003	−0.002	−0.001	0.000
0.250	2.425	**5.374**	1.119	0.308	0.072	−0.006	−0.025	−0.023	−0.013	−0.004	−0.002
0.500	0.126	0.248	0.537	0.437	0.231	0.099	0.034	0.006	−0.002	−0.001	0.000
0.750	0.023	0.047	0.128	0.172	0.143	0.092	0.050	0.023	0.009	−0.009	0.009
0.980	0.001	0.003	0.008	0.012	0.011	0.008	0.005	−0.003	0.001	0.000	0.000
						$\eta = 0.50\,b$					
0.020	0.006	0.012	0.030	0.034	0.023	0.005	0.004	0.001	0.000	0.000	0.000
0.250	0.126	0.248	0.537	0.437	0.231	0.099	0.034	0.006	−0.002	−0.001	0.000
0.500	2.448	**5.421**	1.247	0.480	0.215	0.087	0.025	0.000	−0.003	−0.013	0.007
0.750	0.126	0.248	0.537	0.437	0.231	0.099	0.034	0.006	−0.002	−0.001	0.000
0.980	0.006	0.012	0.030	0.034	0.023	0.005	0.004	0.001	0.000	0.000	0.000

$\xi = 0.25\,a$

y/b \\ x/a	0.020	0.125	0.230	0.250	0.270	0.375	0.500	0.625	0.730	0.750	0.770	0.875	0.980
							$\eta = d$						
0.020	0.003	0.039	−**2.236**	−0.001	**2.234**	−0.042	−0.008	−0.002	−0.001	−0.001	−0.001	0.000	0.000
0.250	−0.032	−0.068	−0.026	−0.002	0.022	0.069	0.030	0.002	−0.002	−0.002	−0.002	−0.001	0.000
0.500	−0.001	−0.004	0.004	0.010	0.001	0.017	0.018	0.012	0.007	0.010	−0.002	0.002	0.004
0.750	0.000	0.002	0.004	0.004	0.005	0.007	0.012	0.006	0.004	0.004	0.004	0.002	0.004
0.980	0.000	0.000	0.000	0.000	0.000	0.001	0.001	0.001	−0.007	0.000	0.008	0.000	0.000

382

Table A.10 (continued)

x/a \ y/b	0.020	0.125	0.250	0.375	0.480	0.500	0.520	0.625	0.750	0.875	0.980
					η = 25 b						
0.020	−0.032	−0.068	−0.026	−0.002	0.022	0.069	0.030	0.002	−0.002	−0.002	−0.001	0.000
0.250	−0.065	−0.503	−4.052	−0.003	4.046	0.525	0.139	0.027	−0.003	−0.002	−0.006	−0.001
0.500	−0.043	−0.161	−0.009	0.048	0.109	0.289	0.219	0.113	0.048	0.042	0.016	−0.004
0.750	0.000	0.007	0.041	0.045	0.051	0.088	0.097	0.075	0.045	0.045	0.021	−0.004
0.980	0.000	0.002	0.004	0.004	0.005	0.007	0.012	0.006	0.004	0.004	0.002	−0.004
					η = 0.50 b						
0.020	−0.001	−0.004	0.004	0.010	0.001	0.017	0.018	0.012	0.010	−0.002	−0.002	0.000
0.250	−0.043	−0.161	−0.009	0.048	0.109	0.289	0.219	0.113	0.048	0.042	0.016	−0.004
0.500	−0.065	−0.496	−4.010	0.042	4.097	0.613	0.236	0.102	0.042	0.043	0.015	0.002
0.750	−0.043	−0.161	−0.009	0.048	0.109	0.289	0.219	0.113	0.048	0.042	0.016	−0.004
0.980	−0.001	−0.004	0.004	0.010	0.001	0.017	0.018	0.012	0.019	−0.002	−0.002	0.000

ξ = 0.50 a

x/a \ y/b	0.020	0.125	0.250	0.375	0.480	0.500	0.520	0.625	0.750	0.875	0.980
					η = d						
0.020	0.000	0.002	0.007	0.041	−2.235	0.000	2.235	−0.041	−0.007	−0.002	0.000
0.250	0.000	−0.005	−0.034	−0.072	−0.025	0.000	0.025	0.072	0.034	0.005	0.000
0.500	−0.010	−0.009	−0.017	−0.015	−0.003	0.000	0.003	0.015	0.017	0.009	0.010
0.750	−0.001	−0.003	0.000	−0.004	−0.001	0.000	0.001	0.004	0.000	0.003	0.001
0.980	0.000	0.000	0.000	0.000	−0.007	0.000	0.007	0.000	0.000	0.000	0.000
					η = 0.25 b						
0.020	0.000	−0.005	−0.034	−0.072	−0.025	0.000	0.025	0.072	0.034	0.005	0.000
0.250	0.000	−0.045	−0.161	−0.543	−4.050	0.000	4.050	0.543	0.161	0.045	0.000
0.500	−0.016	−0.098	−0.213	−0.274	−0.067	0.000	0.067	0.274	0.213	0.098	0.016
0.750	−0.015	−0.049	−0.077	−0.062	−0.018	0.000	0.018	0.062	0.077	0.049	0.015
0.980	−0.001	−0.003	0.000	−0.004	−0.001	0.000	0.001	0.004	0.000	0.003	0.001
					η = 0.50 b						
0.020	−0.010	−0.009	−0.017	−0.015	−0.003	0.000	0.003	0.015	0.017	0.009	0.010
0.250	−0.016	−0.098	−0.213	−0.274	−0.067	0.000	0.067	0.274	0.213	0.098	0.016
0.500	−0.015	−0.094	−0.238	−0.605	−4.068	0.000	4.068	0.605	0.238	0.094	0.015
0.750	−0.016	−0.098	−0.213	−0.274	−0.667	0.000	0.067	0.274	0.213	0.098	0.016
0.980	−0.010	−0.009	−0.017	−0.015	−0.003	0.000	0.003	0.015	0.017	0.009	0.010

383

Appendix

Table A.11
BENDING MOMENTS M_x — COEFFICIENTS k_M^x

$$M_x = \frac{M}{a} k_M^x \qquad \boxed{\xi = c}$$

x/a \\ y/b	0.020	0.040	0.125	0.250	0.375	0.500	0.625	0.750	0.875	0.960	0.98
\multicolumn{12}{c}{$\eta = d$}
0.020	1.137	2.983	0.021	0.001	0.000	0.000	0.000	0.000	0.000	0.000	0.000
0.250	0.040	0.077	0.136	0.058	0.023	0.008	0.003	0.001	0.000	0.000	0.000
0.500	0.005	0.010	0.027	0.033	0.025	0.008	0.007	0.004	0.001	0.000	0.000
0.750	0.001	0.003	0.007	0.011	0.011	0.008	0.005	−0.003	0.001	0.000	0.000
0.980	0.000	0.000	0.000	0.001	0.001	0.001	0.000	0.000	0.000	−0.014	0.010
\multicolumn{12}{c}{$\eta = 0.25\,b$}
0.020	0.040	0.077	0.136	0.058	0.023	0.008	0.003	0.001	0.000	0.000	0.000
0.250	2.990	6.326	1.566	0.545	0.225	0.094	0.039	0.016	0.006	0.002	0.001
0.500	0.112	0.222	0.499	0.454	0.282	0.156	0.082	0.039	0.015	0.005	0.002
0.750	0.021	0.042	0.115	0.160	0.143	0.103	0.064	0.035	0.016	−0.008	0.010
0.980	0.001	0.003	0.007	0.011	0.011	0.008	0.005	−0.003	0.001	0.000	0.000
\multicolumn{12}{c}{$\eta = 0.50\,b$}
0.020	0.005	0.010	0.027	0.033	0.025	0.008	0.007	0.004	0.001	0.000	0.000
0.250	0.112	0.222	0.499	0.454	0.282	0.156	0.082	0.039	0.015	−0.005	0.002
0.500	3.010	6.367	1.681	0.705	0.367	0.197	0.104	0.051	0.022	−0.006	0.011
0.750	0.112	0.222	0.499	0.454	0.282	0.156	0.082	0.039	0.015	0.005	0.002
0.980	0.005	0.010	0.027	0.033	0.025	0.008	0.007	0.004	0.001	0.000	0.000

$$\boxed{\xi = 0.25\,a}$$

x/a \\ y/b	0.020	0.0125	0.0230	0.0250	0.270	0.375	0.500	0.625	0.730	0.750	0.770	0.875	0.980
\multicolumn{14}{c}{$\eta = d$}
0.020	0.000	−0.011	−2.869	0.000	2.869	0.011	0.000	0.000	0.000	0.000	0.000	0.000	0.000
0.250	−0.036	−0.058	−0.017	0.004	0.025	0.071	0.043	0.012	0.005	0.004	0.003	0.001	0.000
0.500	−0.001	−0.001	0.007	0.014	0.001	0.018	0.019	0.013	0.010	0.014	−0.002	0.003	0.000
0.750	0.000	0.002	0.004	0.004	0.005	0.007	0.013	0.006	0.005	0.004	0.004	0.002	0.005
0.980	0.000	0.000	0.000	0.000	0.000	0.000	0.001	0.001	−0.009	0.000	0.010	0.000	0.000

$b = a$
$c = d = 0.02\,a$
$\mu = 1/3$

384

Table A.11 (continued)

x/a \ y/b	0.020	0.125	0.250	0.375				0.625	0.750	0.875	0.980
					η = 0.25 b						
0.020	−0.036	−0.058	−0.017	0.004	0.025	0.071	0.043	0.005	0.004	0.003	0.001	0.000
0.250	−0.083	−0.646	−4.544	0.048	4.638	0.776	0.276	0.050	0.048	0.044	0.017	0.002
0.500	−0.037	−0.115	0.027	0.077	0.129	0.293	0.242	0.087	0.077	0.069	0.031	0.003
0.750	−0.002	−0.014	0.048	0.049	0.055	0.088	0.097	0.052	0.049	0.002	0.051	0.004
0.980	0.000	0.002	0.004	0.004	0.005	0.007	0.013	0.005	0.004	0.005	0.004	0.005

					η = 0.50 b						
0.020	−0.001	−0.001	−0.007	0.014	0.001	0.018	0.019	0.010	0.014	0.002	0.003	0.000
0.250	−0.037	−0.115	−0.027	0.077	0.129	0.293	0.242	0.087	0.077	0.069	0.031	0.003
0.500	−0.081	−0.632	−4.496	0.097	4.693	0.864	0.373	0.103	0.097	0.095	0.042	0.007
0.750	−0.037	−0.115	−0.027	0.077	0.129	0.293	0.242	0.087	0.077	0.069	0.031	0.003
0.980	−0.001	−0.001	−0.007	0.014	0.001	0.018	0.019	0.010	0.014	0.002	0.003	0.000

ξ = 0.50 a

x/a \ y/b	0.020	0.125	0.250	0.375	0.480	0.500	0.520	0.625	0.750	0.875	0.980
					η = d						
0.020	0.000	0.000	0.000	−0.011	−2.369	0.000	2.369	0.011	0.000	0.000	0.000
0.250	−0.001	−0.010	−0.042	−0.070	−0.022	0.000	0.022	0.070	0.042	0.010	0.001
0.500	−0.012	−0.008	−0.015	−0.013	−0.003	0.000	0.003	0.013	0.015	0.008	0.012
0.750	0.000	−0.002	0.002	−0.003	−0.001	0.000	0.001	0.003	−0.002	0.002	0.000
0.980	0.000	0.000	0.000	0.000	−0.009	0.000	0.009	0.000	0.000	0.000	0.000

					η = 0.25 b						
0.020	−0.001	−0.010	−0.042	−0.070	−0.022	0.000	0.022	0.070	0.042	0.010	0.001
0.250	−0.005	−0.089	−0.259	−0.753	−4.593	0.000	4.593	0.753	0.259	0.089	0.005
0.500	−0.017	−0.100	−0.202	−0.246	−0.058	0.000	0.058	0.246	0.202	0.100	0.017
0.750	−0.015	−0.042	−0.065	−0.052	−0.017	0.000	0.017	0.052	0.065	0.042	0.015
0.980	0.000	−0.002	0.002	−0.003	−0.001	0.000	0.001	0.003	−0.002	0.002	0.000

					η = 0.50 b						
0.020	−0.012	−0.008	−0.015	−0.013	−0.003	0.000	0.003	0.013	0.015	0.008	0.012
0.250	−0.017	−0.100	−0.202	−0.246	−0.058	0.000	0.058	0.246	0.202	0.100	0.017
0.500	−0.020	−0.131	−0.324	−0.805	−4.611	0.000	4.611	0.805	0.324	0.131	0.020
0.750	−0.017	−0.100	−0.202	−0.246	−0.058	0.000	0.058	0.246	0.202	0.100	0.017
0.980	−0.012	−0.008	−0.015	−0.013	−0.003	0.000	0.003	0.013	0.015	0.008	0.012

Appendix

Table A.12
BENDING MOMENTS M_x — COEFFICIENTS k_M^x

$$M_x = \frac{M}{a} k_M^x \qquad \begin{array}{l} b = a \\ c = d = 0.01a \\ \mu = 0 \end{array}$$

$\xi = e$

x/a \ y/b	0.040	0.080	0.125	0.250	0.375	0.500	0.625	0.750	0.875	0.920	0.960
						$\eta = d$					
0.040	0.198	1.435	0.133	−0.045	−0.017	−0.008	−0.004	−0.002	−0.001	−0.001	0.000
0.250	0.140	0.231	0.256	0.107	0.014	−0.006	−0.008	−0.006	−0.003	−0.002	−0.001
0.500	0.022	0.041	0.058	0.067	0.044	0.024	0.007	0.002	0.000	0.000	0.000
0.750	0.006	0.011	0.016	0.024	0.023	0.016	0.008	0.006	0.000	0.001	0.001
0.960	0.001	0.001	0.002	0.003	0.003	0.003	0.002	0.001	−0.001	−0.002	−0.002
						$\eta = 0.25\,b$					
0.040	0.140	0.231	0.256	0.107	0.014	−0.006	−0.008	−0.006	−0.003	−0.002	−0.001
0.250	1.098	2.944	1.391	0.335	0.078	−0.002	−0.024	−0.022	−0.012	−0.008	−0.004
0.500	0.219	0.393	0.505	0.431	0.234	0.101	0.034	0.007	−0.001	−0.001	−0.001
0.750	0.045	0.086	0.124	0.169	0.143	0.091	0.050	0.022	0.007	0.002	0.000
0.960	0.006	0.011	0.016	0.024	0.023	0.016	0.008	0.006	0.000	0.001	0.001
						$\eta = 0.50\,b$					
0.040	0.022	0.041	0.058	0.067	0.044	0.024	0.007	0.002	0.000	0.000	0.000
0.250	0.219	0.393	0.505	0.431	0.234	0.101	0.034	0.007	−0.001	−0.001	−0.001
0.500	1.143	3.029	1.515	0.503	0.221	0.089	0.026	0.000	−0.005	−0.006	−0.004
0.750	0.219	0.393	0.505	0.431	0.234	0.101	0.034	0.007	−0.001	−0.001	−0.001
0.960	0.022	0.041	0.058	0.067	0.044	0.024	0.007	0.002	0.000	0.000	0.000

$\xi = 0.25\,a$

x/a \ y/b	0.040	0.125	0.210	0.250	0.290	0.375	0.500	0.625	0.710	0.750	0.790	0.875	0.960
							$\eta = d$						
0.040	0.015	0.012	−1.459	−0.004	1.451	−0.022	−0.023	−0.009	−0.005	−0.004	−0.003	−0.001	0.000
0.250	−0.055	−0.134	−0.093	−0.003	0.087	0.135	0.046	0.004	−0.002	−0.003	−0.004	−0.003	−0.001
0.500	−0.004	−0.009	0.005	0.012	0.019	0.033	0.035	0.021	0.016	0.012	0.007	0.003	0.001
0.750	0.001	0.003	0.006	0.008	0.010	0.012	0.016	0.011	0.010	0.008	0.007	0.004	0.003
0.960	0.000	0.001	0.001	0.001	0.002	0.002	0.002	0.002	0.000	0.001	0.003	0.001	0.000

Table A.12 (continued)

x/a \ y/b	0.040	0.125	0.250	0.375	0.460	0.500	0.540	0.625	0.750	0.875	0.960

η = 0.25 b

0.040	−0.055	−0.134	−0.093	−0.003	0.087	0.135	0.046	−0.002	−0.003	−0.004	−0.003	−0.001
0.250	−0.133	−0.538	−2.326	−0.001	2.324	0.562	0.145	0.002	−0.001	−0.004	−0.006	−0.002
0.500	−0.072	−0.156	−0.060	0.049	0.161	0.288	0.218	0.066	0.049	0.036	0.018	0.002
0.750	0.000	0.007	0.031	0.044	0.060	0.089	0.098	0.054	0.044	0.038	0.020	0.007
0.960	0.001	0.003	0.006	0.008	0.010	0.012	0.016	0.010	0.008	0.007	0.004	0.003

η = 0.50 b

0.040	−0.004	−0.009	0.005	0.012	0.019	0.033	0.021	0.016	0.012	0.007	0.003	0.001
0.250	−0.072	−0.156	−0.060	0.049	0.161	0.288	0.218	0.114	0.066	0.049	0.036	0.018	0.002
0.500	−0.132	−0.531	−2.294	0.043	2.384	0.651	0.242	0.106	0.056	0.043	0.034	0.014	0.004
0.750	−0.072	−0.156	−0.060	0.049	0.161	0.288	0.218	0.114	0.066	0.049	0.036	0.018	0.002
0.960	−0.004	−0.009	0.005	0.012	0.019	0.033	0.035	0.021	0.016	0.012	0.007	0.003	0.001

ξ = 0.50 a

η = d

0.040	0.002	0.006	0.020	0.019	−1.454	0.000	1.454	−0.019	−0.020	−0.006	−0.002
0.250	−0.002	−0.010	−0.053	−0.141	−0.092	0.000	0.092	0.141	0.053	0.010	0.002
0.500	−0.009	−0.018	−0.033	−0.030	−0.012	0.000	0.012	0.030	0.033	0.018	0.009
0.750	−0.002	−0.008	−0.009	−0.008	−0.003	0.000	0.003	0.008	0.009	0.008	0.002
0.960	0.000	−0.001	−0.001	−0.001	−0.002	0.000	0.002	0.001	0.001	0.001	0.000

η = 0.25 b

0.040	−0.002	−0.010	−0.053	−0.141	−0.092	0.000	0.092	0.141	0.053	0.010	0.002
0.250	−0.011	−0.049	−0.167	−0.581	−2.332	0.000	2.332	0.581	0.167	0.049	0.011
0.500	−0.031	−0.098	−0.210	−0.272	−0.126	0.000	0.126	0.272	0.210	0.098	0.031
0.750	−0.019	−0.046	−0.075	−0.060	−0.025	0.000	0.025	0.060	0.075	0.046	0.019
0.960	−0.002	−0.008	−0.009	−0.008	−0.003	0.000	0.003	0.008	0.009	0.008	0.002

η = 0.50 b

0.040	−0.009	−0.018	−0.033	−0.030	−0.012	0.000	0.012	0.030	0.033	0.018	0.009
0.250	−0.031	−0.098	−0.210	−0.272	−0.126	0.000	0.126	0.272	0.210	0.098	0.031
0.500	−0.029	−0.095	−0.242	−0.640	−2.357	0.000	2.357	0.640	0.242	0.095	0.029
0.750	−0.031	−0.098	−0.210	−0.272	−0.126	0.000	0.126	0.272	0.210	0.098	0.031
0.960	−0.009	−0.018	−0.033	−0.030	−0.012	0.000	0.012	0.030	0.033	0.018	0.009

Appendix

Table A.13
BENDING MOMENTS M_x — COEFFICIENTS k_M^x

$M_x = \dfrac{M}{a} k_M^x$
$\xi = e$
$b = a$
$c = d = 0.04\,a$
$\mu = 1/3$

y/b \ x/a	0.040	0.080	0.125	0.250	0.375	0.500	0.625	0.750	0.875	0.920	0.960
\multicolumn{12}{c}{$\eta = d$}
0.040	0.451	1.886	0.446	0.014	0.002	0.000	0.000	0.000	0.000	0.000	0.000
0.250	0.130	0.219	0.252	0.138	0.047	0.017	0.006	0.002	0.001	0.000	0.000
0.500	0.020	0.037	0.053	0.065	0.048	0.031	0.014	0.008	0.003	0.002	0.001
0.750	0.005	0.010	0.014	0.022	0.022	0.017	0.010	0.008	0.002	0.001	0.001
0.960	0.001	0.001	0.002	0.003	0.003	0.003	0.002	0.001	−0.001	−0.002	−0.002
\multicolumn{12}{c}{$\eta = 0.25\,b$}
0.040	0.130	0.219	0.252	0.138	0.047	0.017	0.006	0.002	0.001	0.000	0.000
0.250	1.372	3.428	1.825	0.571	0.231	0.099	0.041	0.017	0.006	0.003	0.002
0.500	0.198	0.360	0.473	0.446	0.284	0.157	0.081	0.040	0.016	0.009	0.005
0.750	0.040	0.077	0.111	0.158	0.143	0.101	0.064	0.034	0.014	0.007	0.002
0.960	0.005	0.010	0.014	0.022	0.022	0.017	0.010	0.008	0.002	0.001	0.001
\multicolumn{12}{c}{$\eta = 0.50\,b$}
0.040	0.020	0.037	0.053	0.065	0.048	0.031	0.014	0.008	0.003	0.001	0.001
0.250	0.198	0.360	0.473	0.446	0.284	0.157	0.081	0.040	0.016	0.009	0.005
0.500	1.411	3.505	1.936	0.729	0.374	0.200	0.105	0.051	0.020	0.011	0.003
0.750	0.198	0.360	0.473	0.446	0.284	0.157	0.081	0.040	0.016	0.009	0.005
0.960	0.020	0.037	0.053	0.065	0.048	0.031	0.014	0.008	0.003	0.002	0.001

$\xi = 0.25\,a$

y/b \ x/a	0.040	0.125	0.210	0.250	0.290	0.375	0.500	0.625	0.710	0.750	0.790	0.875	0.960
\multicolumn{14}{c}{$\eta = d$}
0.040	−0.009	−0.111	1.763	0.000	1.764	0.112	0.005	0.001	0.000	0.000	0.000	0.000	0.000
0.250	−0.051	−0.115	−0.072	0.008	0.088	0.140	0.068	0.022	0.011	0.008	0.006	0.002	0.001
0.500	−0.002	−0.004	−0.009	0.015	0.021	0.034	0.037	0.025	0.020	0.015	0.010	0.005	0.002
0.750	0.001	0.004	0.007	0.008	0.010	0.012	0.015	0.011	0.010	0.008	0.007	0.004	0.003
0.960	0.000	0.001	0.001	0.001	0.002	0.002	0.002	0.002	0.000	0.001	0.002	0.001	0.000

Table A.13 (continued)

η = 0.25 b

x/a \ y/b	0.040	0.125	0.250⁻	0.250	0.250⁺	0.375	0.460	0.500	0.540	0.625	0.750	0.875	0.960
0.040	−0.051	−0.115	−0.072	0.008	0.088	0.140	0.068	0.022	0.011	0.008	0.006	0.002	0.001
0.250	−0.168	−0.669	−2.540	0.049	2.639	0.801	0.280	0.115	0.062	0.049	0.037	0.017	0.005
0.500	−0.055	−0.111	−0.018	0.078	0.176	0.293	0.242	0.148	0.097	0.078	0.060	0.033	0.008
0.750	0.003	0.014	0.038	0.050	0.063	0.090	0.098	0.080	0.059	0.050	0.043	0.024	0.008
0.960	0.001	0.004	0.007	0.008	0.010	0.012	0.015	0.011	0.010	0.008	0.007	0.004	0.003

η = 0.50 b

x/a \ y/b	0.040	0.125	0.250	0.375	0.460	0.500⁻	0.500	0.500⁺	0.540	0.625	0.750	0.875	0.960
0.040	−0.002	−0.004	−0.009	0.015	0.021	0.034	0.037	0.025	0.020	0.015	0.010	0.005	0.002
0.250	−0.055	−0.111	−0.018	0.078	0.176	0.293	0.242	0.148	0.097	0.078	0.060	0.033	0.008
0.500	−0.165	−0.655	−2.502	0.099	2.702	0.891	0.378	0.195	0.121	0.099	0.079	0.041	0.013
0.750	−0.055	−0.111	−0.018	0.078	0.176	0.293	0.242	0.148	0.097	0.078	0.060	0.033	0.008
0.960	−0.002	−0.004	−0.009	0.015	0.021	0.034	0.037	0.025	0.020	0.015	0.010	0.005	0.002

ξ = 0.50 a

η = d

x/a \ y/b	0.040	0.125	0.250	0.375	0.460	0.500⁻	0.500	0.500⁺	0.540	0.625	0.750	0.875	0.960
0.040	0.000	−0.001	−0.005	−0.112	−1.764	0.000	1.764	0.112	0.005	0.001	0.000		
0.250	−0.005	−0.019	−0.066	−0.137	−0.083	0.000	0.083	0.137	0.066	0.019	0.005		
0.500	−0.009	−0.017	−0.029	−0.026	−0.010	0.000	0.010	0.026	0.029	0.017	0.009		
0.750	−0.002	−0.006	−0.007	−0.007	−0.002	0.000	0.002	0.007	0.007	0.006	0.002		
0.960	0.000	−0.001	−0.001	−0.001	−0.002	0.000	0.002	0.001	0.001	0.001	0.000		

η = 0.25 b

x/a \ y/b	0.040	0.125	0.250⁻	0.250	0.250⁺	0.375	0.460	0.500	0.540	0.625	0.750	0.875	0.960
0.040	−0.005	−0.019	−0.066	−0.137	−0.083	0.000	0.083	0.137	0.066	0.019	0.005		
0.250	−0.024	−0.092	−0.264	−0.778	−2.601	0.000	2.601	0.778	0.264	0.092	0.024		
0.500	−0.032	−0.100	−0.202	−0.244	−0.111	0.000	0.111	0.244	0.202	0.100	0.032		
0.750	−0.016	−0.039	−0.063	−0.049	−0.021	0.000	0.021	0.049	0.063	0.039	0.016		
0.960	−0.002	−0.006	−0.007	−0.007	−0.002	0.000	0.002	0.007	0.006	0.006	0.002		

η = 0.50 b

x/a \ y/b	0.040	0.125	0.250	0.375	0.460	0.500⁻	0.500	0.500⁺	0.540	0.625	0.750	0.875	0.960
0.040	−0.009	−0.017	−0.029	−0.026	−0.010	0.000	0.010	0.026	0.029	0.017	0.009		
0.250	−0.032	−0.100	−0.202	−0.244	−0.111	0.000	0.111	0.244	0.202	0.100	0.032		
0.500	−0.040	−0.132	−0.327	−0.828	−2.622	0.000	2.622	0.828	0.327	0.132	0.040		
0.750	−0.032	−0.100	−0.202	−0.244	−0.111	0.000	0.111	0.244	0.202	0.100	0.032		
0.960	−0.009	−0.017	−0.029	−0.026	−0.010	0.000	0.010	0.026	0.029	0.017	0.009		

Appendix

Table A. 14
BENDING MOMENTS M_x — COEFFICIENTS k_M^x

$$M_x = \frac{M}{a} k_M^x \qquad \xi = e \qquad \begin{array}{l} b = a \\ c = d = 0.06\,a \\ \mu = 0 \end{array}$$

x/a \ y/b	0.060	0.120	0.125	0.250	0.375	0.500	0.625	0.750	0.875	0.880	0.940
					$\eta = d$						
0.060	0.117	1.025	0.983	−0.044	−0.030	−0.015	−0.008	−0.004	−0.002	−0.002	−0.001
0.250	0.213	0.327	0.331	0.169	0.027	−0.006	−0.011	−0.009	−0.005	−0.004	−0.002
0.500	0.045	0.079	0.081	0.097	0.067	0.035	0.012	0.004	0.000	0.000	0.000
0.750	0.012	0.023	0.023	0.035	0.033	0.024	0.014	0.008	0.002	0.002	0.001
0.940	0.002	0.005	0.005	0.008	0.008	0.006	0.004	0.002	0.000	0.000	0.002
					$\eta = 0.25\,b$						
0.060	0.213	0.327	0.331	0.169	0.027	−0.006	−0.011	−0.009	−0.005	−0.004	−0.002
0.250	0.655	1.924	1.887	0.385	0.093	0.003	−0.021	−0.021	−0.012	−0.011	−0.006
0.500	0.264	0.440	0.449	0.425	0.238	0.105	0.037	0.008	0.000	−0.001	0.000
0.750	0.062	0.114	0.117	0.163	0.141	0.093	0.050	0.023	0.008	0.007	0.004
0.940	0.012	0.023	0.023	0.035	0.033	0.024	0.014	0.008	0.002	0.002	0.001
					$\eta = 0.50\,b$						
0.060	0.045	0.079	0.081	0.097	0.067	0.035	0.012	0.004	0.000	0.000	0.000
0.250	0.264	0.440	0.449	0.425	0.238	0.105	0.037	0.008	0.000	−0.001	0.000
0.500	0.717	2.037	2.004	0.548	0.234	0.095	0.029	0.002	−0.004	−0.005	−0.002
0.750	0.264	0.440	0.449	0.425	0.238	0.105	0.037	0.008	0.000	−0.001	0.000
0.940	0.045	0.079	0.081	0.097	0.067	0.035	0.012	0.004	0.000	0.000	0.000

$$\xi = 0.25\,a$$

x/a \ y/b	0.060	0.125	0.190	0.250	0.310	0.375	0.500	0.625	0.690	0.750	0.810	0.875	0.940
						$\eta = d$							
0.060	0.003	−0.103	−1.043	−0.007	1.033	0.085	−0.028	−0.015	−0.010	−0.007	−0.005	−0.003	−0.001
0.250	−0.111	−0.200	−0.180	−0.004	0.174	0.204	0.066	0.007	−0.001	−0.004	−0.005	−0.004	−0.002
0.500	−0.010	−0.011	0.000	0.015	0.035	0.052	0.053	0.034	0.026	0.015	0.009	0.006	0.003
0.750	0.002	0.005	0.008	0.012	0.016	0.020	0.021	0.019	0.016	0.012	0.009	0.006	0.004
0.940	0.001	0.001	0.002	0.003	0.004	0.004	0.005	0.004	0.002	0.003	0.004	0.002	0.000

Table A.14 (continued)

y/b \ x/a	0.060	0.125	0.250	0.375	0.440	0.500	0.560	0.625	0.750	0.875	0.940
\multicolumn{12}{c}{$\eta = 0.25\,b$}											
0.060	−0.111	−0.200	−0.180	−0.004	0.174	0.204	0.066	−0.001	−0.004	−0.005	−0.004	−0.002
0.250	−0.214	−0.577	−1.580	−0.003	1.583	0.605	0.151	0.008	−0.003	−0.005	−0.005	−0.003
0.500	−0.099	−0.153	−0.098	0.049	0.203	0.282	0.221	0.075	0.049	0.031	0.016	0.006
0.750	0.001	0.007	0.024	0.047	0.068	0.087	0.096	0.059	0.047	0.033	0.021	0.010
0.940	0.002	0.005	0.009	0.012	0.016	0.020	0.021	0.016	0.012	0.009	0.006	0.004
\multicolumn{12}{c}{$\eta = 0.50\,b$}											
0.060	−0.010	−0.011	0.000	0.015	0.035	0.052	0.053	0.034	0.026	0.015	0.009	0.006	0.003
0.250	−0.099	−0.153	−0.098	0.049	0.203	0.282	0.221	0.113	0.075	0.049	0.031	0.016	0.006
0.500	−0.213	−0.570	−1.555	0.044	1.651	0.692	0.248	0.106	0.067	0.044	0.028	0.016	0.007
0.750	−0.099	−0.153	−0.098	0.049	0.203	0.282	0.221	0.113	0.075	0.049	0.031	0.016	0.006
0.940	−0.010	−0.011	0.000	0.015	0.035	0.052	0.053	0.034	0.026	0.015	0.009	0.006	0.003

$\xi = 0.50\,a$

y/b \ x/a	0.060	0.125	0.250	0.375	0.440	0.500	0.560	0.625	0.750	0.875	0.940
\multicolumn{12}{c}{$\eta = d$}											
0.060	0.005	0.010	0.023	−0.090	−1.039	0.000	1.039	0.090	−0.023	−0.010	−0.005
0.250	−0.006	−0.016	−0.075	−0.212	−0.181	0.000	0.181	0.212	0.075	0.016	0.006
0.500	−0.016	−0.028	−0.049	−0.045	−0.026	0.000	0.026	0.045	0.049	0.028	0.016
0.750	−0.005	−0.009	−0.016	−0.011	−0.005	0.000	0.005	0.011	0.016	0.009	0.005
0.940	−0.001	−0.002	−0.003	−0.002	−0.002	0.000	0.002	0.002	0.003	0.002	0.001
\multicolumn{12}{c}{$\eta = 0.25\,b$}											
0.060	−0.006	−0.016	−0.075	−0.212	−0.181	0.000	0.181	0.212	0.075	0.016	0.006
0.250	−0.019	−0.049	−0.173	−0.622	−1.594	0.000	1.594	0.622	0.173	0.049	0.019
0.500	−0.045	−0.097	−0.217	−0.266	−0.173	0.000	0.173	0.266	0.217	0.097	0.045
0.750	−0.025	−0.048	−0.075	−0.061	−0.034	0.000	0.034	0.061	0.075	0.048	0.025
0.940	−0.005	−0.009	−0.016	−0.011	−0.005	0.000	0.005	0.011	0.016	0.009	0.005
\multicolumn{12}{c}{$\eta = 0.50\,b$}											
0.060	−0.016	−0.028	−0.049	−0.045	−0.026	0.000	0.026	0.045	0.049	0.028	0.016
0.250	−0.045	−0.097	−0.217	−0.266	−0.173	0.000	0.173	0.266	0.217	0.097	0.045
0.500	−0.044	−0.097	−0.248	−0.683	−1.628	0.000	1.628	0.683	0.248	0.097	0.044
0.750	−0.045	−0.097	−0.217	−0.266	−0.173	0.000	0.173	0.266	0.217	0.097	0.045
0.940	−0.016	−0.028	−0.049	−0.045	−0.026	0.000	0.026	0.045	0.049	0.028	0.016

Appendix

Table A. 15
BENDING MOMENTS M_x — COEFFICIENTS k_M^x

$$M_x = \frac{M}{a} k_M^x \qquad \begin{array}{l} b = a \\ c = d = 0.06\,a \\ \mu = 1/3 \end{array}$$

$\boxed{\xi = c}$

x/a \ y/b	0.060	0.120	0.125	0.250	0.375	0.500	0.625	0.750	0.875	0.880	0.940
η = d											
0.060	0.285	1.325	1.284	0.071	0.009	0.002	0.000	0.000	0.000	0.000	0.000
0.250	0.211	0.330	0.336	0.215	0.075	0.028	0.010	0.003	0.001	0.001	0.000
0.500	0.040	0.072	0.074	0.094	0.073	0.046	0.022	0.012	0.005	0.004	0.002
0.750	0.011	0.020	0.021	0.032	0.032	0.025	0.016	0.010	0.003	0.003	0.002
0.940	0.002	0.004	0.004	0.007	0.007	0.006	0.004	0.003	0.001	0.000	0.002
η = 0.25 b											
0.060	0.211	0.330	0.336	0.215	0.075	0.028	0.010	0.003	0.001	0.001	0.000
0.250	0.842	2.255	2.222	0.619	0.246	0.104	0.044	0.018	0.006	0.006	0.003
0.500	0.246	0.417	0.426	0.440	0.286	0.161	0.084	0.041	0.017	0.015	0.008
0.750	0.056	0.103	0.106	0.153	0.141	0.102	0.064	0.035	0.014	0.013	0.007
0.940	0.011	0.020	0.021	0.032	0.032	0.025	0.016	0.010	0.003	0.003	0.002
η = 0.50 b											
0.060	0.040	0.072	0.074	0.094	0.073	0.046	0.022	0.012	0.005	0.004	0.002
0.250	0.246	0.417	0.426	0.440	0.286	0.161	0.084	0.041	0.017	0.015	0.008
0.500	0.897	2.357	2.328	0.772	0.387	0.206	0.108	0.053	0.021	0.019	0.010
0.750	0.246	0.417	0.426	0.440	0.286	0.161	0.084	0.041	0.017	0.015	0.008
0.940	0.040	0.072	0.074	0.094	0.073	0.046	0.022	0.012	0.005	0.004	0.002

$\boxed{\xi = 0.25\,a}$

x/a \ y/b	0.060	0.125	0.190	0.250	0.310	0.375	0.500	0.625	0.690	0.750	0.810	0.875	0.940
η = d													
0.060	−0.054	−0.246	−1.243	0.001	1.244	0.249	0.023	0.003	0.001	0.001	0.000	0.000	0.000
0.250	−0.102	−0.175	−0.147	0.012	0.175	0.213	0.099	0.035	0.021	0.012	0.007	0.004	0.002
0.500	−0.005	−0.004	0.008	0.021	0.038	0.053	0.055	0.040	0.032	0.021	0.014	0.009	0.005
0.750	0.002	0.005	0.009	0.013	0.016	0.020	0.020	0.019	0.016	0.013	0.009	0.007	0.004
0.940	0.001	0.001	0.002	0.003	0.004	0.004	0.005	0.004	0.002	0.003	0.004	0.002	0.000

392

Table A.15 (continued)

x/a y/b	\multicolumn{9}{c}{$\eta = 0.25\,b$}

x/a \ y/b	0.060	0.125	0.250	0.375	0.440	0.500	0.560	0.625	0.750	0.875	0.940
					$\eta = 0.25\,b$						
0.060	−0.102	−0.175	−0.147	0.012	0.175	0.213	0.099	0.035	0.021	0.012	0.007	0.004	0.002
0.250	−0.263	−0.686	−1.703	0.047	**1.807**	0.820	0.284	0.116	0.073	0.047	0.031	0.018	0.008
0.500	−0.075	−0.109	−0.053	0.077	0.213	0.288	0.247	0.147	0.107	0.077	0.053	0.032	0.014
0.750	−0.005	0.014	0.031	0.052	0.071	0.088	0.097	0.078	0.064	0.052	0.038	0.024	0.011
0.940	0.002	0.005	0.009	0.013	0.016	0.020	0.020	0.019	0.016	0.013	0.009	0.007	0.004
					$\eta = 0.50\,b$						
0.060	−0.005	−0.004	0.008	0.021	0.038	0.053	0.055	0.040	0.032	0.021	0.014	0.009	0.005
0.250	−0.075	−0.109	−0.053	0.077	0.213	0.288	0.247	0.147	0.107	0.077	0.053	0.032	0.014
0.500	−0.259	−0.672	−1.672	0.099	**1.878**	0.908	0.381	0.194	0.137	0.099	0.069	0.042	0.019
0.750	−0.075	−0.109	−0.053	0.077	0.213	0.288	0.247	0.147	0.107	0.077	0.053	0.032	0.014
0.940	−0.005	−0.004	0.008	0.021	0.038	0.053	0.055	0.040	0.032	0.021	0.014	0.009	0.005

x/a \ y/b	0.060	0.125	0.250	0.375	0.440	0.500	0.560	0.625	0.750	0.875	0.940
					$\xi = 0.50\,a$						
					$\eta = d$						
0.060	−0.001	−0.003	−0.023	−0.249	−1.244	0.000	1.244	0.249	0.023	0.003	0.001
0.250	−0.013	−0.030	−0.096	−0.209	0.167	0.000	0.167	0.209	0.096	0.030	0.013
0.500	−0.016	−0.026	−0.044	−0.039	−0.022	0.000	0.022	0.039	0.044	0.026	0.016
0.750	−0.004	−0.007	−0.013	−0.009	−0.004	0.000	0.004	0.009	0.013	0.007	0.004
0.940	−0.001	−0.002	−0.002	−0.002	−0.002	0.000	0.002	0.002	0.002	0.002	0.001
					$\eta = 0.25\,b$						
0.060	−0.013	−0.030	−0.096	−0.209	−0.167	0.000	0.167	0.209	0.096	0.030	0.013
0.250	−0.039	−0.092	−0.267	−0.796	−1.773	0.000	1.773	0.796	0.267	0.092	0.039
0.500	−0.047	−0.098	−0.209	−0.240	−0.153	0.000	0.153	0.240	0.209	0.098	0.047
0.750	−0.022	−0.041	−0.063	−0.051	−0.028	0.000	0.028	0.051	0.063	0.041	0.022
0.940	−0.004	−0.007	−0.013	−0.009	−0.004	0.000	0.004	0.009	0.013	0.007	0.004
					$\eta = 0.50\,b$						
0.060	−0.016	−0.026	−0.044	−0.039	−0.022	0.000	0.022	0.039	0.044	0.026	0.016
0.250	−0.047	−0.098	−0.209	−0.240	−0.153	0.000	0.153	0.240	0.209	0.098	0.047
0.500	−0.061	−0.133	−0.331	−0.847	−**1.302**	0.000	**1.302**	0.847	0.331	0.133	0.061
0.750	−0.047	−0.098	−0.209	−0.240	−0.153	0.000	0.153	0.240	0.209	0.098	0.047
0.940	−0.016	−0.026	−0.044	−0.039	−0.022	0.000	0.022	0.039	0.044	0.026	0.016

Appendix

Table A. 16
TORSIONAL MOMENTS M_{xy} — COEFFICIENTS k_M^{xy}

$$M_{xy} = \frac{M}{a} k_M^{xy}$$

$\xi = c$

$b = a$
$c = d = 0.02\,a$
$\mu = 0$

x/a y/b	0.000	0.020	0.040	0.125	0.250	0.375	0.500	0.625	0.750	0.875	0.980	1.000
\multicolumn{13}{c}{$\eta = d$}
0.000	−2.779	−2.730	−0.881	−0.384	−0.107	−0.051	−0.031	−0.022	−0.018	−0.015	−0.014	−0.014
0.020	−1.899	−1.660	−0.759	−0.297	−0.101	−0.050	−0.031	−0.022	−0.017	−0.015	−0.014	−0.014
0.040	−0.412	−0.214	−0.484	−0.108	−0.085	−0.047	−0.030	−0.022	−0.017	−0.015	−0.014	−0.014
0.250	0.102	0.096	0.082	−0.014	−0.042	−0.017	−0.001	0.005	0.007	0.008	0.008	0.008
0.500	0.031	0.031	0.030	0.019	0.000	−0.009	−0.011	−0.007	−0.004	−0.003	−0.002	−0.002
0.750	0.017	0.017	0.017	0.014	0.007	0.000	−0.004	−0.007	−0.008	−0.008	−0.008	−0.008
0.980	0.014	0.014	0.014	0.013	0.002	−0.002	−0.002	−0.006	−0.008	−0.009	−0.009	−0.010
\multicolumn{13}{c}{$\eta = 0.25\,b$}
0.020	−1.171	−1.147	−1.080	−0.529	−0.072	0.221	0.222	0.195	0.172	0.157	0.152	0.152
0.230	−1.644	−1.618	−0.688	−0.004	−0.026	0.041	0.083	0.098	0.102	0.101	0.101	0.101
0.250	−0.238	−0.237	−0.232	−0.187	−0.080	0.010	0.061	0.084	0.091	0.093	0.093	0.093
0.270	1.166	1.143	−0.223	−0.369	−0.133	−0.020	0.041	0.069	0.080	0.083	0.084	0.084
0.500	0.486	0.475	0.443	0.185	−0.083	−0.123	−0.092	−0.059	−0.035	−0.024	−0.020	−0.020
0.750	0.238	0.237	0.232	0.187	0.080	−0.010	−0.061	−0.084	−0.091	−0.093	−0.093	−0.093
0.980	0.193	0.192	0.190	0.165	0.099	−0.023	−0.038	−0.078	−0.099	−0.110	−0.113	−0.113
\multicolumn{13}{c}{$\eta = 0.50\,b$}
0.020	−0.478	−0.475	−0.466	−0.373	−0.159	0.021	0.124	0.168	0.182	0.185	0.185	0.185
0.250	−0.678	−0.666	−0.633	−0.350	−0.015	0.099	0.131	0.137	0.137	0.134	0.133	0.133
0.480	−1.397	−1.372	−0.448	−0.185	−0.049	0.024	0.017	0.014	0.013	0.012	0.012	0.011
0.500	0.000	0.000	0.000	0.000	0.000	0.000	0.000	0.000	0.000	0.000	0.000	0.000
0.520	1.397	1.372	0.448	−0.185	−0.049	−0.024	−0.017	−0.014	−0.013	−0.012	−0.012	−0.011
0.750	0.678	0.666	0.633	0.350	0.015	−0.099	−0.131	−0.137	−0.137	−0.134	−0.133	−0.133
0.980	0.478	0.475	0.466	0.373	0.159	−0.021	−0.124	−0.168	−0.182	−0.185	−0.185	−0.185

Table A.16 (continued)

$\xi = 0.25\,a$

x/a \ y/b	0.000	0.020	0.125	0.230	0.250	0.270	0.375	0.500	0.625	0.750	0.875	0.980	1.000
η = a													
0.000	0.105	0.107	0.207	−1.379	−3.240	−1.381	0.193	0.061	0.033	0.023	0.019	0.018	0.017
0.020	0.099	0.101	0.171	−0.933	−1.546	−0.935	0.157	0.058	0.032	0.023	0.019	0.017	0.017
0.040	0.084	0.085	0.085	−0.183	−0.426	−0.185	0.072	0.051	0.031	0.022	0.018	0.017	0.017
0.250	−0.046	−0.045	−0.017	0.050	0.053	0.051	−0.006	−0.019	−0.004	0.004	0.007	0.007	0.007
0.500	−0.001	0.000	0.005	0.008	0.012	0.012	0.006	−0.002	−0.006	−0.005	−0.005	−0.004	−0.004
0.750	0.007	0.007	0.007	0.007	0.007	0.006	0.004	−0.001	−0.004	−0.006	−0.007	−0.008	−0.010
0.980	0.008	0.008	0.008	0.006	0.006	0.006	0.003	0.000	−0.003	−0.006	−0.008	−0.008	−0.008
η = 0.25 b													
0.020	0.078	0.072	−0.151	−0.473	−0.487	−0.476	−0.165	0.125	0.190	0.187	0.176	0.172	0.172
0.230	−0.027	−0.026	0.013	−0.781	−1.731	−0.780	0.042	0.038	0.071	0.091	0.100	0.102	0.102
0.250	−0.080	−0.080	−0.088	−0.091	−0.086	−0.086	−0.052	0.006	0.052	0.079	0.088	0.091	0.091
0.270	−0.132	−0.133	−0.189	0.600	1.553	0.608	−0.145	−0.026	0.032	0.063	0.076	0.080	0.080
0.500	−0.090	−0.083	0.030	0.194	0.202	0.200	0.062	−0.064	−0.073	−0.056	−0.041	−0.035	−0.039
0.750	0.080	0.080	0.088	0.091	0.088	0.086	0.052	−0.006	−0.052	−0.079	−0.088	−0.091	−0.091
0.980	0.099	0.099	0.094	0.081	0.077	0.073	0.044	−0.001	−0.044	−0.076	−0.094	−0.099	−0.101
η = 0.50 b													
0.020	−0.159	−0.159	−0.176	−0.181	−0.178	−0.172	−0.103	0.012	0.103	0.154	0.176	0.182	0.182
0.250	−0.009	−0.015	−0.124	−0.275	−0.280	−0.273	−0.105	0.062	0.117	0.132	0.136	0.137	0.134
0.480	0.048	0.049	0.100	−0.690	−1.639	−0.690	0.095	0.031	0.018	0.016	0.013	0.013	0.013
0.500	0.000	0.000	0.000	0.000	0.000	0.000	0.000	0.000	0.000	0.000	0.000	0.000	0.000
0.520	−0.048	−0.049	−0.100	0.690	1.639	0.690	−0.095	−0.031	−0.018	−0.016	−0.013	−0.013	−0.013
0.750	0.009	0.015	0.124	0.275	0.280	0.273	0.105	−0.062	−0.117	−0.132	−0.136	−0.137	−0.134
0.980	0.159	0.159	0.176	0.181	0.178	0.172	0.103	−0.012	−0.103	−0.154	−0.176	−0.182	−0.182

Appendix

Table A.16 (continued)

	ξ = 0.50 a												
x/a \ y/b	0.000	0.020	0.125	0.250	0.375	0.480	0.500	0.520	0.625	0.750	0.875	0.980	1.000

η = a

y/b													
0.000	0.031	0.031	0.036	0.061	0.190	1.388	3.248	1.388	0.190	0.061	0.036	0.031	0.031
0.020	0.031	0.031	0.036	0.058	0.154	0.942	1.581	0.942	0.154	0.058	0.036	0.031	0.031
0.040	0.030	0.030	0.034	0.051	0.069	0.192	0.418	0.192	0.069	0.051	0.034	0.030	0.030
0.250	−0.001	−0.001	−0.005	−0.019	−0.005	0.055	0.058	0.055	−0.005	−0.019	−0.005	−0.001	−0.001
0.500	−0.007	−0.009	−0.008	−0.002	0.008	0.014	0.014	0.014	0.008	−0.002	−0.008	−0.009	−0.007
0.750	−0.004	−0.004	−0.003	−0.001	0.003	0.005	0.005	0.005	0.003	−0.001	−0.003	−0.004	−0.004
0.980	−0.002	−0.002	−0.002	0.000	0.002	0.003	0.003	0.003	0.002	0.000	−0.002	−0.002	−0.002

η = 0.25 b

y/b													
0.020	0.222	0.222	0.209	0.124	−0.184	−0.509	−0.522	−0.509	−0.184	0.124	0.209	0.222	0.222
0.230	0.082	0.081	0.070	0.038	−0.043	−0.775	−1.705	−0.775	−0.043	0.038	0.070	0.081	0.082
0.250	0.059	0.062	0.047	0.006	−0.047	−0.073	−0.07	−0.073	−0.047	0.006	0.047	0.062	0.059
0.270	0.043	0.041	0.025	−0.026	−0.138	0.628	1.557	0.628	−0.138	−0.026	0.025	0.041	0.043
0.500	−0.092	−0.092	−0.091	−0.064	0.080	0.233	0.239	0.233	0.080	−0.064	−0.091	−0.092	−0.092
0.750	−0.059	−0.062	−0.047	−0.006	0.047	0.073	0.074	0.073	0.047	−0.006	−0.047	−0.062	−0.059
0.980	−0.039	−0.038	−0.028	−0.001	0.028	0.040	0.040	0.040	0.028	−0.001	−0.028	−0.038	−0.039

η = 0.50 b

y/b													
0.020	0.125	0.122	0.095	0.012	−0.095	−0.146	−0.148	−0.146	−0.095	0.012	0.095	0.122	0.125
0.250	0.131	0.131	0.119	0.066	−0.107	−0.273	−0.279	−0.273	−0.107	0.066	0.119	0.131	0.131
0.480	0.017	0.017	0.019	0.031	0.094	−0.696	−1.626	−0.696	0.094	0.031	0.019	0.017	0.017
0.500	0.000	0.000	0.000	0.000	0.000	0.000	0.000	0.000	0.000	0.000	0.000	0.000	0.000
0.520	−0.017	−0.017	−0.019	−0.031	−0.094	0.696	1.626	0.696	−0.094	−0.031	−0.019	−0.017	−0.017
0.750	−0.131	−0.131	−0.119	−0.066	0.107	0.273	0.279	0.273	0.107	−0.066	−0.119	−0.131	−0.131
0.980	−0.125	−0.122	−0.095	−0.012	0.095	0.146	0.148	0.146	0.095	−0.012	−0.095	−0.122	−0.125

Appendix

Table A. 17
TORSIONAL MOMENTS M_{xy} — COEFFICIENTS k_M^{xy}

$$M_{xy} = \frac{M}{a} k_M^{xy} \qquad \boxed{\xi = c} \qquad \begin{array}{l} b = a \\ c = d = 0.04a \\ \mu = 0 \end{array}$$

x/a \ y/b	0.000	0.040	0.080	0.125	0.250	0.375	0.500	0.625	0.750	0.875	0.960	1.000
\multicolumn{13}{c}{$\eta = d$}											
0.000	−1.364	−1.336	−0.431	0.451	0.201	0.100	0.062	0.044	0.035	0.030	0.029	0.029
0.040	−0.938	−0.830	−0.370	0.112	0.158	0.091	0.059	0.043	0.034	0.030	0.029	0.028
0.080	−0.200	−0.105	−0.234	−0.300	0.063	0.067	0.051	0.039	0.032	0.029	0.027	0.027
0.250	0.166	0.138	0.068	−0.016	−0.083	−0.033	−0.002	0.011	0.015	0.016	0.016	0.016
0.500	0.060	0.057	0.049	0.037	0.000	−0.018	−0.018	−0.013	−0.009	−0.006	−0.005	−0.005
0.750	0.034	0.034	0.032	0.028	0.015	0.001	−0.008	−0.013	−0.015	−0.016	−0.016	−0.016
0.960	0.029	0.028	0.027	0.025	0.016	0.005	−0.005	−0.012	−0.016	−0.018	−0.019	−0.019
\multicolumn{13}{c}{$\eta = 0.25\,b$}											
0.040	−1.097	−1.021	−0.817	−0.525	0.054	0.210	0.216	0.192	0.170	0.156	0.151	0.151
0.210	−0.952	−0.926	−0.453	0.023	0.021	0.069	0.100	0.112	0.112	0.110	0.109	0.109
0.250	−0.233	−0.227	−0.212	−0.184	−0.081	0.009	0.060	0.083	0.091	0.092	0.092	0,092
0.290	0.483	0.468	0.028	−0.391	−0.180	−0.050	0.020	0.054	0.068	0.074	0.075	0.075
0.500	0.445	0.412	0.322	0.191	−0.073	−0.120	−0.092	−0.059	−0.036	−0.024	−0.020	−0.020
0.750	0.233	0.227	0.212	0.184	0.081	−0.009	−0.060	−0.083	−0.091	−0.092	−0.092	−0.092
0.960	0.191	0.188	0.179	0.164	0.098	−0.023	−0.038	−0.078	−0.099	−0.109	−0.112	−0.112
\multicolumn{13}{c}{$\eta = 0.50\,b$}											
0.040	−0.469	−0.458	−0.425	−0.368	−0.159	0.019	0.122	0.166	0.181	0.183	0.183	0.183
0.250	−0.635	−0.598	−0.501	−0.354	−0.025	0.095	0.129	0.136	0.136	0.135	0.133	0.133
0.460	−0.702	−0.683	−0.229	0.213	0.093	0.048	0.033	0.028	0.025	0.024	0.024	0.023
0.500	0.000	0.000	0.000	0.000	0.000	0.000	0.000	0.000	0.000	0.000	0.000	0.000
0.540	0.702	0.683	0.229	−0.213	−0.093	−0.048	−0.033	−0.028	−0.025	−0.024	−0.024	−0.023
0.750	0.635	0.598	0.501	0.354	0.025	−0.095	−0.129	−0.136	−0.136	−0.135	−0.133	−0.133
0.960	0.469	0.458	0.425	0.368	0.159	−0.019	−0.122	−0.166	−0.181	−0.183	−0.183	−0.183

Appendix

Table A.17 (continued)

$\xi = 0.25\,a$												
x/a \ y/b	0.000	0.040	0.125	0.210	0.250	0.290	0.375	0.500	0.625	0.750	0.875	0.960	1.000

η = a

0.000	0.191	0.201	0.286	—0.653	—1.565	—0.660	0.259	0.113	0.064	0.045	0.037	0.035	0.035
0.040	0.155	0.158	0.149	—0.436	—0.773	—0.443	0.125	0.095	0.060	0.043	0.036	0.034	0.034
0.080	0.071	0.063	—0.078	—0.072	—0.224	—0.077	—0.092	0.051	0.048	0.039	0.034	0.032	0.032
0.250	—0.087	—0.083	—0.034	0.079	0.105	0.085	—0.014	—0.036	—0.007	0.008	0.013	0.015	0.015
0.500	—0.001	0.000	0.010	0.020	0.022	0.022	0.012	—0.005	—0.012	—0.012	—0.010	—0.009	—0.009
0.750	0.015	0.015	0.015	0.014	0.013	0.012	0.007	0.000	—0.008	—0.012	—0.015	—0.015	—0.015
0.960	0.016	0.016	0.015	0.013	0.012	0.011	0.007	0.000	—0.007	—0.012	—0.015	—0.016	—0.016

η = 0.25 b

0.040	0.078	0.054	—0.142	—0.438	—0.492	—0.444	—0.155	0.124	0.186	0.184	0.174	0.170	0.169
0.210	0.018	0.021	0.050	—0.424	—0.883	—0.417	0.072	0.065	0.089	0.105	0.111	0.112	0.112
0.250	—0.080	—0.081	—0.088	—0.092	—0.089	—0.081	—0.051	0.006	0.052	0.077	0.088	0.091	0.091
0.290	—0.174	—0.180	—0.226	0.238	0.704	0.254	—0.175	—0.053	0.014	0.048	0.064	0.068	0.069
0.500	—0.086	—0.073	0.030	0.171	0.199	0.183	0.062	—0.061	—0.073	—0.056	—0.041	—0.036	—0.036
0.750	0.080	0.081	0.088	0.092	0.089	0.081	0.051	—0.006	—0.052	—0.077	—0.088	—0.091	—0.091
0.960	0.098	0.098	0.094	0.085	0.077	0.068	0.043	—0.001	—0.044	—0.076	—0.095	—0.099	—0.100

η = 0.50 b

0.040	—0.156	—0.159	—0.176	—0.185	—0.179	—0.163	—0.104	0.013	0.103	0.154	0.175	0.181	0.181
0.250	—0.013	—0.025	—0.125	—0.256	—0.276	—0.251	—0.106	0.062	0.116	0.131	0.135	0.136	0.136
0.460	0.088	0.093	0.135	—0.328	—0.789	—0.330	0.126	0.057	0.035	0.028	0.026	0.025	0.025
0.500	0.000	0.000	0.000	0.000	0.000	0.000	0.000	0.000	0.000	0.000	0.000	0.000	0.000
0.540	—0.088	—0.093	—0.135	0.328	0.789	0.330	—0.126	—0.057	—0.035	—0.028	—0.026	—0.025	—0.025
0.750	0.013	0.025	0.125	0.256	0.276	0.251	0.106	—0.062	—0.116	—0.131	—0.135	—0.136	—0.136
0.960	0.156	0.159	0.176	0.185	0.179	0.163	0.104	—0.013	—0.103	—0.154	—0.175	—0.181	—0.181

Table A.17 (continued)

$\xi = 0.50\,a$

y/b \ x/a	0.000	0.040	0.125	0.250	0.375	0.460	0.500	0.540	0.625	0.750	0.875	0.960	1.000
$\eta = d$													
0.000	0.061	0.062	0.071	0.113	0.252	0.673	−1.581	−0.673	0.252	0.113	0.071	0.062	0.061
0.040	0.058	0.059	0.066	0.095	0.119	−0.454	−0.788	−0.454	0.119	0.095	0.066	0.059	0.058
0.080	0.051	0.051	0.053	0.051	−0.097	−0.086	0.213	−0.086	−0.097	0.051	0.053	0.051	0.051
0.250	−0.001	−0.002	−0.009	−0.036	−0.012	0.091	0.113	0.091	−0.012	−0.036	−0.009	−0.002	−0.001
0.500	−0.018	−0.018	−0.016	−0.005	0.016	0.027	0.029	0.027	0.016	−0.005	−0.016	−0.018	−0.018
0.750	−0.009	−0.008	−0.007	0.000	0.007	0.009	0.009	0.009	0.007	0.000	−0.007	−0.008	−0.009
0.960	−0.005	−0.005	−0.004	0.000	0.004	0.005	0.005	0.005	0.004	0.000	−0.004	−0.005	−0.005
$\eta = 0.25\,b$													
0.040	0.217	0.216	0.205	0.124	−0.173	−0.473	−0.525	−0.473	−0.173	0.124	0.205	0.216	0.217
0.210	0.102	0.100	0.090	0.065	0.071	−0.416	−0.880	−0.416	0.071	0.065	0.090	0.100	0.102
0.250	0.061	0.060	0.047	0.006	−0.047	−0.070	−0.073	−0.070	−0.047	0.006	0.047	0.060	0.061
0.290	0.022	0.020	0.004	−0.053	−0.166	0.273	0.730	0.273	−0.166	−0.053	0.004	0.020	0.022
0.500	−0.092	−0.092	−0.090	−0.061	0.079	0.213	0.235	0.213	0.079	−0.061	−0.090	−0.092	−0.092
0.750	−0.061	−0.060	−0.047	−0.006	0.047	0.070	0.073	0.070	0.047	−0.006	−0.047	−0.060	−0.061
0.960	−0.039	−0.038	−0.028	−0.001	0.028	0.039	0.041	0.039	0.028	−0.001	−0.028	−0.038	−0.039
$\eta = 0.50\,b$													
0.040	0.124	0.122	0.094	0.013	−0.095	−0.143	−0.149	−0.143	−0.095	0.013	0.094	0.122	0.124
0.250	0.130	0.129	0.118	0.062	−0.107	−0.251	−0.275	−0.251	−0.107	0.062	0.118	0.129	0.130
0.460	0.033	0.033	0.037	0.057	0.124	−0.334	−0.795	−0.334	0.124	0.057	0.037	0.033	0.033
0.500	0.000	0.000	0.000	0.000	0.000	0.000	0.000	0.000	0.000	0.000	0.000	0.000	0.000
0.540	−0.033	−0.033	−0.037	−0.057	−0.124	0.334	0.795	0.334	−0.124	−0.057	−0.037	−0.033	−0.033
0.750	−0.130	−0.129	−0.118	−0.062	0.107	0.251	0.275	0.251	0.107	−0.062	−0.118	−0.129	−0.130
0.960	−0.124	−0.122	−0.094	−0.013	0.095	0.143	0.149	0.143	0.095	−0.013	−0.094	−0.122	−0.124

Appendix

Appendix

Table A.18

TORSIONAL MOMENTS M_{xy} — COEFFIENTS k_M^{xy}

$$M_{xy} = \frac{M}{a} k_M^{xy} \qquad \overline{\xi = c} \qquad \begin{array}{l} b = a \\ c = d = 0.06\,a \\ \mu = 0 \end{array}$$

x/a \ y/b	0.000	0.060	0.120	0.125	0.250	0.375	0.500	0.625	0.750	0.875	0.940	1.000
η = d												
0.000	−0.901	−0.880	−0.277	−0.124	0.257	0.144	0.091	0.066	0.052	0.045	0.044	0.043
0.060	−0.614	−0.539	−0.237	−0.203	0.143	0.115	0.082	0.061	0.050	0.044	0.042	0.042
0.120	−0.127	−0.059	−0.145	−0.200	−0.046	0.050	0.057	0.050	0.044	0.040	0.038	0.038
0.250	0.166	0.126	0.015	0.005	−0.111	−0.048	−0.003	0.016	0.022	0.024	0.024	0.024
0.500	0.083	0.076	0.056	0.054	0.002	−0.025	−0.027	−0.020	−0.013	−0.009	−0.008	−0.007
0.750	0.050	0.048	0.042	0.042	0.022	−0.002	−0.012	−0.020	−0.023	−0.024	−0.024	−0.024
0.940	0.042	0.041	0.038	0.037	0.024	−0.007	−0.007	−0.018	−0.024	−0.027	−0.028	−0.028
η = 0.25 b												
0.060	−0.965	−0.853	−0.557	−0.529	0.032	0.193	0.206	0.187	0.167	0.153	0.150	0.149
0.190	−0.720	−0.694	−0.349	−0.268	0.048	0.092	0.118	0.124	0.121	0.118	0.116	0.116
0.250	−0.224	−0.213	−0.182	−0.179	−0.082	0.005	0.058	0.082	0.090	0.092	0.092	0.092
0.310	0.265	0.262	−0.017	−0.091	−0.206	−0.076	−0.001	0.038	0.056	0.063	0.065	0.065
0.500	0.387	0.336	0.208	0.195	−0.055	−0.113	−0.090	−0.059	−0.037	−0.024	−0.021	−0.020
0.750	0.224	0.213	0.182	0.179	−0.082	−0.005	−0.058	−0.082	−0.090	−0.092	−0.092	−0.092
0.940	0.187	0.181	0.163	0.161	0.097	−0.023	−0.037	−0.076	−0.098	−0.108	−0.110	−0.111
η = 0.50 b												
0.060	−0.455	−0.431	−0.366	−0.359	−0.158	0.016	0.117	0.162	0.178	0.181	0.181	0.181
0.250	−0.572	−0.516	−0.370	−0.355	−0.043	0.088	0.126	0.135	0.135	0.133	0.133	0.132
0.440	−0.470	−0.461	−0.157	−0.082	0.117	0.068	0.049	0.042	0.038	0.036	0.036	0.035
0.500	0.000	0.000	0.000	0.000	0.000	0.000	0.000	0.000	0.000	0.000	0.000	0.000
0.560	0.470	0.461	0.157	0.082	−0.117	−0.068	−0.049	−0.042	−0.038	−0.036	−0.036	−0.035
0.750	0.572	0.516	0.370	0.355	0.043	−0.088	−0.126	−0.135	−0.135	−0.133	−0.133	−0.132
0.940	0.455	0.431	0.366	0.359	0.158	−0.016	−0.117	−0.162	−0.178	−0.181	−0.181	−0.181

Table A.18 (continued)

$\xi = 0.25\,a$

y/b \ x/a	0.000	0.060	0.125	0.190	0.250	0.310	0.375	0.500	0.625	0.750	0.875	0.940	1.000
\multicolumn{14}{c}{$\eta = d$}
0.000	0.245	0.257	0.252	−0.397	−1.013	−0.413	0.216	0.148	0.091	0.066	0.055	0.052	0.051
0.060	0.157	0.143	0.054	−0.260	−0.482	−0.272	0.027	0.103	0.078	0.061	0.052	0.050	0.049
0.120	−0.011	−0.046	−0.158	−0.034	0.166	−0.037	−0.162	0.016	0.048	0.048	0.045	0.044	0.043
0.250	−0.122	−0.111	−0.050	0.075	0.148	0.088	−0.020	−0.050	−0.010	0.012	0.020	0.022	0.022
0.500	−0.002	0.002	0.014	0.027	0.033	0.029	0.018	−0.008	−0.018	−0.017	−0.014	−0.013	−0.013
0.750	0.022	0.022	0.022	0.021	0.020	0.016	0.011	−0.001	−0.011	−0.019	−0.022	−0.023	−0.023
0.940	0.024	0.024	0.023	0.021	0.018	0.015	0.010	0.000	−0.010	−0.018	−0.023	−0.024	−0.025
\multicolumn{14}{c}{$\eta = 0.25\,b$}
0.060	0.079	0.032	−0.134	−0.376	−0.494	−0.383	−0.144	0.122	0.179	0.178	0.169	0.167	0.166
0.190	0.048	0.048	0.032	−0.304	−0.613	−0.298	0.047	0.085	0.106	0.118	0.121	0.121	0.121
0.250	−0.079	−0.082	−0.088	−0.092	−0.088	−0.075	−0.051	0.006	0.051	0.076	0.088	0.090	0.091
0.310	−0.200	−0.206	−0.206	0.119	0.433	0.145	−0.150	−0.071	−0.003	0.035	0.052	0.056	0.057
0.500	−0.082	−0.055	0.031	0.140	0.192	0.156	0.062	−0.059	−0.071	−0.055	−0.041	−0.037	−0.035
0.750	0.079	0.082	0.088	0.092	0.088	0.075	0.051	−0.006	−0.051	−0.076	−0.088	−0.090	−0.091
0.940	0.098	0.097	0.094	0.087	0.077	0.063	0.044	−0.001	−0.044	−0.075	−0.094	−0.098	−0.100
\multicolumn{14}{c}{$\eta = 0.50\,b$}
0.060	−0.152	−0.158	−0.173	−0.185	−0.180	−0.153	−0.103	0.013	0.103	0.152	0.173	0.178	0.179
0.250	−0.016	−0.043	−0.125	−0.227	−0.269	−0.219	−0.105	0.060	0.115	0.130	0.135	0.135	0.136
0.440	0.111	0.117	0.115	−0.209	−0.516	−0.214	0.103	0.074	0.051	0.042	0.039	0.038	0.038
0.500	0.000	0.000	0.000	0.000	0.000	0.000	0.000	0.000	0.000	0.000	0.000	0.000	0.000
0.560	−0.111	−0.117	−0.115	0.209	0.516	0.214	−0.103	−0.074	−0.051	−0.042	−0.039	−0.038	−0.038
0.750	0.016	0.043	0.125	0.227	0.269	0.219	0.105	−0.060	−0.115	−0.130	−0.135	−0.135	−0.136
0.940	0.152	0.158	0.173	0.185	0.180	0.153	0.103	−0.013	−0.103	−0.152	−0.173	−0.178	−0.179

Appendix

Table A.18 (continued)

$\xi = 0.50\,a$

x/a \ y/b	0.000	0.060	0.125	0.250	0.375	0.440	0.500	0.560	0.625	0.750	0.875	0.940	1.000
η = a														
0.000	0.089	0.091	0.101	0.148	0.206	−0.429	−1.037	−0.429	0.206	0.148	0.101	0.091	0.089
0.060	0.080	0.082	0.087	0.103	0.019	−0.286	−0.503	−0.286	0.019	0.103	0.087	0.082	0.080
0.120	0.058	0.057	0.053	0.016	−0.167	−0.045	0.157	−0.045	−0.167	0.016	0.053	0.057	0.058
0.250	0.000	−0.003	−0.013	−0.050	−0.017	0.095	0.160	0.095	−0.017	−0.050	−0.013	−0.003	0.000
0.500	−0.028	−0.027	−0.024	−0.008	0.024	0.038	0.043	0.038	0.024	−0.008	−0.024	−0.027	−0.028
0.750	−0.013	−0.012	−0.010	−0.001	0.010	0.013	0.014	0.013	0.010	−0.001	−0.010	−0.012	−0.013
0.940	−0.008	−0.007	−0.006	0.000	0.006	0.007	0.008	0.007	0.006	0.000	−0.006	−0.007	−0.008
η = 0.25 b														
0.060	0.209	0.206	0.196	0.122	−0.161	−0.408	−0.524	−0.408	−0.161	0.122	0.196	0.206	0.209
0.190	0.120	0.118	0.110	0.085	−0.043	−0.302	−0.615	−0.302	−0.043	0.085	0.110	0.118	0.120
0.250	0.061	0.058	0.047	0.006	−0.047	−0.066	−0.073	−0.066	−0.047	0.006	0.047	0.058	0.061
0.310	0.004	−0.001	−0.015	−0.071	−0.139	0.165	0.464	0.165	−0.139	−0.071	−0.015	−0.001	0.004
0.500	−0.091	−0.090	−0.089	−0.059	0.079	0.183	0.228	0.183	0.079	−0.059	−0.089	−0.090	−0.091
0.750	−0.061	−0.058	−0.047	−0.006	0.047	0.066	0.073	0.066	0.047	−0.006	−0.047	−0.058	−0.061
0.940	−0.040	−0.037	−0.029	−0.001	0.029	0.038	0.041	0.038	0.029	−0.001	−0.029	−0.037	−0.040
η = 0.50 b														
0.060	0.124	0.117	0.096	0.013	−0.095	−0.137	−0.151	−0.137	−0.095	0.013	0.096	0.117	0.124
0.250	0.129	0.126	0.116	0.060	−0.107	−0.220	−0.267	−0.220	−0.107	0.060	0.116	0.126	0.129
0.440	0.048	0.049	0.053	0.074	0.100	0.218	0.522	0.218	0.100	0.074	0.053	0.049	0.048
0.500	0.000	0.000	0.000	0.000	0.000	0.000	0.000	0.000	0.000	0.000	0.000	0.000	0.000
0.560	−0.048	−0.049	−0.053	−0.074	−0.100	0.218	0.522	0.218	−0.100	−0.074	−0.053	−0.049	−0.048
0.750	−0.129	−0.126	−0.116	−0.060	0.107	0.220	0.267	0.220	0.107	−0.060	−0.116	−0.126	−0.129
0.940	−0.124	−0.117	−0.096	−0.013	0.095	0.137	0.151	0.137	0.095	−0.013	−0.096	−0.117	−0.124

References

1 Theory of elasticity. Theory of structures. Theory of plates

1. Nádai, A. — *Die elastischen Platten*, Julius Springer, Berlin, 1925 (Nachdruck 1968).
2. Marcus, H. — *Die Theorie elastischer Gewebe und ihre Anwendung auf die Berechnung biegsamer Platten*, Julius Springer, Berlin, 1932 (1924).
3. Galerkin, B. G. — *Uprugie tonkie pliti*, Gosstroizdat, Moskow, 1933 (Sobranie socinenyi, 1953).
4. Love, A. E. H. — *A Treatise on the Mathematical Theory of Elasticity*, Dover Publications, New York, 1944 (Cambridge University Press, 1934).
5. Westergaard, H. M. — *Theory of Elasticity and Plasticity*, Harvard University Press, 1952.
6. Timoshenko, S P., Goodier, J. N. — *Theory of Elasticity*, 2nd edition, McGraw-Hill Book Company, New York, Toronto, London, 1954 (1936); Béranger, Paris, Liège, 1961 (1948).
7. Timoshenko, S. P., Woinowski-Krieger, S. — *Theory of Plates and Shells*, 2nd edition, McGraw-Hill Book Company, New York, London, 1959 (1940); Béranger, Paris, 1961 (1951); Editura tehnică, București, 1968.
8. Mazilu, P. — *Statica construcțiilor (II)*, Editura tehnică, București, 1959.
9. Girkmann, K. — *Flächentragwerke*, 6. Auflage, Springer Verlag, Wien, 1963 (1959, 1956, 1946).
10. Ping-Chun Wang — *Numerical and Matrix Methods in Structural Mechanics*, John Wiley & Sons, New York, London, Sidney, 1966; Editura tehnică, București, 1970.
11. Zienkiewicz, O. C. — *The Finite Element Method in Engineering Science*, 2nd edition, Mc Graw-Hill Book Company, London, 1971.
12. Massonnet, Ch., Deprez, G., Maquoi, R. a.o. — *Calcul des structures sur ordinateur*, Eyrolles et Masson, Paris, 1972 (1968); Editura tehnică, București, 1974.
13. Ifrim, M. — *Analiza dinamică a structurilor și inginerie seismică*, Editura didactică și pedagogică, București, 1973 (1962).
14. Frey, F., Rondal, J. — Fourneaux, N. — *Application de la méthode des éléments finis à la résolution des problèmes de structure*, Université de Liège, Centre d'études, de recherches et d'essais scientifiques du génie civil, Liège, 1974, **48,** 7.
15. Gheorghiu, Al. — *Statica, stabilitatea și dinamica construcțiilor*, Editura didactică și pedagogică, București, 1974 (1968).
16. Szilard, R. — *Theory and Analysis of Plates*, Prentice-Hall, Englewood Cliffs, New Jersey, London, Toronto, Tokyo, 1974.
17. Clough, R. W., Penzien, J. — *Dynamics of Structures*, McGraw-Hill Book Company, New York, 1975.
18. Sandi, H. — *Metode matriceale in mecanica structurilor*, Editura tehnică, București, 1975.

References

19. Stiglat, K., Wippel, H. — *Platten*, 3. Auflage, Wilhelm Ernst & Sohn, Berlin, München, 1983 (1973, 1966).
20. Voinea, R. a.o. — *Elasticitate și plasticitate (I, II)*, Institutul Politehnic București, 1976.
21. Soare, M. — *Plăci plane. Manual pentru calculul construcțiilor (I)*. Section VI, Editura tehnică, București, 1977 (1959).
22. Gheorghiu, A. — *Statica construcțiilor (II, III)*, Editura tehnică, București, 1965, 1980.
23. Teodorescu, P. P., Ille, V. — *Teoria elasticității și introducere în mecanica solidelor deformabile (I, II, III)*, Editura Dacia, Cluj-Napoca, 1976, 1979, 1980 (1961, 1966).
24. Stiglat, K., Wippel, H. — *Massive Platten. Besondere Kapitel der Schnittkraftermittlung und Bemessung* — Beton-Kalender (I), Wilhelm Ernst & Sohn, Berlin, München, Düsseldorf, 1971, 1973, 1975, 1977, 1979, 1981, 1983.

2 Flat slabs. Elastic analysis of slab structures

2.1 Classical literature. Flat slabs with capitals

25. Westergaard, H., Slater, W. —Moments and Stresses in Slabs, "Proceedings of the American Concrete Institute", 1921, **XVII**, *12*, 415.
26. Nádai, A. — Über die Biegung durchlaufender Platten und der rechteckigen Platten mit freien Rändern, "Zeitschrift für angewandte Mathematik und Mechanik", Berlin, 1922, **II**, *1*, 1 ("Der Bauingenieur", 1921, **II**, *11)*.
27. Lewe, V. — Strenge Lösungen der elastischen Probleme endlich ausgedehnter Pilzdecken und anderer Platten vermittels Fourierscher Reihen, "Der Bauingenieur", Berlin, 1922, **III**, *11* (1922, **III**, *4*, 111 ; 1920, **I**, *22*, 631 and "Beton und Eisen", 1915, **VII**, *8)*.
28. Lewe, V. — *Pilzdecken und andere trägerlose Eisenbetondecken*, Julius Springer, Berlin, 1929 (1926).
29. Turneaure, F., Maurer, E. — *Principles of Reinforced Concrete Constructions*, 4th edition, John Wiley & Sons, New York, 1932 (1924).
30. Tölke, F. — Über Spannungszustände in dünnen Rechteckplatten, *Ingenieur Archiv*, Berlin, 1934, **V**, 187.
31. Woinowsky-Krieger, S. — Beitrag zur Theorie der Pilzdecken, "Zeitschrift für angewandte Mathematik und Mechanik", Berlin, 1934, **XIV**, *1*, 13.
32. Marcus, H. — Caractéristique et méthode de calcul des planchers-champignons, "La Technique moderne", Paris, 1937, **10** ("Beton und Eisen", Berlin, 1933, **14** and 1925, **15**; "Der Bauingenieur", 1924, 20).
33. Grein, K. — *Pilzdecken. Theorie und Berechnung*, 3. Auflage, Wilhelm Ernst & Sohn, Berlin, 1948 (1941).
34. Müller, W. — Die Momentenfläche der elastischen Platte oder Pilzdecke und die Bestimmung der Durchbiegung aus den Momenten, *"Ingenieur Archiv*, Berlin, 1953, **21**, 63 (1952, **20**, 278).
35. Woinowsky-Krieger, S. — On Bending of a Flat Slab Supported by Square-shaped Columns and Clamped, "Journal of Applied Mechanics", 1954, **XXI**, *9*, 263.

2.2 Modern literature. Flat slabs without capitals

36. Devars du Mayne, R., Turin, M. a.o. — Les planchers-dalles sans champignons, *Annales de l'Institut technique du bâtiment et des travaux publics*, Paris, 1951, **1**.

References

37. Escher, A. — Pilzdecken ohne Pilz, "Schweizeriche Bauzeitung", Zürich, 1960, **78**, *18*.
38. Duddeck, H., Berger, F., von Gunten, H. — Praktische Berechnung der Pilzdecken ohne Sützenkopfverstärkung (Flachdecke), "Beton- und Stahlbetonbau", Berlin, 1963, **58**, *3*, 56.
39. Rabe, J. — *Beitrag zur Berechnung der Eckfelder von Flachdecken mit Randträgern*, Dissertation, Technische Hochschule Karlsruhe, 1963.
40. Sozen, M. A., Siess, C. P. — Investigation of Multiple-panel Reinforced Concrete Floor Slabs. Design Methods. Their Evolution and Comparison, "Journal of the American Concrete Institute", Detroit, 1963, **35**, *8*, 999.
41. Bretthauer, G., Seiler, F. — Platten mit einen und zwei freien Rändern unter beliebiger Last (I, II), "Beton- und Stahlbetonbau", Berlin, 1964, **59**, *6*, 137 and *9*, 205.
42. Andersson, J. L. — Inspänningsmoment i kantpelare vid plattor utan kantbalkar, "Nordisk Betong", Stokholm, 1965, **1**, *61*.
43. Mehmel, A. — Flachdecken — *Vorträge Betontag*, Verlag Deutscher Betonverein, 1965, 360.
44. Baader, W., Krebs, A. — Ein Beitrag zur Untersuchung des freien punktgestützten Randes von Flachdecken unter besonderer Berücksichtigung einer Deckenauskragung, *Mitteilungen aus dem Institut für Massivbau*, Technische Hochschule Darmstadt, 1966, 13 (Baader, W. — Dissertation, Darmstadt, 1965).
45. Bretthauer, G., Seiler, F. — Die Plizdecke ohne verstärkte Säulenköpfe (Flachdecke) bei verschiedenen Randbedingungen, "Beton- und Stahlbetonbau", Berlin, 1966, **61**, *9*, 229 and *11*, 279.
46. Nisenbaum, M. I. — Les moments fléchissants dans les planchers-dalles à panneaux rectangulaires des immeubles d'habitation, *Comité Européen du Béton, Bulletin d'information*, Paris, 1966, **56**, *8*.
47. Franz, G., Rabe, J. — Der Stützenbereich von Flachdecken (Flat Plates) ans Stahlbeton, *Comité Européen du Béton, Bulletin d'information*, Paris, 1966, **58**, *1* ("Beton- und Stahlbetonbau", 1965, **60**, *1* and 1964, **59**, *6*).
48. Pfaffinger, D., Thürliman, B. — *Tabellen für unterzugslose Decken*, Verlags-AG der akademischen technischen Vereine, Zürich, 1967.
49. Bretthauer, G., Nötzold, F. — Zur Berechnung von Pilzdecken, "Beton- und Stahlbetonbau", Berlin, 1968, **63**, *10*, 221 ; *11*, 251 and *12*, 277.
50. Davidovici, V. E., Jalil, W. A. — Planchers-dalles, *Annales de l'Institut Technique du Bâtiment et des Travaux Publiques*, Paris, 1969, **264**, *12*, 1909.
51. Rabe, J. — Ein neunfeldriges, randträgerversteiftes Flachdecken-system unter gleichmässig verteilter Vollast, Theorie und Praxis des Stahlbetonbaues, *Festschrift Franz*, Wilhelm Ernst & Sohn, Berlin, München, 1969, 143.
52. Szerdahely, D. — Beeinflussung der Momentenverteilung durch Aussparungen in einer Flachdecke, "Schweizerische Bauzeitung", Zürich, 1970, *8*, 147.
53. Pfaffinger, D., Heck, B., Pestalozzi, U. — *Statische Analyse unterzugsloser Decken*, 2 Version (Programm SAUD 2), FIDES Rechenzentrum, Zürich, 1972.
54. Reiss, M., Sokal, J. — Design of Ribbed Flat Slabs, "The Structural Engineer," London, 1972, **50**, *8*, 303.
55. Glahn, H., Trost, H. — Zur Berechnung von Pilzdecken, "Der Bauingenieur", Berlin, 1974, **49**, *4*, 122 (Glahn, H. — Dissertation, Aachen, 1973).
56. Agent, R. — Construcții din beton armat (II). Chap. 9. Plăci cu reazeme concentrate, Institutul de Construcții București, 1979.
57. Andrä, H. P., Baur, H., Stiglat, K. — Zum Tragverhalten, Konstruieren und Bemessen von Flachdecken *(I, II, III)*, "Beton- und Stahlbetonbau", Berlin, 1984, **10**, 258 ; **11**, 303 and **12**, 328.

References

2.3 Surface couple. Effective plate width. Elastic analysis of slab structures

58. Mohammed, I. A., Popov, E. P. — Effective Stiffness of a Plate Subjected to a Local Edge Moment, *Proceedings of the Second U.S. National Congress of Applied Mechanics*, 1954, 423 ("Proceedings of the American Society of Civil Engineers", 1954, *80*, 529—1).
59. Tsuboi, Y., Kawaguchi, M. — On Earthquake-Resistant Design of Flat Slabs and Concrete Shell Structures, *Proceedings of the 2nd World Conference on Earthquake Engineering*, Tokyo, 1960; Science Council of Japan, Tokyo, 1960, 1963.
60. Stiglat, K. — Der Plattenstreifen unter dem Angriff von Flächenmomenten (Flachdecken). Mitwirkende Plattenbreiten (I, II), "Die Bautechnik", Berlin, 1963, **40**, *4*, 113, and 1969, **46**, *4*, 127.
61. Blakey, F. A. — The Deflection of Flat Plate Structures, "Civil Engineering and Public Works Review", 1963, *9*.
62. Carpenter, J. E. — *Flexural Characteristics of Flat Plate Floors in Buildings Subjected to Lateral Loads*, PhD Dissertation, Purdue University, 1965.
63. Krebs, A., Kruse, W. — Über die Steifigkeit von Rahmenrigeln in Flachdecken, *Festschrift Mehmel*, Beton Verlag, Düsseldorf, 1967, 115.
64. Neguțiu, R. — Plăci plane acționate prin cupluri de suprafață (I), "Buletinul științific al Institutului de Construcții București", 1969, **XII**, *2*, 175.
65. Neguțiu, R. — *Efectul cuplului de suprafață asupra stării de deformație a plăcilor plane*, Sesiunea științifică jubiliară de mecanică, Universitatea București, December 12—14, 1969.
66. Scholz, G. — Flachdeckenrahmen bei Horizontalbelastung, "Stahlbetonbau, Berichte aus Forschung und Praxis", *Festschrift Rüsch*, Wilhelm Ernst & Sohn, Berlin, München, 1969, 265.
67. Corley, W. G., Jirsa, J. O. — Equivalent Frame Analysis for Slab Design, "Journal of the American Concrete Institute", Detroit, 1970, **67**, *11*, 875 ("Structural Research", Series 217, Department of Civil Engineering, University of Illinois, Urbana, 1961).
68. Neguțiu, R. — Plaques planes actionnées par des couples de surface (II). Étude des déplacements élastiques, "Buletinul științific al Institutului de Construcții București", 1970, **XIII**, *3*, 115.
69. Neguțiu, R. — *Contribuții la studiul structurilor cu planșee-dală*, Sesiunea științifică a cadrelor didactice, Institutul de petrol, gaze și geologie, București, April 16—17, 1971.
70. Neguțiu, R. — Distribution des efforts sectionnels dans les plaques planes actionnées par des couples de surface, "Revue roumaine des sciences techniques — Série de Mécanique appliquée", Bucarest, 1971, **XVI**, *6*, 1307.
71. Pfaffinger, D. — Column-Plate Interaction in Flat Slab Structures, "Journal of the Structural Division", Proceedings of the American Society of Civil Engineers, 1972, **98**, *ST1*, 8670, 307.
72. Aalami, B. — Moment-Rotation Relation between Column and Slab, "Journal of the American Concrete Institute", Detroit, 1972, **69**, *5*, 263.
73. Neguțiu, R. — L'effet des groupements de couples de surface sur les plaques planes, "Revue roumaine des sciences techniques — Série de Mécanique appliquée", Bucarest, 1973, **XVIII**, *1*, 179.
74. Neguțiu, R. — *Modelarea efectului legăturilor rigide la studiul static al structurilor cu planșee-dală*, Sesiunea științifică a cadrelor didactice, Institutul de Construcții București, April 26—28, 1973.
75. Neguțiu, R.— *Efectul legăturilor rigide la calculul structurilor cu planșee-dală*, Doctoral thesis, Institutul de Construcții București, July 7, 1973.
76. Mehrain, M., Aalami, B. — Rotational Stiffness of Concrete Slabs, "Journal of the American Concrete Institute", Detroit, 1974, **71**, *9*, 429.

77. Negruțiu, R. — *Legături rigide de tip bară-placă plană la calculul organelor de mașini și al structurilor de rezistențe metalice, cu sau fără simetrii*, Symposium "Utilaj petrolier — Direcții de modernizare și perfecționare", Institutul de petrol și gaze, Ploiești, November 8—9, 1974.
78. Negruțiu, R. — *Metodă imbunătățită de calcul al structurilor cu planșee-dală, prin introducerea forțelor de legătură-cupluri pe două direcții ale planului*, Symposium "Concepții și metode de calcul pentru optimizarea soluțiilor in domeniul construcțiilor", Institutul de Construcții București, April 25—27, 1975.
79. Pecknold, D. A. — Slab Effective Width for Equivalent Frame Analysis, "Journal of American Concrete Institute", Detroit, 1975, **72**, *4*, 135..
80. Allen, F. H., — *Lateral Load Characterics of Flat Plate Structures*, MSc Thesis, Monash University, Melbourne, 1976 *(Proceedings of the Fifth Australian Conference on the Mechanics of Structures and Materials*, Melbourne, 1975, University of Melbourne and Monash University, 1975, 1).
81. Allen, F., Darvall, P. — Lateral Load Equivalent Frame, "Journal of the American Concrete Institute", Detroit, 1977, **74**, 7, 294.
82. Negruțiu, R. — Plăci plane sub acțiunea unui cuplu de suprafață. Noi modele matematice ale suprafeței mediane deformate, "Buletinul științific al Institutului de Construcții București", 1978, **XXI**, *3—4*, 175.
83. Negruțiu, R. — Introducere in analiza dinamică a structurilor cu dale (I, II), "Studii și cercetări de mecanică aplicată", București, 1980, **39**, *2*, 261 (1980, **39**, *1*, 83).
84. Brotchie, J. F. — Some Australian Research on Flat Plate Structures, "Journal of the American Concrete Institute", Detroit, 1980, **77**, *1*, 3 (1964, **61**, *8*, 959; 1959, **56**, *8* and 1957, **54**, 7, 31).
85. Ioani, A. M.—Calculul spațial al structurilor alcătuite din planșee și stîlpi, Doctoral Thesis, Institutul Politehnic Cluj-Napoca, December 19, 1981.
86. Negruțiu, R. — *Unele aspecte noi in analiza elastică a structurilor cu dale*, Symposium "Structuri spațiale", Institutul Politehnic Cluj-Napoca, May 16—18, 1985.

Selective bibliography on related fields

3 Rectangular plates in the elastic range

3.1 Continuous plates

87. Woinowsky-Krieger, S. — Beitrage zur Theorie der durchlaufenden Platten, *Ingenieur Archiv*, 1938, **IX**, *5*, 396.
88. Bechert, H. — Über die Stützmomente durchlaufender Platten bei vollständiger Lastumordnung, "Die Bautecknik", 1961, **38**, *11*.
89. Morrison, G. D. — *Solutions for nine-panel continuous plates with stiffening beams*, MSC Thesis, Department of Civil Engineering, University of Illinois, 1961.
90. Stiglat, K., Wippel, H. — Punktgestützte Rechteckplatten, "Schweizerische Bauzeitung", Zürich, 1962, **80**, *20*.
91. Woodring, R. E., Siess, C. P. — An Analytical Study of the Moments in Continuous Slabs Subjected to Concentrated Loads "Civil Engineering Studies", Structural Research, Series 264, University of Illinois, Urbana, 1963, 5.

References

92. Lang, E. — Die Schnittmomente der durchlaufenden Platte mit gleichen Feldweiten, ermittelt mit Hilfe des Superpositionsgesetzes, "Die Bautechnik", 1965, **42**, 12.
93. Pieper, K., Martens, P. — Durchlaufende vierseitig gestützte Platten im Hoch bau, "Beton- und Stahlbetonbau", Berlin, 1966, **61**, 6.
94. Franz, G. — Neue Erkenntnisse über rand- und punktgestützten Platten, Österreichischer Beton-Verein, 1967, 38.
95. Hees, G. — Balken und Platten unter Momentenbelastung, "Die Bautechnik", Berlin, 1968, **45**, 4.
96. Stiglat, K. — Einfluss von biege- und torsionssteifen Unterstützungen bei Platten, "Die Bautechnik", Berlin, 1970, *10*, 355 and *11*, 381.
97. Zöphel, J. — Der Plattenstreifen mit unterbrochener Mittels-stützung unter Gleichlast, "Die Bautechnik", Berlin, 1972, **49**, 5, 169.
98. Zöphel, J. — Plattenstrefen mit äquidistanten Querwänden (Stichwänden) unter Gleichlast, "Die Bautechnik", Berlin, 1973, **50**, 6, 208.
99. Eisenbiegler, G. — Rechteckplatten mit randparallelen Liniengelenken, "Die Bautechnik", Berlin, 1974, **51**, 4, 133 (1973, **50**, 3, 92).
100. Eisenbiegler, G. — Stützenquerkräfte isotropen Zweifeldplatten infolge von Punkt-, Linien- und Rechtecklasten, "Beton- und Stahlbetonbau", Berlin, 1978, *9*, 215 and *10*, 252.

3.2 Cantilever plates. Clamped plates

101. Girkmann, K., Tungl, E. — Näherungsweise Berechnung eingespannter Rechteckplatten, "Österreichische Bauzeitschrift", Wien, 1953, **VIII**, 3, 47.
102. Lardy, P. — Sur une méthode nouvelle de résolution du problème des dalles rectangulaires encastrées, *Association Internationale des Ponts et Charpentes, Mémoires*, Zürich, 1953, XIII.
103. Neurwirth, H. — Berüksichtigung teilweiser Randeinspannung bei der Untersuchung von Rechteckplatten mittels Differenzenrechnung, "Beton- und Stahlbetonbau", Berlin, 1958, **53**, 11.
104. Mehmel, A. — Zweiseitig über Eck gelagerte Platten, *Vortrag auf dem Deutschen Betontag*, München, 1959.
105. Stiglat, K., Wippel, H. — Die an zwei benachbarten Rändern eingespannte (gelagerte) Platte unter Gleichlast, "Beton- und Stahlbetonbau", Berlin, 1959 **54**, 7 and 1960, **55**, 4.
106. Salvadori, G. M., Reggini, H. C. — Simply Supported Corner Plate, *Proceedings of the American Society of Civil Engineers*, Journal of the Structural Division", 1960, **86**, ST 11.
107. Bechert, H. — Die vierseitig starr eingespannte Platte auf elastischen Trägern unter Gleichlast, "Beton- und Stahlbetonbau", Berlin, 1961, **56**, 1.
108. Franz, G. — Um eine Ecke laufende Kragplatten, "Beton- und Stahlbetonbau", Berlin, 1968, **63**, 3, 64.
109. Stiglat, K. — Vierseitig gelagerte Platten mit Tielrandmoment (Kragplatten), "Die Bautechnik", 1968, **45**, 3, 97 and 4, 127.
110. Werner, H., Schikora, K. — Der eingespannte Plattenstreifen mit einem freien Randstück unter Gleichlast, "Der Bauingenieur", Berlin, 1969, **46**, 7, 258.
111. Davies, J. — Analysis of Corner Supported Rectangular Slabs, "The Structural Engineer", London, 1970, **48**, 2.
112. Ludwig, W. — Berechnung der um eine Ecke laufenden Kragplatten nach dem Differenzverfahren, "Beton- und Stahlbetonbau", Berlin, 1971, **66**, 6, 55.
113. Zöphel, J. — Die Quadratplatte mit teilweise frei drehbaren und teilweise starr eingespannten Rändern unter Gleichlast, "Die Bautechnik", Berlin, 1974, **51**, 9.

3.3 Elastic-edge plates. Free-edge plates

114. Fuchs. S. J. — Plates with boundary conditions of elastic support, "Proceedings of the American Society of Civil Engineers", 1953, **79**, *9*, 199.
115. Olsen, H., Reinitzhuber, F. — *Die zweiseitig gelagerte Plate* (I, II), Wilhelm Ernst & Sohn, Berlin, 1959 (1951, 1944).
116. von Gunten, H. — Platten mit freien Rändern, *Mitteilungen aus dem Institut für Baustatik an der Eidgenössischen Technischen Hochschule Zürich*, Verlag Leemann, Zürich, 1960.
117. Appleton, J. — Reinforced Concrete Floor Slabs on Flexible Beams, *Civil Engineering Studies, Structural Research*, Series 223, University of Illinois, Urbana, 1961.
118. Stiglat, K. — *Rechteckige und Schiefe Platten mit Randbalken*, Wilhelm Ernst & Sohn, Berlin, München, 1962.
119. Ertürk, J. — *Zwei-, drei, und vierseitig gestutzte Rechteckplatten*, Wilhelm Ernst & Sohn, Berlin, München, Düsseldorf, 1965.
120. Stiglat, K. — *Einflussfelder rechteckiger und schiefer Platten mit Randbalken*, Wilhelm Ernst & Sohn, Berlin, München, Düsseldorf, 1965.
121. Werner, H. — *Anwendung der Funktionentheorie auf die Berechnung von Platten mit freier Umrand*, Dissertation, Technische Hochschule Hanover, 1965.
122. Özden, K. — Berechnung einer an einen Rand und zwei Ecken frei drehbar gestützten Rechteckplatte unter gleichmässig verteilter Belastung, "Beton- und Stahlbetonbau", 1968, **63**, *2*.
123. Eibl, J., Ivanyi, G. — Momententafeln für dreiseitig gelagerte Platten bei angreifenden Randmomenten, "Beton- und Stahlbetonbau," Berlin, 1969, **64**, *11*.
124. Molkenthin, A. — *Einflussfelder zweifeldriger Platten mit freien Längsrändern*, Springer Verlag, Berlin, Heidelberg, New York, 1971.

3.4 Plates of variable thickness. Plates with openings

125. Dedič, O. — Spannungszustand einer quadratischen Platte mit kreisrunder Öffnung, Beiträge zur angewandten Mechanik. *Federhofer-Girkmann Festschrift*, Deutige Verlag, Wien, 1950.
126. Höland, G. — Ein Beitrag zur Berechnung örtlich dickerer Platten, "Beton- und Stahlbetonbau", Berlin, 1959, **54**, *3*.
127. Fluhr, N., Ang, N., Siess, C. P. — Theoretical Analysis of the Effects of Openings on the Bending Moments in Square Plates with Fixed Edges, *Civil Engineering Studies, Structural Research*, Series 203, University of Illinois, Urbana, 1960.
128. Buchholz, E. — *Beitrag zur Berechnung der Schnittkräfte und Durchbiegungen von umfangsgelagerten Rechteckplatten mit einer Spannrichtung veränderlichem Querschnitt*, Dissertation, Technische Hochschule Karlsruhe, 1963.
129. Reinizhuber, F., Krug, H. — Platten mit Verstärkungen, "Die Bautechnik", 1963, **40**, *9*.
130. Homberg, H., Ropers, W. — *Fahrbahnplatten mit veränderlicher Dicke*, Springer Verlag, Berlin, Heidelberg, New York, 1965.
131. Frenzel, D. — *Zweiseitig gestützte, quadratische Platten mid einer grösseren Öffnung bei Beanspruchung durch eine gleichmässig verteilte Flächenlast*, Dissertation, Technische Hochschule Braunschweig, 1966.
132. Homberg, H., Ropers, W. — Kragplatten mit veränderlicher Dicke, "Beton- und Stahlbetonbau," Berlin, 1867, **62**, *3*.
133. Zuber, E. — *Ein Beitrag zur Berechnung elastischer Platten mit Rechtecköffnungen*, Dissertation, Technische Hochschule Darmstadt, 1967.
134. Beck, H., Zuber, E. — Näherungsweise Berechnung von Stahlbetonplatten mit Rechtecköffnungen unter Gleichflächenlast, "Die Bautechnik", Berlin, 1969, **42**, *12*, 397.

References

135. Bergfelder, J. — Berechnung von Platten veränderlicher Steifigkeit nach dem Differenzenverfahren, *Konstruktiver Ingeneurbau Berichte*, Vulkan Verlag, Essen, 1969, 4, "Beton- und Stahlbetonbau", 1967, **62**, *12*; Dissertation, Technische Hochschule Hannover, 1966.
136. Buchholz, E. — Umfangsgelagerte Platten mit in einer Spannrichtung veränderlichem Querschnitt, "Die Bautechnik", Berlin, 1971, **48**, *5*, 171, and *6*, 201.
137. Eisenbiegler, G. — Plattentragwerke mit abschnittsweise unterschiedlichen Dicken und versetzten Mittelflächen, "Die Bautechnik", Berlin, 1972, **49**, *12*, 425.
138. Eisenbiegler, G. — *Dreiseitig gelagerte isotrope Rechteckplatten mit linear veränderlicher Dicke*, Abhandlungen, Internationale Vereinigung für Brückenbau und Hochbau, Zürich, 1974, **34**, 2.

3.5 Tables with computation values

139. Sonier, P. — *Tables pour le calcul rationnel des planchers sans nervures et des dalles rectangulaires*, Dunod, Paris, 1929 (1922).
140. Pucher, A. — *Einflussfelder elastischer Platten*, 3. Auflage, Springer Verlag, Wien, New York, 1964 (1958).
141. Rüsch, H. — *Berechnungstafeln für rechtwinklige Fahrbahnplatten von Strassenbrücken*, 6. Auflage. Wilhelm Ernst & Sohn, Berlin, München, Düsseldorf, 1965.
142. Bareš, R. — *Tables pour le calcul des dalles et des parois*, Dunod, Paris, 1969 (Wiesbaden, 1969).
143. Czerny, F. — *Tafeln für vierseitig und dreiseitig gelagerte Rechteckplatten*, Beton-Kalender (I), Wilhelm Ernst & Sohn, 1974 (1972, 1970).

4 Rectangular plates and flat slabs in the plastic range

144. Johansen, K. — *Pladeformler*, Polyteknisk Forening, Köbenhavn, 1956 (1949).
145. László, N. — Planșee-dală fără capitel, "Revista construcțiilor și a materialelor de construcții", București, 1959, *3*.
146. Albigès, M., Frederiksen, M. — Calcul à la rupture des dalles par la théorie de Johansen, *Annales de l'Institut Thechnique du Bâtiment et des Travaux Publics*, Paris, 1960, **54**, 1.
147. Johansen, K. W. — *The Application of the Yield-Line Theory to Calculation of the Flexural Strength of Slabs and Flatslab Floors*, Cement and Concrete Association, London, 1962.
148. Sawczuk, A., Jäger, Th. — *Grenztragfähigkeitstheorie der Platten*, Julius Springer, Berlin, Göttingen, Heidelberg, 1963.
149. Fergusson, M. P. — *Reinforced Concrete Fundamentals with Emphasis on Ultimate Strength*, John Wiley & Sons, New York, 1965.
150. Kwiecinski, M. W. — Yield Criterion for Initially Isotropic Reinforced Slab, "Magazine of Concrete Reserarch", 1965, **XVII**, *51*.
151. Petcu, V. — Mecanisme de rupere exacte și aproximative în calculul plastic al plăcilor dreptunghiulare de beton armat "Studii și cercetări de mecanică aplicată", București, 1966, **XVII**, -*3* (1965, **XVI**, -*1*, *2*, *5* and *6*; "Standardizarea", București, 1962, *4* and *5*).
152. Lenschow, R., Sozen, M. A. — A Yield Criterion for Reinforced Concrete Slabs, "Journal of the American Concrete Institute", Detroit, 1967, **39**, *5*, 266.
153. Petcu, V. — *Calculul structurilor de beton armat in domeniul plastic*, Editura tehnică, București, 1972.

References

5 Code requirements. Standards

154. * * * — *Règles techniques de conception et de calcul des ouvrages et constructions en béton armé, CCBA 68*, Société de diffusion des techniques du bâtiment et de travaux publics, Paris, 1970 (1963).
155. * * * — *Beton- und Stahlbetonbau, Bemessung und Ausführung — DIN 1045*, Deutscher Ausschuss für Stahlbeton, Berlin, Köln, 1972 (1959).
156. * * * — *Normes pour le calcul, la construction et l'exécution des ouvrages en béton, en béton armé et en béton précontraint, SIA 162*, Societété suisse des ingénieurs et des architectes, Zürich, 1972 (1968, 1956).
157. * * * — *Stroitelnyie normy i pravila. Betonnyie i jelezobetonnyie konstruktzii. Normy proektirovania, SNiP II, B. 1—73*, Gostroiizdat, Moskow, 1973 (1962, 1949).
158. * * * — *Code of Practice for the Structural Use of Reinforced Concrete — CP 110*: Part 1: 1972, British Standard Institute, London, 1974 (1969, 1957).
159. * * * — *Construcţii civile şi industriale. Calculul şi alcătuirea elementelor din beton, beton armat şi beton precomprimat — Stas 10107/0—76—*, Oficiul de Stat pentru Standarde, Bucureşti, 1976.
160. * * * — *Building Code Requirements for Reinforced Concrete, ACI Standard 318-77*, American Concrete Institute, Detroit, 1977 (1971, 1963, 1956).
161. * * * — *Recommandations internationales pour le calcul et l'exécution des ouvrages en béton armé (I, II) — CEB—FIP 1978 —*, Comité Européen du Béton, Paris, 1978, (1972, 1964).
162. * * * — *Normativ privind calculul şi alcătuirea planşeelor din beton armat şi beton precomprimat*, Chap. 7, Planşee cu dale groase. Planşee-dală. Planşee-ciuperci. Proiect INCERC, Institutul de Construcţii Bucureşti, 1979.
163. * * * — *Normativ pentru proiectarea antiseismică a construcţiilor — P 100-81*, Institutul central de cercetare, proiectare şi directivare în construcţii, Bucureşti, 1981 (1978, 1970).

Author index

Aalami, B. — [72], [76], 8, 64(2), 69
Agent, R. — [56]
Albigès, M. — [146], 11
Alembert, J. R. d' — 320, 329
Allen, F. H. — [80], [81], 11, 55, 64(2), 69(5), 70(2), 84
Andersson, J. L. — [42], 13, 26, 27, 30, 40(2), 41(3), 42, 43, 44, 51, 83
Andrä, H. P. — [57], 59
Ang, N. — [127], 11
Appleton, J. — [117], 11, 30

Baader, W. — [44], 30
Bareš, R. — [142], 11
Baur, H. — [57], 59
Bechert, H. — [88], [107], 11(2)
Beck, H. — [134], 11
Berger, F. — [38], 30
Bergfelder, J.— [135], 8, 11
Blakey, F. A — [61], 64
Bretthauer, G. — [41], [45], [49], 30(3), 44(3), 45(5), 46, 48(3), 49, 52(2), 54(2), 82, 83
Brotchie, J. F. — [84], 64, 69
Buchholz, E.— [128], [136], 11(2)

Carpenter, J. E. — [62], 64, 69
Clough, R. W. — [17], 319, 320, 325, 328

Corley, W. G. — [67], 64
Czerny, F. — [143], 11

Darvall, P. — [81], 55, 64, 69(4), 70(2), 84
Davidovici, V. E. — [50], 8, 9, 12
Davies, J. — [111], 11
Dedič, O. — [125], 11
Deprez, G. — [12]
Devars du Mayne, R. — [36]
Duddeck, H. — [38], 30

Eibl, J. — [123], 11
Eisenbiegler, G. — [99], [100], [137], [138], 11(4)
Ertürk, J. — [119], 11
Escher, A. — [37]

Fergusson, M. P. — [149], 11
Fluhr, N. — [127], 11
Fourier, J. — 21, 66, 80, 101, 190, 219
Fourneaux, N. — [14], 10
Franz, G. — [47], [94], [108], 11(2), 30(2)
Frederiksen, M. — [146], 11
Frenzel, D. — [131], 11
Frey, F. — [14], 10
Fuchs, S. J. — [114], 11, 30, 32(2), 35

Note : — The number in square brackets denotes the indicated author's references.
— The number in parentheses denotes the number of the indicated author's references on the same page.

Author index

Galerkin, B. G. — [3], 8(2), 9, 12, 13, 356
Germain, S. — 6, 55, 84, 94(2), 100, 137, 356
Gheorghiu, Al. — [15], [22], 5(2), 89(2), 196, 319, 320, 325, 328
Girkmann, K. — [9], [101], 5, 8(3), 9, 11, 12, 13, 22, 23(2), 24, 25(2), 26(4), 27(2), 28(5), 29(8), 30(3), 37, 41, 46, 51(2), 54, 55(5), 56(2), 58, 59, 66, 77, 89, 90, 96(2), 97, 100, 101, 111, 196
Glahn, H. — [55], 261
Goodier, J. N. — [6], 5
Goursat, E. — 45
Grein, K. — [33], 12, 23, 25, 26(2), 46, 51
Gunten, H. von — [38], [116], 11, 30

Heck, B. — [53], 74
Hees, G. — [95], 11
Höland, G. — [126], 11
Homberg, H. — [130], [132], 11(2)
Hooke, R. — 10

Ifrim, M. — [13], 319, 320, 325, 328
Ille, V. — [23], 5
Ioani, A. M. — [85], 10, 64
Ivanyi, G. — [123], 11

Jäger, Th. — [148], 11
Jalil, W. A. — [50], 8, 9 12
Jirsa, J. O. — [67], 64
Johansen, K. — [144], [147], 11(2)

Kawaguchi, H. — [59], 64, 69
Krebs, A. — [44], [63], 30, 64
Krug, H. — [129], 11
Kruse, W. — [63], 64
Kwiecinski, M. W. — [150], 11

Lagrange, L. de — 6, 10, 13, 55, 84, 94(2), 100, 137, 356
Lang, E. — [92], 11
Laplace, P.-S. de — 91
Lardy, P. — [102], 11
Laszlo, N. — [145], 11
Lenschow, R. — [152] 11,

Lévy, M. — 66
Lewe, V. — [27], [28], 8(2), 12(2), 20(4), 21, 22(3), 28, 81, 82
Love, A. E. H. — [4], 5
Ludwig, W. — [112], 8, 11

Maquoi, R. — [12]
Marcus, H. — [2], [32], 8(2), 9(3), 12(4), 31(2), 35(2), 40(2), 82
Martens, P. — [93], 11
Massonnet, Ch. — [12]
Maurer, E. — [29], 12
Maxwell, J. C. — 196(2), 197, 250
Mazilu, P. — [8], 5, 89, 196
Mehmel, A. — [43], [104], 11
Mehrain, M. — [76], 64
Mohammed, I. A. — [58], 54, 56, 57(3), 58(2), 59(4), 61(2), 62, 64(2), 66, 249
Mohr, O. — 250
Molkenthin, A. — [124], 11
Morrison, G. D. — [89], 11, 30, 31
Müller, W. — [34]

Nádai, A. — [1], [26], 8, 12(2), 13(4), 16, 17, 20, 26, 28(2), 29, 82, 196
Navier, L. M. H. — 8, 110, 111, 115
Negruţiu, R. — [64], [65], [68], [69], [70], [73], [74], [75], [77], [78], [82], [83], [86], 4(13), 8, 11(6), 17(2), 30, 44, 48(2), 54(7), 56, 61, 62(2), 63(3), 64(3), 67, 68(2), 69, 70(6), 73(2), 74(2), 84(3), 92(3), 95, 98(2), 99, 101, 103, 104(2), 105(2), 107, 108, 110(4), 112, 114, 115, 116(3), 118(4), 137, 141(5), 152(2), 157(2), 167, 179(2), 185(2), 190, 196(2), 200(4), 201, 203, 206(2), 225(2), 236(6), 242, 249(3), 257, 262(7), 263(5), 265(4), 274(6), 290(2), 298(3), 303, 319(2), 328(3), 334(2), 346
Neuwirth, H. — [103], 8, 11
Niesenbaum, M. I. — [46]
Nötzold, F. — [49], 30, 44, 48, 49, 52, 54(2)

Author index

Olsen, H. — [115], 11
Özden, K. — [122], 11

Pecknold, D. A. — [79], 11, 55, 64(2), 66(7), 67(4), 69(4), 70(3), 84
Penzien, J. — [17], 319, 320, 325, 328
Pestalozzi, U. — [53], 74
Petcu, V. — [151], [153], 11(2)
Pfaffinger, D. — [48], [53], [71], 11, 30, 35(2), 38(2), 40, 70, 74(6), 77(4), 78(3), 79, 80(5), 81(2), 82, 84(3), 90, 110(3), 192, 264, 303, 355
Pieper, K. — [93], 11
Popov, E. P. — [58], 54, 56, 57(3), 58(2), 59(4), 61(2), 62, 64(2), 66, 249
Pucher, A. — [140], 11

Rabe, J. — [39], [47], [51], 30(4), 31(2), 32(3), 33, 35(5), 78, 81, 82
Rayleigh, J. W. S. — 8, 9
Reggini, H. C. — [106], 11
Reinitzhuber, F. — [115], [129], 11(2)
Reiss, M. — [54], 261
Ritz, W. — 8, 356
Rondal, J. — [14], 10
Ropers, W. — [130], [132], 11(2)
Rüsch, H. — [141], 11

Salvadori, G. H. — [106], 11
Sandi, H. — [18]
Sawczuk, A. — [148], 11
Schikora, K. — [110], 11
Scholz, G. — [66], 11, 70(3), 71(3), 72(2), 73(4), 74(5), 83, 84, 192, 200(3)
Seiler, F. — [41], [45], 30(2), 44(2), 45(2), 46, 48, 54
Siess, C. P. — [40], [91], [127], 11(2), 12
Slater, W. — [25], 8, 12
Soare, M. — [21], 8(2), 9, 89
Sokal, J. — [54], 261
Sonier, P. — [139], 11, 196
Sozen, M. A. — [40], [152], 11, 12

Stiglat, K. — [19], [24], [57], [60], [90], [96], [105], [109], [118], [120], 11(7), 54(3), 59(10), 61(5), 62(5), 63(3), 64(4), 66(2), 68, 70, 84, 98, 110, 192, 200, 249(2), 254, 257, 354
Szerdahely, D. — [52]
Szilard, R. — [16], 5, 8(2), 9, 55, 89, 111

Teodorescu, P. P. — [23], 5
Thürlimann, B. — [48], 30, 35(2), 38(2), 40, 74(2), 82, 90, 192
Timoshenko, S. P. — [6], [7], 5(2), 8(4), 9, 12, 13, 14, 15, 16(2), 17(2), 18(2), 19, 20(3), 22(2), 26(2), 28, 31, 33, 41, 42, 55, 57, 58, 66, 78, 82, 89, 90, 96, 111, 142
Tölke, F. — [30], 12
Trost, H. — [55], 261
Tsuboi, Y. — [59], 64, 69
Tungl, E. — [101], 11, 196
Turin, H. — [36]
Turneaure, F. — [29], 12

Voinea, R. — [20], 5, 8(2), 9, 89

Wang, Ping-Chun — [10]
Werner, H. — [110], [121], 11(2)
Westergaard, H. M. — [5], [25], 5, 8(2), 12
Wippel, H. — [19], [24], [90], [105], 11(2)
Woinowski-Krieger, S. — [7], [31], [35], [87], 5, 8(2), 9, 11, 12, 13, 16(3), 17, 18, 31, 33, 55, 57, 66, 78, 89, 90, 96, 111, 142
Woodring, R. E. — [91], 11

Zienkiewicz, O. C. — [11], 10
Zöphel, J. — [97], [98], [113], 11(3)
Zuber, E. — [133], [134], 11(2)

Subject index

Airy stress function *(see* Differential equation, biharmonic; *see also* Function)

Amplitude, acceleration *(see* Dynamic magnitude; Matrix; Vibration)

Angular elastic displacement *(see* Infinitely long plate; Rectangular plate; Square plate; *see also* Slab structure, comparative values)

Approximate elastic analysis, theoretical methods and/or arocedures for:
 of elastic plates and/or flat slabs:
  elastic network method, 9, 12—13
  energy method, 8—9, 74—84, 356
  finite difference method, 8—10, 12—13, 68—69
  finite element method, 9—10, 68—69, 184
  membrane analogy method, 9, 13
 of flat slabs with capitals:
  Girkmann's procedure, 26—30, 41, 51
  Grein-Girkmann's procedure, 22—26, 51
  Lewe's procedure, 20—22, 26, 82
  Marcu's procedure, 12—13, 31, 36—40, 82
  Nádai-Timoshenko's procedure, 13—17, 26—27, 82
  Timoshenko's procedure, 18—20, 40—42, 82
 of flat slabs without capitals:
  Andersson's procedure, 26—27, 40—44, 51, 83

  Bretthauner's procedure, 44—54, 82—83
  Pfaffinger-Thürlimann's procedure, 35—40, 82
  Rabe's procedure, 30—37, 82
 of slab structures *(see also* Elastic analysis of slab structures):
  Negruțiu's procedure, 73—74, 84, 86, 192, 200—201, 312—317
  Scholz' procedure, 70—74, 84, 192, 200—201

Assumption:
 basic *(see also* Fundamental hypotheses), 5, 8, 30, 81, 89—90, 204, 320
  continuity, homogeneity, isotropy, 74, 89ff.
 of linear elastic behaviour *(see also* Behaviour), 5, 10—11, 70, 74, 84, 89ff.
 of small displacements, of straight normals, of strain linear function, 74, 89—90ff.
 of small vibrations, 320
 of superposition of effects *(see also* Principle), 85, 90, 320
 particular/simplifying:
  in dynamic analysis, 320, 322, 328—331, 335—336, 346—347
  in static analysis, 12—13, 18—24, 27, 31, 44—45, 51—54, 57—58, 64—66, 69—72, 74—76, 82—84, 249, 261, 298, 303—304, 309
 of free horizontal translation, 298, 303, 335, 346

Subject index

Bar, elastic, 1—2, 9, 60, 86, 196, 319
Bar-plate connection *(see* Connection)
Basic system *(see also* Fundamental static scheme, system), 37, 52, 71—72, 75—80ff.
Beam ;
   continuous, 23—24, 46
   discrete, 3, 56—61, 245
   edge *(see* Edge beam)
   equivalent frame, 2, 56—59, 62—68, 248—249
Beam-shearwall connection *(see* Connection)
Bearing element *(see* Structural element/member)
Bearingwall *(see* Shearwall)
Behaviour of structural member, system :
   in elastic (linear-) range *(see also* Response) :
     of plates, 10—12, 85, 95—115, 119—137, 141—201
     of slab structures, 4—5, 70, 74, 84—87
       similar and/or identical, of plates, slab structures, 137, 142, 145, 152, 167, 174, 178, 184, 190, 201, 205, 212, 219, 233, 236, 244
   in plastic /post-elastic range, 11, 184
Bending moment *(see* Infinitely long plate ; Moment ; Rectangular plate ; Square plate ; *see also* Slab structure, comparative values)
Boundary conditions *(see also* Edge ; Equation, compatibility ; Mathematical modelling), 4—11, 14—16, 19, 24, 31—38, 45, 75—78, 92—93, 96, 102—103, 108—109, 115, 118, 191
Boundary reactions *(see also* Infinitely long plate ; Rectangular plate), 4, 19, 33, 78, 85, 91—94, 103, 115—118ff.

Cartesian coordinates *(see also* Numerical computation grid) :
   of column position 6—7, 36, 63, 87
   of contact area, line, point:
     concentrated and/or linear local couple, 56, 62, 100, 113—115ff.
     surface couple, 62—63, 73, 93, 97—99, 105—107, 112—114, 138—201, 205ff.
     transverse force/load: 6—7, 55—57, 64—65, 93—96, 102—104, 110—111, 205ff.
   of discretized mass, 320—321, 329—332
   of disturbance, 329—331
Column :
   axial force-reaction *(see also* Reaction, force), 13, 20—21, 24—40, 45—54, 76—80ff.
   cross-section :
     area, size of, 6—7, 15—16, 26, 34—38, 48, 52, 62—68, 75—83, 87, 138—141ff.
     rectangular, 16, 20—22, 30—40, 61—70, 75—80, 87, 138—141ff.
     square 16, 21—22, 34—40, 62—63, 67—70, 80—83, 138—141ff.
   couple-reaction *(see also* Reaction, couple-), 26—30, 41—44, 51—54, 66, 72—73, 76—80
   edge- *(see also* Support, edge-), 18—20, 40—44, 63, 69, 138—140
   elastic displacement *(see* Elastic displacement)
   equidistant *(see also* Row), 13—54, 63—70, 140ff.
   interior/intermediate, 12—17, 20—40, 45—54, 61—83, 303—311
   moment in *(see* Moment)
   position of *(see* Cartesian coordinates)
     arbitrary, 36—40, 62
   rigidity *(see* Rigidity ; *see also* Relative rigidity)
   row of *(see also* Row), 13—54, 63—70
   with capital, 1—3, 12—30, 68, 75—83
   without capital, 1—3, 30—54, 61—84
Column-slab connection *(see* Connection)

Subject index

Compatibility condition *(see* Continuity condition; *see also* Equation)

Connection:
  bar-plate, 2—4, 11, 60, 84, 93—95, 137ff.
  beam-shearwall, 3, 11, 56—61, 95
  column-slab, 3, 11—54, 61—87, 92—95, 137ff.
  non-point-like, 2, 4, 13, 20—22, 31—40, 54, 56—87, 92—95, 137ff.
  non-rigid, 12—26, 30—40, 45—51, 80
  point-like, 12—20, 23—30, 45—56, 81
  rigid, 2—4, 11, 13, 26—30, 41—44, 51—87, 92—95, 137ff.

Contact area and/or loaded area:
  edge of, 57—58, 64—66
  position of *(see* Cartesian coordinates)
  sizes of, 6—7
    beam-shearwall, 57, 60, 64
    capital-slab, 3, 16, 26
    column-slab *(see also* Column cross-section), 15—16, 31, 35, 40, 48, 54, 64—69, 85, 110, 138—141ff.
    locally distributed couple, force, 85—87, 93—96, 110, 118, 137, 142ff.
    mass-plate *(see* Mass)

Continuity condition at slab-column nodes *(see also* Equation, continuity): 12—14, 17, 19—21, 26—28, 30—32, 38, 40—42, 45, 49, 52, 72—73, 77, 82—83ff.

Couple:
  concentrated, 26—30, 41—44, 51—56, 60—62, 66, 83, 100, 113—115
  distributed:
    continuously, edge, 18, 41—44
    linearly, and/or linear edge, 51—66
    locally *(see also* Surface couple), 56—60, 62—64ff.
  disturbing *(see* Dynamic load)
  effect on plate *(see* Effect)
  end-, of beam, column, 28—29, 41—44, 51—52, 58—59
  given/external, 30, 54—70

  model *(see* Mathematical modelling
  moment of, 18, 41—44, 56—57, 60—63ff.
  position of *(see* Cartesian coordinates)
  -reaction/-unknown *(see* Reaction)
  unit *(see* Dynamic load; Generalized interacting force; Reaction)

Cylindrical bending, 23, 27, 101

Cylindrical rigidity of plate *(see* Rigidity)

Deflected/deformed/ elastic middle surface of plate, 6, 9, 14, 18, 21—24, 28—29, 38, 55, 58, 66, 77—79, 89—92, 95—115, 142—146, 152—157, 207—223, 225—237, 240—247, 284—285, 288, 294—295, 302, 305, 308
  differential equation of *(see* Differential equation of plates)
  equation of *(see* Solution of fundamental equation)

Deformation *(see* Strain)

Differential equation:
  in dynamic analysis:
    acceleration, 323
    inertia force, 320
    motion of mass, 320—321, 323, 328
  in static analysis
    biharmonic, 102
    of equilibrium, 8—9
    of minimum complementary, minimum total energy, 8, 80
    of plates *(see also* Lagrange — Sophie Germain equation):
      fundamental, 5—8, 12, 91—92
      homogeneous, 6, 18, 33, 37
      partial, of fourth order, 8—9, 13, 92
      partial, of second order, 9
      solution of *(see* Solution of fundamental and/or homogeneous equation of plates)

Dimensionless coefficients used in Elastic analysis of slab structures *(see*

Subject index

Infinitely long plate; Rectangular plate; Square plate; *see also* Relative rigidity):
  expression of, 116—118, 137, 140—141
    for infinitely long plate, 120—124
    for rectangular plate simply supported along entire boundary, 133—136
    for rectangular plate with two parallel free edges, 126—132
    for relative rigidity factor, column-slab, 274—275, 279, 299—300
  numerical data/value of *(see also* Numerical computation), 5—6, 87, 94, 137
    for column-slab relative rigidity factor, 73, 274—275, 280, 282—289, 292—298, 300—302, 305—311, 338—341, 347—353
    for elastic displacements;
      angular/rotation, 63, 147—151, 184, 190—200, 366—380
      transverse, 15, 142—157, 166, 174—175, 178—179, 184—199, 360—365
    for moments;
      bending, 15, 17, 157—180, 184—196, 200—201, 382—393
      torsional, 157, 180—185, 190—191, 195, 394—402
    for shearing forces and/or boundary reactions, 157, 191

Displacement:
  dynamic *(see* Dynamic magnitude)
  elastic, static *(see* Elastic displacement)
  given/of known magnitude (see Support failure)
  small *(see* Assumption, basic)

Disturbance/disturbing couple, force *(see* Dynamic load)

Dynamic analysis of slab structures, Negruțiu's method for, *(see also* Dynamic characteristics, load, magnitude), 2, 70, 84—87, 118, 245, 319—353

for forced vibration study, 319, 327—335, 341—346, 351
for free vibration study, 319—327, 335—341, 346—353

Dynamic characteristics of moving mass:
  degree of freedom of mass, 321, 324, 329—335, 342
  inertia of mass:
    rotatory, 322, 331—332, 342, 347—348
    translatory, 332
  motion of mass, 87, 319, 346
    law of, 320—321, 323, 328—329, 342
    rotation, 321—322, 329—334, 342
    translation, horizontal, 334—336, 341, 346, 348—351
    translation, vertical, 321—322, 329—336, 341, 348—351
  oscilation/vibration of mass *(see* Vibration)

Dynamic characteristics of slab structure:
  dynamic diagram/scheme, elastic system *(see also* Vibrating system), 320, 323—325, 327—329, 335—337, 341—342, 346—348
  degree of freedom of:
    direction of, 320—321, 324, 329—338, 341—343, 348—350
    number of, 320—321, 324—325, 327, 335—337, 341—343, 347—350, 353
  mass of column, 334—335
  mass of plate, slab *(see* Mass)
  oscillation/vibration of *(see* Vibration)
  dynamic deflected shape, 344—345, 350
  eigenvalue *(see also* Vibration), 324—327
  eigenfrequency/natural frequency, 324—325
  natural circular frequency, 323—327, 335—343, 347—353

period, 324—326
spectrum, 325
general shape, 320
inertia force *(see also* Dynamic magnitude), 320, 329
law of variation of, 320, 329
natural shape, 320, 325—327
ordinate of, 325—327
orthogonality property of, 327
natural vibration mode, 325—328

Dynamic load:
disturbance, system of, 328—329, 333,
frequency of, 328—329, 333, 342
harmonic, 328—329, 342
position of *(see* Cartesian coordinates)
disturbing locally distributed force, 322, 325, 327—335, 342
grouping of/set of associated, 330—331
inplane/horizontal component, 336—337, 342—350
transverse/vertical component, 336—337, 342—350
unit, 321—322, 329—332, 335—337, 348
disturbing surface couple, 322, 328—335
grouping of/set of associated, 330—331
unit, 321—322, 330—332, 335
initial pulse-excitation, 320
seismic lateral load, 325—326

Dynamic magnitude/quantity:
acceleration of moving mass, 319—321, 323
amplitude value:
of disturbing locally distributed force, 328—329, 333—334
of disturbing surface couple, 328—329, 333
of inertia force, 329, 333—334, 342—343
of moving mass, 323—326, 329, 333
ratio of, 325—327

coefficients of earthquake-resistant design, 326
comparative value of:
natural circular frequency, 338—314, 349—353
unit dynamic displacement, 337—338, 349—352
dynamic coefficient, 343—346
dynamic displacement:
angular/rotation, 320—322, 329—334, 344—345, 348
grouping of/set of associated, 327, 330—332, 334
inplane/horizontal, 329, 334—338, 341, 345—346, 348
transverse/vertical, 320—322, 329—337, 344—346, 348
unit *(see also* Unit elastic displacement), 321—338, 342—343, 348
dynamic moment, stress
in column, 344—346
in plate, slab, 344—346

Edge of panel, plate, slab *(see* Mathematical modelling of boundary conditions; *see also* Plate);
cantilever free, 92
clamped, 11—12, 14—16, 36—38, 45, 78, 81—83, 92
elastic, 11, 31—38, 78, 92
free (parallel), 11, 18—20, 36—38, 40—48, 51—52, 78, 81—83, 85, 92, 101—110
indefinite, 13—17, 20—23, 44—54, 64—70
moment *(see* Moment in plate)
simply supported, 12, 23—30, 35—40, 45, 51, 55—64, 71—84, 85, 92—93, 95—115
— support *(see* Support, edge)
Edge-beam of plate, slab *(see also* Mathematical modelling);
elastic, 31—36, 75—80, 92
elastically/ partially and/or rigidly clamped, 78, 92

Subject index

rigidity *(see* Rigidity)
Effect *(see also* Connection; Continuity; Interaction; Rigidity);
   column cross-section, 15—17, 20, 26, 54, 82
   column rigidity, 12, 14, 23, 37, 54, 74, 84
   combined working/ structural, 1—2ff.
   continuity, of mass *(see also* Mass), 320, 335
   rigid connection, 4, 11, 13, 22, 29, 35, 41, 51, 54, 66, 74, 81—82, 84, 95, 137, 261—262
   rotatory inertia, of mass *(see* Dynamic characteristics)
   space interaction, 2, 261
   superposition of *(see* Superposition of effects)
   support failure, on plate, slab *(see also* Support failure), 1—2, 302—311
     local perturbation characteristic of, 305—308
   surface couple, on plate, slab, 54, 60—70, 74, 85, 90—94, 97—100, 104—107, 111—115, 137—201, 328
     local perturbation, characteristic of, 70, 137, 158—159, 170, 174—189, 206—212, 214—218, 244—245
Effective plate width *(see also* Mathematical modelling), 54—70, 84—86, 248—259, 314
   expression of, 59—60, 66—67, 251, 257, 314
   value of, 59, 62, 68—70, 252—256
   average/mean, 59—60, 64, 257—258
Elastic analysis of slab structures:
   Negruțiu's method, 2—8, 11, 17, 48, 54—56, 62—64, 68—70, 73, 84—87, 94—95, 115—118, 137, 146, 203—205, 261—277ff.
     under dynamic load *(see* Dynamic analysis), 319—353
     under failure of support *(see* Support failure), 302—311
     under horizontal/inplane load, 298—302, 312—317
     under transverse locally distributed load, 287—289
     under uniformly distributed load, 277—287, 289—298
   Pfaffinger's method, 70, 74—84, 241, 302—303
Elastic displacement formula *(see* Method, Maxwell-Mohr)
Elastic displacement of slab structure members:
   distribution of, 4—5, 11, 22, 48, 52, 85—86, 137, 141ff.
   experimentally determined, 6, 190
   magnitude of, 5—6, 48, 52, 85—87, 110, 137, 141ff.
   of column and/or column end;
     angular/rotation, 4, 28, 42, 52—53, 63, 72, 287—289, 307—310
     axial/longitudinal, 13, 28, 77—83, 303
   of equivalent beam, angular, 58—59, 62, 65—66
   of mass *(see* Dynamic characteristic, magnitude)
   of plate, slab *(see* Infinitely long plate; Rectangular plate; Square plate; *see also* Slab structure, comparative values):
     algorithm for *(see* Dimensionless coefficients; *see also* Numerical computation)
     angular/rotation, 4—5, 14, 22—24, 28—30, 42, 52—53, 58—66, 70—73, 78, 81—82, 86, 91—92, 115—118, 140—141ff.
     transverse, 4—5, 12, 15, 24, 28—30, 34—35, 38, 49, 52, 78, 90—92, 115—118, 140—141ff.
   unit *(see* Unit elastic displacement; *see also* Matrix, square)
Energy *(see* Principle; *see also* Theorem):
   complementary, 74, 78—80
   condition of minimum complementary, minimum total, 8, 74, 80

deformation *(see* Energy, strain)
potential, total, 8
strain, total strain, 8—9, 74
Equation ;
  algebraic, 8—9, 267—310, 323—326, 333, 342
  compatibility, for boundary conditions *(see* Boundary reactions ; Moment ; Shearing force ; Transverse displacement ; *see also* Mathematical modelling), 14, 19, 24, 33, 78, 103
  continuity/compatibility, at slab-column nodes *(see also* Continuity condition), 203, 213, 236, 303, 327
    for rotations, 28, 42, 52, 72—73, 276, 299—300, 306, 337
    for transverse displacements, 12, 20, 28, 38, 49, 261, 263, 275—277, 280, 304—305
    for transverse and angular displacements, system of, 30, 263—268, 275—276, 279, 290—291, 299, 307, 309—310, 337
  differential *(see* Differential equation)
  homogeneous, 323—325
  linear, 33, 38, 267—310, 323—326, 333, 342
  of equilibrium, 8—9, 24, 50
  of natural circular frequencies/secular *(see also* Matrix, square), 324—325, 327, 337—338, 348—351
  of plate *(see* Solution of fundamental equation)
  system of, 14, 19, 24, 30, 33, 38, 49, 72—73, 78—80, 263—310
  system of motion *(see also* Differential equation), 323—327, 333, 342
  transcedental, 33, 103
Equivalence criterion, equation, 51—52, 59, 62, 65—66, 77, 250—251
Equivalent frame *(see also* Method, approximate analysis), 71—74, 248—249, 314—317
Experimental research *(see* Model, experimental)

Failure of support *(see* Support failure)
Field, displacement-, force-, strain-, stress-, 8—9
Flat slab *(see* Approximate elastic analysis ; *see also* Method, approximate analysis) :
  geometrical and/or mechanical, topological parameters of, 23, 31—37, 42—44, 49, 52—53, 59ff.
  one-level, 12—54
  physical parameters of *(see also* Modulus ; Poisson's ratio), 15—16, 35, 42, 52, 54, 59ff.
  reinforced concrete, 1, 38ff.
  topological symmetrical, 12—35, 38—60ff.
    two-fold, 14—17, 20—22, 25—27, 31—44, 49—54
  topological unsymmetrical, 36—38, 56—58
  with capitals, 1—3, 12—30ff.
  without capitals, 1—3, 12—13, 30—54, 56—60ff.
Flexibility :
  factor *(see* Relative flexibility)
  matrix *(see* Matrix ; *see also* Method)
Force *(see also* Load) :
  concentrated, 13, 44—50, 55—56, 100
  distributed :
    linearly *(see* Load, linearly distributed)
    locally, 37, 55, 61, 85, 93—97, 101—104, 110—111, 119—121, 124—127, 132—134ff.
  disturbing *(see* Dynamic load)
  equidistant *(see* Row of concentrated and/or locally distributed forces)
  generalized interacting *(see* Generalized interacting force)
  horizontal/inplane, 1—2, 68—74, 80
  inertia *(see* Differential equation ; *see also* Dynamic characteristic, magnitude)
  position of *(see* Cartesian coordinates)
  -reaction/-unknown *(see* Reaction)
  seismic *(see* Load, seismic)

423

Subject index

set of associated *(see* Grouping; *see also* Set)
shearing *(see* Shearing force)
system *(see* Loading system)
transverse/vertical, 8—10, 13, 37, 44—50, 55—56, 61, 85, 93—97, 100—104, 108—111, 115ff.
unit *(see* Dynamic load; Generalized interacting force; Reaction)
-unknown *(see* Reaction, force-)

Forced and/or free vibration *(see* Dynamic analysis of slab structures, *see also* Vibration)

Force-reflection procedure *(see also* Approximate elastic analysis, Bretthauer's procedure), 45—46

FORTRAN IV language *(see also* Numerical computation), 94, 137

Fourier series *(see also* Series), 21—22, 66—67, 78—80, 101, 145, 190, 219

Frequency *(see* Dynamic characteristic, magnitude; *see also* Vibration)

Function:
 Airy *(see also* Differential equation, biharmonic), 102
 complementary energy *(see also* Energy), 74, 79—80
 exponential *(see also* Series, simple), 8, 22, 37, 55—59, 78, 96—108, 118, 145, 193—195
 harmonic, 323, 328—329
 hyperbolic *(see also* Series, simple), 15—18, 22—24, 28, 32, 42, 58, 67, 78, 96—100, 102—110, 118, 145, 193, 219
 load *(see* Load function)
 of complex variable *(see also* Approximate elastic analysis; Bretthauer's procedure), 45—48
 strain linear *(see* Assumption, basic)
 trigonometric *(see also* Series, double and/or simple), 8, 15—16, 18, 21—24, 28, 32, 37, 42, 55—58, 66—67, 78, 96—115, 118, 145, 193—195, 219

Fundamental equation of plates *(see* Differential equation)

Fundamental hypotheses *(see also* Assumption, basic):
 of Theory of Elasticity, 5, 81
 of Theory of (hyperstatic) Structures, 5, 89, 320
 of Theory of Plates, 8, 30, 89, 90, 203—204, 320

Fundamental static scheme/system in Elastic analysis of slab structures: basic structural system:
 for dynamic analysis (dynamic scheme), 320—321, 328—331, 335—337, 341—342, 346—348
 for static analysis, 203—204, 264—268, 278—279, 298—299, 303, 336

 infinitely long plate under:
  locally distributed load, 95—97, 119—121
  surface couple, 97—100, 119, 121—123, 138—149, 152—169, 174—178, 180—190
  uniformly distributed load, 100—101, 119, 124

 rectangular plate simply supported along entire boundary under:
  locally distributed load, 110—111, 132—134
  surface couple, 111—114, 132—133, 134—135, 138—146, 149—157, 160, 166—174, 178—185, 189—190, 360—402
  uniformly distributed load, 115, 132—133, 135—136

 rectangular plate with two parallel free edges under:
  locally distributed load, 101—104, 124—127
  surface couple, 104—107, 124—125, 128—131
  uniformly distributed load, 108—110, 125—126, 131—132

424

Subject index

Generalized interacting force *(see* Fundamental static scheme; *see also* Reaction):
  concentrated:
    couple, 26—30, 51—54
    force, 24, 26—30, 45
  distributed:
    linearly, couple, 51—54
    linearly, force, 30—31, 45
    locally, force, 20—22, 30—31, 72, 76—80, 85, 94—95, 118, 196—199, 203—205, 303—310
    surface couple, 64—66, 70—73, 76—80, 84—85, 94—95, 118, 137, 197—199, 203—205, 262, 302—304, 306—311
    triangularly, 65—66, 76—78
    uniformly, 30—31, 51—54, 64, 66, 76—78, 94ff.
  unit, 196—199
Grouping *(see* Set)

Inertia:
  force *(see* Dynamic magnitude)
  matrix *(see* Matrix, inertia)
  moment of *(see* Moment of inertia)
  rotatory and/or translatory *(see* Dynamic characteristics)
Infinitely long plate as structural member *(see also* Dimensionless coefficients; Fundamental static scheme), 90—92, 115—118
  elastic displacements of:
    angular/rotation, 62—63, 66, 73, 120—124, 146—149, 184, 190—193, 196—200, 256—259, 287, 289
    transverse, 95—101, 119—124, 142—149, 152—157, 166, 174—175, 184—187, 190—199, 205—218, 225—232, 237—243, 284—286
  moments of:
    bending, 96, 120—124, 157—169, 174—178, 184—196, 200—201, 205—218, 225—233, 239—244, 284—286, 302
      extreme value of, 141, 158—169, 174—177, 180, 184—188, 192—196, 205—218, 226—232, 239—243, 284, 302
    torsional, 96, 120—124, 180—185, 190—191, 195
      extreme value of, 180—182, 184
  shearing forces and/or boundary reactions of, 120—124, 157, 190—191
Influence surface, line, 44—46, 184, 263
Integral *(see* Solution of fundamental and/or homogeneous equation of plates)
Integration of fundamental and/or homogeneous equation of plates:
  constants, 18—21, 24, 33, 38, 41—42, 57, 78, 101—104
Interacting force *(see* Generalized interacting force)
Interaction *(see* Connection; *see also* Effect, structural):
  bar-plate, column-slab, 2, 26, 35, 44, 48, 81 ff.
  space *(see* Space interaction)
Iteration:
  computation scheme, distribution coefficients, transmission factor of, 315
  convergence of, 315

Lagrange-Sophie Germain equation of plates *(see also* Differential equation, fundamental and/or homogeneous), 6, 8, 10, 13, 20, 26, 55, 85, 94, 100, 137
Laplace operator *(see* Operator)
Law, constitutive, Hooke's, 10
Load on plate, slab, slab structure:
  arbitrary *(see also* Load, unsymmetrical), 2, 84—86
  concentrated (point-like), 12, 36, 44—45
  concentrated couple as, 30, 41—42, 54—56, 100, 113, 114—115
  distributed (non-point-like):
    continuously, 12—54, 80—83

425

Subject index

in strips, 21, 23—30, 36
in chess board fashion, 21, 36
linearly, 20, 24—25, 33, 44—54, 57—58, 64—66
locally, 6—7, 36—38, 57, 61, 77, 80—83, 85—87, 93—97, 101—104, 110—111, 120—121, 126—127, 133—134, 142, 146, 157, 191—193
uniformly, 57—58, 61, 66, 77, 80—83
uniformly, throughout floor, plate, slab, 13—22, 25—27, 30—54, 80—82, 85, 87, 93, 95, 100, 101, 108—110, 115, 124, 131—132, 135—136, 157, 191
dynamic *(see* Dynamic load)
equivalent, 51—52, 65—66
function, 6—8, 19—23, 27, 66, 67
given/design/external/service, 5—7, 8—54, 71—84, 85, 89, 93—95, 118
gravity *(see also* Load, vertical) 319, 327
horizontal/inplane/lateral, 1—2, 37, 61—74, 77, 84, 86—87, 302, 319,
intensity of, 6, 65—66
position of *(see* Cartesian coordinates)
seismic, 64, 84, 224, 248
set of associated *(see* Set)
sign convention, 90
surface couple as, 30, 54, 56, 60—70, 77—78, 85—87, 92—95, 97—100, 104—107, 111—114, 121—123, 128—131, 134—135, 137—201, 203—224, 236—246
symmetrical and/or antisymmetrical, 12—22, 25—26, 30—54, 60—74, 80—83, 93, 100—101, 108—110, 115, 124, 131—132, 135—136, 205—236
transverse/normal to mid-plane, 2, 5—8, 12—58, 61, 64—66, 68, 75—84, 85—86, 203—205, 224—247
type of, 5—7, 13—23, 41—47, 60, 70—71, 93—95, 115—119, 124—126, 132—133, 303

unit *(see* Couple ; Force ; *see also* Dynamic load; Reaction), 32—33, 38, 44, 52—53, 79, 196—199
unsymmetrical, 12, 23—24, 27—29, 36—38, 55—74, 77—80, 93, 95—100, 101—107, 110—115, 120—123, 126—131, 133—135, 137—201, 244—248
vertical/gravity, 1—2, 12—54, 68, 75, 87, 90, 302
Loaded area *(see* Contact area)
Loading :
assumption/case, 304—306, 309
system :
of horizontal forces, 66, 70—74, 77, 80, 86, 319
of set of associated couples and/or forces, 204—248
Local perturbation characteristic *(see* Effect, support failure, surface couple)

Mass of plate, slab *(see also* Dynamic characteristics):
continuous, 320, 335
discretized/concentrated, 319—320, 327, 334, 346—347, 353
distribution of, 320, 325, 327, 346
arbitrary, 320
grouping of/set of associated, 327, 334
magnitude of, 319, 347
-plate, -slab contact area, 87, 349
position of *(see* Cartesian coordinates)
sizes of discrete, 320—321, 334, 336, 342, 347—349
large, 322, 331
unbalanced, 319
Mathematical modelling :
of boundary conditions *(see also* Equation, compatibility):
arbitrary, 38, 78
for elastic edge, 33, 36, 78
for free edge, 19, 35—36, 44, 103
for intermediate support, 24
for rigidly clamped edge, 14, 35—36

for simply supported edge, 24, 35—36
of couple effect on plate, slab:
for concentrated couple:
Girkmann's procedure, 54—56, 60, 66, 97, 100
Negruțiu's procedure, 56, 100, 112—113, 114—115
for linear edge couple:
Mohammed-Popov's procedure, 54, 56—58, 60, 66
for surface couple:
Negruțiu's procedure, 54, 61, 64, 70, 85, 90, 93—95, 98—99, 104—107, 111—114ff.
Pecknold's procedure, 64—68, 69—70
Pfaffinger's procedure, 76—78, 110, 192
Stiglat's procedure, 54, 60—64, 66, 73, 97—98, 110, 192, 200—201
of effective plate width:
Allen-Darvall's procedure, 55, 69—70, 84
Mohammed-Popov's procedure, 55, 58—60, 61—62, 64
Negruțiu's procedure, 62—63, 64, 67—70, 248—259
Pecknold's procedure, 64, 66—68, 84
Stiglat's procedure, 62—64, 84
Matrix:
column-/vector-:
of accelerations, 323
of amplitudes of inertia forces, 333
of amplitudes of mass, 323—325
of constants of integration, 78
of general interacting forces/reactions/statically indeterminate unknowns, 38, 78—80, 267—268, 303—304
of dynamic displacements, 323, 333
of static displacements
by given load, 78—79, 267—272

by support failure, 267—268, 303—304
diagonal, of masses *(see* Matrix, inertia)
flexibility *(see also* Matrix, square), 77—80, 267—272, 303—304, 323—327, 333
inertia, 323
load, rectangular, topological, 78—80
square:
determinant of, 324, 348—349
of unit dynamic displacements, 323—327, 337, 348—349
of unit static displacements, 38, 72, 267—272, 303—304
Maxwell theorem *(see* Theorem of unit displacement reciprocity)
Method:
approximate analysis of flat slabs, design method for *(see also* Approximate elastic analysis):
empirical method, 5, 261, 263
equivalent frame method, 5, 64, 68, 84, 86, 261, 263, 314—317
dynamic analysis:
of reticular structures, 8, 319—320, 351, 353
of slab structures *(see* Dynamic analysis of slab structures)
elastic, static analysis, 8, 10, 44
of plates *(see* Solution of fundamental equation; *see also* Theory of Plates)
of reticular structures *(see also* Theory of Structures), 1—2, 203, 248
general displacement method, 2, 10, 70, 73, 86, 224, 249, 263, 312—317
general force method, 2, 70—71, 85—86, 204—205, 263
of slab structures *(see* Elastic analysis of slab structures)
flexibility matrix, 70, 86, 324, 334—335
Maxwell-Mohr's, 250—251

427

Subject index

plastic/post-elastic analysis *(see* Theory, Johansen's yield line)
Model of slab structures:
  experimental, in elastic range, 6, 190
    full-scale, 4, 6
    reduced-scale steel, 4
    mathematical *(see also* Mathematical modelling), 1—2, 4, 12, 92, 320
Modulus:
  of elasticity/Young, 7, 23, 31—36, 42—44, 52, 59, 72, 76—78, 80—83, 89, 250—251, 274—275 ff.
  shear, 31, 33, 76—78
Moment, dynamic *(see* Dynamic magnitude)
Moment in column, bending, 38, 73, 76—78

Moment in plate, slab *(see also* Infinitely long plate; Rectangular plate; Square plate):
  algorithm for *(see* Dimensionless coefficients; *see also* Numerical computation)
  bending, 5, 13—19, 21—27, 32—40, 45—48, 52, 61, 70, 73—74, 78—83, 90—92, 96, 103, 115—118, 137, 140—141, 157 ff.
  bending edge, 18—19, 41—42
  distribution *(see* Stress distribution)
  torsional, 5—6, 13—14, 34—35, 38, 45—46, 52, 70, 74, 79, 90—92, 96, 115—118, 137, 140—141, 157 ff.
Moment of couple *(see* Couple)
Moment, sectional *(see* Moment in plate)
Moment, twisting *(see* Moment, torsional)
Moment vector of couple, surface couple, 29, 56, 61, 93, 95—100, 104—107, 111—115, 118, 119, 124, 132, 142 ff.
Moment, work of, 79
Moment of inertia:
  of column bending, 42—44, 52—53, 72—73, 274—275 ff.
  of edge beam:
    bending, 31, 33—36, 75—78
    torsional, 31, 33—37, 75—78

  of equivalent beam, 59, 250—251
Motion *(see* Dynamic characteristic; Mass; Vibration)

Node *(see also* Connection):
  bar-plate, 3—4, 11
  beam-shearwall, 11, 56, 61
  column-slab, 11—14, 19, 26—28, 51, 66, 69, 72, 74—76, 80, 82, 85ff.
  rotation of *(see* Relative rotation)
Numerical computation in Elastic analysis of slab structures *(see* Dimensionless coefficients; *see also* Relative rigidity):
  accuracy-condition of, 87, 94—95, 167, 190—196, 200, 263, 287—289, 308, 311
  algorithm for, 4—5, 85—86, 94—95, 115—137, 191, 203 ff.
  grid, 137—141, 184
  parameter:
    dimensionless, 62—63, 68—69, 137, 141, 148—151, 191ff.
    geometrical, 6—7, 62—63, 68—69, 85—87, 89, 93—94, 116—117, 138—201, 204—205ff.
    mechanical, 4, 7, 85—87, 94, 138—141, 274—275, 280—289, 292—298, 300—302, 305—311, 327
    physical *(see* Modulus; *see also* Poisson's ratio)
    program *(see also* FORTRAN IV language), 94, 137, 167, 190—192, 263
Numerical experiment, 85—86, 137—201

Oblique, parallelogram-like panel, plate, slab *(see* Panel; *see also* Row, offset)
Operator:
  differential, 8, 91
  finite difference, 8
  Laplace, 91
  variational, 9

Subject index

Orthogonality condition, property *(see* Dynamic characteristic of slab structure, natural shape)
Oscillating system *(see* Vibrating system)
Oscillation *(see* Vibration)

Panel of floor, plate, slab:
  corner, 31—40 ff.
  edge, 12, 17, 18—20, 23—48 ff.
  interior/central, 12—22, 25—27, 30 40, 46, 49—54, 66—70 ff.
  oblique, parallelogram-like, rhombic, 237—241, 244—246
  rectangular, 12—22, 23—54, 66—70ff.
  square, 16—17, 21, 26—27, 43—44, 67—70 ff.
Period of vibration *(see* Dynamic characteristic; magnitude; *see also* Vibration)
Plate, category of *(see also* Fundamental static scheme; Infinitely long plate; Rectangular plate; Square plate):
  cantilever, 11, 92
  clamped, 11, 12, 36, 45, 78, 81—83, 92
  continuous, 11
  corner-supported, 31—35
  elastic, as structural member, 1—2, 6—7, 26, 29—30, 54, 70, 74, 77, 85—87, 92—115, 203—248
  elastic boundary, 11, 31—35, 36, 75—80, 92
  finite boundary, 30—40, 62, 64, 85, 92—94, 101—115ff.
  free boundary, 11, 18—20, 36, 40—48, 51—52, 81—83, 85, 92—94, 101—110ff.
  half-plane, 44—48, 52
  infinitely extended *(see* Plate of indefinite sizes)
  infinitely long *(see also* Infinitely long plate), 18—20, 23—30, 37, 40—48, 51, 55—64, 71—74, 77—78, 84, 85, 92—101ff.
  of indefinite sizes, 13—17, 20—22, 23, 44—54, 64—70
  of variable thickness and/or with openings, 11
  rectangular, 30—40, 46, 57—58, 74—85, 92—94, 101—115ff.
  symmetrical, topological, 142—143, 327
  simply supported *(see also* Infinitely long plate; Rectangular plate; Square plate), 12, 23—30, 35—40, 45, 51, 55—64, 71—84, 85, 92—115ff.
  square *(see also* Square plate), 31, 35, 40, 87ff.
Plate, topological characteristics of:
  boundary, edge of *(see* Boundary; *see also* Edge)
  cylindrical/flexural rigidity of *(see* Rigidity)
  edge beam of *(see* Edge beam)
  effective, width *(see* Effective plate width)
  inplane sizes of, 6—7, 30—36, 57, 87, 93, 138—139ff.
  loaded area of *(see* Contact area)
  middle plane of, 6—7, 36, 86, 90, 93, 117ff.
  span of 3, 5, 7, 20, 23, 59, 71, 93, 96—98, 102, 105—107, 111—114, 138—141ff.
  support of *(see* Support)
  thickness, 4, 7, 11, 59, 72, 75, 89—90, 93ff.
Poisson's ratio, 4, 7, 15—17, 34—35, 38, 42, 54, 59, 62, 68—69, 73—74, 80—83, 85—91, 94, 116, 140—141, 158—183, 185—190, 205—206, 210—212, 218—224, 229—235, 246, 251—257, 322
Principle:
  d'Alembert's, 320, 329
  energy variational, 8, 84
  of minimum complementary energy, 74, 77—80

429

Subject index

of superposition of effects *(see also* Superposition of effects), 85, 90, 303, 320

Reaction and/or statically indeterminate unknown *(see also* Generalized interacting forces):
boundary-, *(see* Boundary reactions)
couple-:
concentrated, 26—30, 41—44, 51—54
linear, 51—54
surface, 61, 65—66, 72—73, 76—80, 85, 118, 203—205, 303—304, 306—311
unit, 52—53, 79
force-:
concentrated, axial, 13, 24—30, 36, 45—50, 203—204
distributed:
linearly, 24, 30—36, 45—54
locally, 13, 20—22, 30—38, 76—80, 85, 118, 203—205, 303—305
inplane, 203—204
unit, 32—33, 38, 79
Rectangular plate simply supported along entire boundary as structural member *(see also* Dimensionless coefficients; Fundamental static scheme), 90—92, 115—118
elastic displacements of:
angular/rotation, 133—135, 149—151, 184, 190—193
transverse, 110—115, 132—135, 142—146, 149—157, 166, 174, 184—185, 190—193, 205—206, 225, 233
moments of:
bending, 133—136, 157—158, 160, 166—169, 174, 178, 184—185, 190—196, 205—206, 225, 233
extreme value of, 141, 166—169, 184, 192—196

torsional, 133—136, 184—185, 190—191, 195
extreme value of, 184
shearing forces and/or boundary reactions of, 134—136, 157, 190—191
Rectangular plate with two parallel free edges as structural member *(see also* Dimensionless coefficients; Fundamental static scheme), 90—92, 115—118
elastic displacements of:
angular/rotation, 126—131
transverse, 101—110, 124—131
moments of:
bending, 126—132
torsion, 126—132
shearing forces and/or boundary reactions of, 126—132
Relative flexibility, slab-column:
factor of, 279, 281, 291, 300, 337
value of, 281
Relative rigidity:
column-slab, 3, 17, 37, 61, 70, 82—83, 86—87, 262, 299, 335
factor of, 26, 43, 52, 72—73, 274—275, 279, 299—300
value of, 43—44, 52, 73, 274—275, 280, 282—289, 292—298, 300—302, 305—311, 338—341, 347—353
slab-edge beam,
factor of, 31, 34—37, 78
value of, 34—37
Relative rotation
at bar-plate node, 3—4
at column-slab node, 13, 30
Response of elastic structural element and/or of slab structures:
steady-state, to harmonic disturbance, 328
symmetrical, to unsymmetrical load/ tendency toward symmetrization, 145, 152, 174—176, 179—180, 184
to dynamic and/or seismic action, force, load, 86—87, 319, 325
to support failure, 86—87, 302—311

Rigid connection *(see* Connection)
Rigidity:
  column bending, 12, 14, 27, 42–44, 52–53, 69–70, 74ff.
  edge beam:
    bending, 31, 33–36, 75–78
    torsional, 31–37, 75–78, 94
  equivalent beam bending, 56, 59, 64, 67–70, 86
  plate, slab cylindrical/flexural, 7, 23, 31–36, 42–44, 52, 55, 63, 73, 76–78, 91, 94, 116ff.
Rotation:
  dynamic *(see* Dynamic characteristic; magnitude)
  elastic *(see* Elastic displacement, angular)
  given/of known magnitude *(see* Support failure)
Row *(see also* Set):
  of column *(see* Column)
  of concentrated forces (-reactions), 13, 30, 44–54
  of concentrated couples (-rections), 26–30, 41–44, 51–54
  of locally distributed forces (-reactions):
    simple, 225–229, 233
    double, 229–235, 237–238, 242, 244, 247
    offset/zig-zag distributed, 236–238
    triple (multi-), 237–238, 242–243
  of surface couples (-reactions):
    simple, 62–63, 67, 70, 206–209, 229
    double, 210–212, 216–224, 237–242
    offset/zig-zag distributed, 237–241, 244–246
    triple(multi-), 213–215, 237–238

Series:
  double, of trigonometric functions, 8, 21–22, 111–115, 118, 145, 168–173, 181, 190–196, 219

  finite sum of, 8, 22, 101, 195–196
  last summed term of, 38, 43–44, 73, 80, 162–165, 168–173, 181, 191–196
  simple, of exponential, hyperbolic and/or trigonometric functions, 8, 14–16, 18, 21–24, 28, 32, 37, 42–43, 55–56, 58–59, 66–67, 78–80, 96–110, 118, 145, 162–165, 168–169, 190–196, 219
Set of associated/grouping of:
  couples (-reactions):
    concentrated, 29, 41–44
    surface, 62–63, 67, 70, 86, 107–108, 117, 204–224, 236–248
  disturbing couples, forces *(see* Dynamic load)
  forces (-reactions):
    concentrated, 28–30
    locally distributed, 31–40, 107–108, 117, 204–205, 224–238, 242, 248
  generalized interacting forces and/or statically indeterminate unknowns *(see also* Set of couple-reactions, force-reactions), 85–87, 203–205, 303–311
  symmetrical and/or antisymmetrical, 204–236
  unsymmetrical, 236–246, 327
Shear:
  effect, 4
  modulus *(see* Modulus)
Shearing force *(see also* Infinitely long plate; Rectangular plate), 14, 18, 24, 52, 78, 90–92, 115–118ff.
Shearwall/bearingwall, 3–4, 11, 56–61, 236, 245
Slab *(see* Flat slab; Plate; Slab structure)
Slab structure, 1–2, 5, 11–12, 30, 57, 60–87, 92ff.
  analysis of *(see* Elastic analysis; Dynamic analysis; *see also* Approximate elastic analysis)

431

Subject index

comparative values:
  of staticaly indeterminate unknowns, 280—283, 291—293, 300—301, 304—307, 309—311
  of static magnitudes *(see also* Infinitely long plate; Square plate), 4—5, 117, 141—201, 205—248
    elastic displacement, angular/rotation, 287—289, 308, 310—311
    elastic displacement, transverse, 284—289, 294—297, 305—306, 308—309, 311
    moment, bending, 284—289, 294—298, 302, 305—306, 308—309, 311
degree of freedom of, eigenvalue of *(see* Dynamic characteristic, magnitude; Vibration)
dynamic behaviour of *(see* Dynamic analysis; *see also* Response)
elastic behaviour of *(see* Behaviour in elastic range)
geometrical and/or mechanical characteristic of *(see also* Numerical computation parameter), 1—4, 6—7, 12, 72—73, 76—77, 85—91ff.
hyperstatic/multiple statically indeterminate, 2, 5, 11, 64, 203—205, 263
metal/steel *(see also* Model, experimental), 4, 237, 245
multistoried, 71—74, 84—86, 261, 303, 312—314, 332
  one-level, 80—83, 274—311, 335—346
  four-level, 72—74, 314—317
  two-level, 346—353
oblique/ parallelogram-like, 5, 236—241, 244—246
physical parameter of *(see* Modulus; Poisson's ratio; *see also* Numerical computation)
reinforced concrete, 1—5, 11—12, 38, 69, 261, 306

symmetry and/or antisimmetry of:
  loading, 80—83, 205—236, 304—306, 327
  mass, 327
  topological, 60—86, 103, 110, 203—236, 303—311, 327, 330
  two-fold, 245, 274, 303—305, 335, 341, 346
unsymmetry of:
  loading, 60—74, 77—80, 244—248, 262, 306
  topological, 5, 85—86, 236—248
with capitals, 1—3, 75—83
without capitals, 1—3, 60—84ff.
Solution of fundamental equation of plates:
  approximate *(see also* Approximate elastic analysis of plates), 8—9, 10, 16, 31, 356
  classical *(see also* Solution, rigorous), 8, 10—11, 12, 13—15, 18—20, 21—22, 23—25, 26, 28—29, 37, 55—56, 57—58, 61, 66, 77, 81, 96
  modern *(see also* Solution, rigorous), 5, 7—8, 11, 32—33, 37—38, 42, 56, 61, 73, 77—79, 81, 84—85, 91—94
  particular, 19—20, 28—29, 32—33, 37—38, 77—78, 102—104, 108—110, 323
  rigorous, used in Elastic analysis of slab structures *(see also* Fundamental static scheme), 5, 7—8, 84—85, 91—94, 115—118, 137
  for locally distributed load:
    Girkmann's solution, 96—97, 191
    Navier's solution, 111, 132—133
    Negruțiu's solution, 104, 124—125
  for surface couple:
    Negruțiu's solutions, 99, 105, 107, 112, 114, 119, 124—125, 132—133

Subject index

Stiglat's solution, 97—98, 119
  for uniformly distributed load:
  Girkmann's solution, 101, 119
  Negruţiu's solution, 110, 125—126
  Navier's, 115, 132—133
  superimposed, 19—20, 23—25, 28—29, 32—33, 37—38, 42, 78—79
Solution of homogeneous differential equation of plates, 18, 33, 37, 45, 77—78, 102—104, 108—110
Space interaction, bidimensional, three-dimensional (see also Effect), 1—2
Square plate as structural member, 90—92, 115—118
  elastic displacements of:
    angular/rotation, 146, 149—151, 184, 192—199, 236, 287, 289, 308, 310—311, 366—380
    transverse, 146, 149—157, 174, 178—179, 184, 189—199, 205—207, 219—223, 225—228, 233—236, 244—247, 284—289, 293—296, 305—306, 308—309, 311, 360—365
  moments of:
    bending, 157—158, 160, 166—174, 178—179, 184—185, 189—196, 205—207, 219—228, 233—236, 244—247, 284—289, 292, 294—298, 302, 305—306, 308—309, 311, 382—393
    extreme value of, 141, 166—173, 178—179, 184, 189, 192—196, 206—207, 219—228, 234—235, 246—247, 284—285, 288, 292, 294—297, 302, 305, 308
    torsional, 180—185, 190—191, 195, 394—402
    extreme value of, 183—184
  shearing forces and/or boundary conditions (see Rectangular plate)
  under support failure, 303—311
Stiffness (see Rigidity)
Strain/deformation, 7—9, 74, 78—80
  energy (see Energy)

state of, 4, 7, 14, 27—29, 89, 116, 158, 204, 261
tensor, 8, 91
Stress (see also Moment; Shearing force; Tension):
  algorithm for (see Dimensionless coefficients; see also Numerical computation)
  distribution:
    in capital, 75—78
    in column, 4, 38, 48, 65—66, 75—76
    in plate, slab, 4—5, 11, 13, 22—23, 44, 48, 75—76, 85—87, 90—92, 115—118, 140—141, 157—190, 262
  dynamic (see Dynamic magnitude)
  experimentally determined, 6, 190
  extreme value of (see Infinitely long plate; Rectangular plate; Square plate)
  normal, 7, 9, 74—75, 89—90, 203—204
    linear uniform, 75—76
    triangular, 65—66, 76—78
  redistribution of, 70
  sectional (see Stress in plate)
  sign convention, 90
  state of, 4, 7, 11, 14, 27—29, 51, 75, 82, 89, 116, 157—158, 204, 261
  tangential, 7, 89—90
  tensor, 8, 91
Structural analysis, Structural Mechanics (see Method; see also Theory of Structures), 1, 4, 9, 30, 41, 54, 64, 70, 84—87, 94—95ff.
Structural effect (see Effect)
Structural element/member (see also Bar; Beam; Bearingwall; Column; Plate; Shearwall; Slab), 1—2, 7, 10, 11, 26, 29—30, 54, 70, 77, 85—86, 92—115, 203—248ff.
Structure/structural system:
  beamless, 1—2ff.
  caisson, cellular, shearwall, 3—4, 11, 56—61, 236, 245
  hyperstatic/multiple statically indeterminate, 2, 5, 11, 203—205

Subject index

mixed, consisting of bars and plates, 1—3, 86, 196, 261, 319
reticular, 1—2, 203—204, 319, 346, 351
slab *(see* Slab structure)
symmetrical, unsymmetrical *(see* Flat slab; *see also* Slab structure)
Supperposition of effects *(see also* Principle):
  in dynamic analysis, 320, 327
  in static analysis, 19—20, 24, 28—32, 37, 42, 45—46, 51, 56—58, 61—67, 85, 97—98, 105—106, 112—113, 204—248, 256—259
Support of plate, slab *(see also* Connection)
  edge *(see also* Edge):
    corner, 31—35
    discrete, 18—20, 31—35, 40—47
    elastic, 31—35, 36, 78, 92
    elastically deformable, axial, 13, 77—83
    equidistant *(see also* Row), 13—30, 40—54, 62—63, 65—84ff.
    failure of known magnitude of *(see* Support failure)
    hinged, 2, 26, 30, 45, 52, 77, 80
    interior/intermediate:
      discrete, 2, 12, 13—17, 20—40, 44—54
      linear, continous, 23—25, 28, 45—54
      with arbitrary position, 36—40
    linear, 31—35, 36, 42—43
    non-point-like, 2, 16—17, 20—22, 31—40, 54, 56—87, 92—95, 137ff.
    point-like, 2, 12—20, 23—30, 31—36, 40—54, 81—82
    rigidly connected, 26—30, 41—44, 51—87, 92—95, 137ff.
    undeformable, axial, 12—54, 61—74, 77—83
Support failure of known magnitude, slab structure under, 1—2, 302—311
  behaviour/response under *(see* Response)

effect of *(see* Effect)
static magnitude under *(see also* Slab structure, comparative value), 305—306, 308—309, 311
type of:
  rotation, 302—306
  translation, 302—311
Surface couple:
  as load *(see* Load)
  as generalized interacting force *(see* Generalized interacting force)
  disturbing *(see* Dynamic load)
  effect on plate *(see* Effect)
  model *(see* Mathematical modelling)
  position of *(see* Cartesian coordinates)
  -reaction *(see* Reaction)

Tendency toward symmetrization *(see* Response)
Tension, normal and/or tangential *(see also* Stress), 7, 9, 74—75, 89—90
Tensor *(see* Strain; Stress)
Test/testing *(see* Model, experimental)
Theorem:
  of minimum total energy *(see also* Principle), 8
  of unit displacement reciprocity/ Maxwell's, 196—199
Theory:
  Johansen's yield-line, 11
  of Elasticity, 2, 5, 81, 89
  of Plates, 2, 6—8, 11, 17, 20, 27, 30, 35, 54, 84, 89—90, 137, 261, 320
  of Structures, hyperstatic, 2, 5, 35, 89, 203, 261, 320, 346
Torsional moment *(see* Infinitely long plate; Rectangular plate; Square plate; *see also* Moment)
Transverse elastic displacement *(see* Infinitely long plate; Rectangular plate; Square plate; *see also* Slab structure, comparative values)

Subject index

Unknown:
  in dynamic analysis:
    amplitude as, 324—325
    amplitude ratio as, 326
    inertia force as, 333
    natural circular frequency as, 323—327, 337, 349
    statically indeterminate, 336—338, 345—346, 349—352
  in static analysis:
    statically indeterminate *(see* Reaction; *see also* Generalized interacting force)*, 29—35, 38, 49—54, 72—73, 76—80, 303—311

Unit elastic displacement *(see also* Matrix, flexibility/square)*:
  in dynamic analysis:
    angular/rotation, 321—324, 326—327, 329—332
    grouping of/set of associated, 327, 330—332
    horizontal/inplane, 329
    vertical/transverse, 321—324, 326—327, 329—332
  in static analysis:
    angular/rotation, 52—53, 72—73, 196—199, 205, 236
    grouping of/set of associated, 117, 205
    vertical/transverse, 38, 49, 196—199, 205, 236

Vector:
  -matrix *(see* Matrix, column)
  moment- *(see* Moment vector)

Vibration *(see also* Dynamic analysis of slab structures; Dynamic characteristic, magnitude):
  analysis method of *(see* Method, flexibility matrix)
  damping influence on, 320, 328
  free, 86, 319—328, 334—341, 346—353
    amplitude of, 323—329, 333
    direction of, 320—321, 324, 332, 334, 336—339, 347—350
    frequency of:
      eigen- *(see* Dynamic characteristic)
      fundamental, 324, 335, 337—341, 343, 347, 351—353
      natural circular *(see also* Equation of, secular), 323—327, 335—343, 347—353
    order of, 324—327, 335, 347, 351, 353
    law of motion of, 320—321, 323
    natural mode of, 325—328
    period of, 324—326
  forced, 86, 319—327, 335, 341—346, 351
    amplitude of, 329
    direction of, 329—334, 342—345
    frequency of, 328—329, 333
    law of motion of, 328—329
    steady state of, 328—329, 332—333, 342—343
  initial phase angle of, 323, 328

Vibrating system *(see also* Dynamic diagram):
  elastic rectangular, square plate as, 320, 328—332, 334
  slab structure as, 320, 325, 327—338, 341—353

Zero moment line, point, 31, 71, 312

PRINTED IN ROMANIA